BOLYAI SOCIETY
MATHEMATICAL STUDIES

10

Springer
*Berlin*
*Heidelberg*
*New York*
*Barcelona*
*Hong Kong*
*London*
*Milan*
*Paris*
*Tokyo*

BOLYAI SOCIETY MATHEMATICAL STUDIES



1. **Combinatorics, Paul Erdős is Eighty, Vol. 1**
   D. Miklós, V.T. Sós, T. Szőnyi (Eds.)
2. **Combinatorics, Paul Erdős is Eighty, Vol. 2**
   D. Miklós, V.T. Sós, T. Szőnyi (Eds.)
3. **Extremal Problems for Finite Sets**
   P. Frankl. Z. Füredi, G. Katona, D. Miklós (Eds.)
4. **Topology with Applications**
   A. Császár (Ed.)
5. **Approximation Theory and Function Series**
   P. Vértesi, L. Leindler, Sz. Révész, J. Szabados, V. Totik (Eds.)
6. **Intuitive Geometry**
   I. Bárány, K. Böröczky (Eds.)
7. **Graph Theory and Combinatorial Biology**
   L. Lovász, A. Gyárfás, G. Katona, A. Recski (Eds.)
8. **Low Dimensional Topology**
   K. Böröczky, Jr., W. Neumann, A. Stipsicz (Eds.)
9. **Random Walks**
   P. Révész, B. Tóth (Eds.)
10. **Contemporary Combinatorics**
    B. Bollobás (Ed.)

Béla Bollobás (Ed.)

# Contemporary Combinatorics

Springer

JÁNOS BOLYAI MATHEMATICAL SOCIETY

Béla Bollobás
University of Memphis
Memphis, TN
USA
e-mail: bollobas@msci.memphis.edu

Mathematics Subject Classification (2000): 05-02, 94A24, 91A24, 57M27, 90C27, 05C15, 05C80

Cataloging-in-Publication Data applied for
Die Deutsche Bibliothek – CIP-Einheitsaufnahme
Contemporary combinatorics / János Bolyai Mathematical Society. Béla Bollobás (ed.). –
Berlin; Heidelberg; New York; Barcelona; Hong Kong; London; Milan; Paris; Tokyo: Springer, 2002
(Bolyai Society mathematical studies; 10)
ISBN 3-540-42725-2

ISSN 1217-4696
ISBN 3-540-42725-2 Springer-Verlag Berlin Heidelberg New York
ISBN 963 8022 99 János Bolyai Mathematical Society, Budapest



Springer-Verlag Berlin Heidelberg New York
a member of BertelsmannSpringer Science+Business Media GmbH
http://www.springer.de


Printed in Hungary



Cover design: Erich Kirchner, Heidelberg
Printed on acid-free paper SPIN 10845177 44/3142/db – 5 4 3 2 1 0

# Contents

# PREFACE

This volume is a collection of survey papers in combinatorics that have grown out of the lectures given in the workshop on Probabilistic Combinatorics at the Paul Erdős Summer Research Center in Mathematics in Budapest. The papers, reflecting the many facets of modern-day combinatorics, will be appreciated by specialists and general mathematicians alike: assuming relatively little background, each paper gives a quick introduction to an active area, enabling the reader to learn about the fundamental results and appreciate some of the latest developments. An important feature of the articles, very much in the spirit of Erdős, is the abundance of open problems.

In 1956, Shannon initiated the study of the amount of information a communication channel can transmit without error. The main invariant that arose from this study was the Shannon capacity $c(G)$ of the graph $G = G(C)$ associated with a channel $C$. Roughly, $c(G)$ measures the largest number of inputs that $C$ can communicate without error in a large number of uses.

Combinatorially, the invariant $c(G)$ is defined in terms of one of the many graph powers, the so-called AND power of $G$. In his article on graph powers, Alon surveys the major results concerning the Shannon capacity, together with AND powers of graphs, and goes on to present properties of numerous other graph powers, including OR powers, sparse powers and tensor powers.

The reader will greatly benefit from the large number of conjectures Alon poses in his article; these beautiful and important conjectures are bound to stimulate much exciting research.

In 1960, Erdős and Rényi wrote a fundamental paper on the evolution of random graphs: they studied the growth of a 'typical' random graph as it evolves by acquiring more and more edges at random. In his paper in this volume on the amount of dependence in the prime factorization of a uniform random integer, Arratia starts with a question analogous to the Erdős-Rényi notion of 'growing' a random graph: "can we grow a random integer?" After a fascinating tour of many exciting ideas, Arratia formulates this question precisely, and shows that the answer is in the affirmative.

Among many deep results, Arratia shows that, roughly speaking, $2 + o(1)$ changes in the exponents are necessary and sufficient to pass from a uniformly distributed random integer from 1 to $n$ to a random integer $\prod_{p \le n} p^{Z_p}$. Here $p$ is a prime and $Z_2, Z_3, \dots$ are independent geometric random variables, with $\mathbb{E}Z_p = 1/(p-1)$. This beautiful result is *much* stronger than the classical theorem of Hardy and Ramanujan on the number of prime factors, and the fundamental central limit theorem of Erdős and Kac.

The paper has numerous tantalizing conjectures; taking his cue from Erdős, Arratia even attaches monetary rewards to some of his conjectures.

In recent years, combinatorial game theory has greatly benefited from surprising applications of probabilistic methods. To be precise, Beck has used his randomization-derandomization approach to tackle classical positional games, and obtained many deep results. Although positional games are deterministic and so have deterministic optimal strategies, Beck has used statistical analysis of the randomized game to construct optimal strategies with the aid of potential functions.

Beck's article is an excellent survey of these exciting developments. His starting point is the traditional game of tic-tac-toe; from there, he goes on to discuss strong positional games which are far-reaching generalizations of tic-tac-toe, describes the use of diverse branches of combinatorics to study games: Ramsey theory, extremal set theory and probability theory, and shows how his quasi-probabilistic method can be used to analyze complex games that resist other attempts.

Polynomials of knots have been studied for almost eighty years, but their study greatly intensified in 1985, when Vaughn Jones constructed a polynomial invariant for knots with the aid of von Neumann algebras. It was soon discovered that this polynomial could be defined in a purely combinatorial way; in particular, the Jones polynomial of an alternating knot is a variant of the classical Tutte polynomial of the associated graph.

Inspired by the Jones polynomial, several important knot polynomials were discovered, including the one- and two-variable Kauffman polynomial, and the HOMFLY polynomial of Freyd, Hoste, Lickorish, Millett, Ocneanu, Przytycki, Traczyk and Yetter. The many proofs of the existence of the HOMFLY polynomial are considerably harder than those of other polynomials, and contain many more pitfalls. In their survey in this volume, Bollobás, Pebody and Riordan give an entirely combinatorial state space definition of the HOMFLY polynomial, somewhat resembling the much simpler state space definition Kauffman gave for his own polynomial. Although the state space definition does not simplify the proof of existence and uniqueness too much, it does have the advantage that it streamlines some of the arguments.

The well-known Max Cut problem of combinatorial optimization asks for a largest *cut* of a graph $G$, i.e., for the maximal number of edges contained in a bipartite subgraph of $G$. This problem is $NP$-hard, and has been extensively studied, both in the algorithmic context and from the extremal perspective of combinatorics. The article of Bollobás and Scott in this volume gives a survey of many of the major results in both areas, and also contains numerous new results.

¿From the extremal point of view, the basic object of study is the function $f(m)$, the maximal number of edges that can be guaranteed in every graph with $m$ edges. Erdős observed that $f(m) \geq m/2$, and in 1973 Edwards proved a lower bound for $f(m)$ which is exact for infinitely many values of $m$.

The paper presents the recent extensions of the inequality of Edwards due to Alon and Halperin, and goes on to determine $f(m)$ for a rather large set of integers $m$, together with the corresponding extremal graphs. Concerning fast approximation algorithms for Max Cut, the paper concentrates on graphs of size $m$ whose maximal cut is not much larger than $f(m)$. The final part of the paper concerns the Max $k$-Cut problem for $k > 2$.

Gibbs measures on lattices have been studied for several decades by physicists and probabilists, but combinatorialists have only recently discovered this fruitful and important area. In their paper, Brightwell and Winkler study a particularly pleasant class of Gibbs measures, the class of simple semi-invariant Gibbs measures on the space of proper colourings of the Cayley tree or Bethe lattice $T_r$, that is, the connected and acyclic infinite graph in which each vertex has degree $r + 1$.

In the simplest case, the space of all proper colourings is endowed with the measure derived from the uniform distribution on finite sets. Brightwell and Winkler generalize this rather simple probability space in two important ways. First, given a 'constraint graph' $H$ with vertex set $\{1, 2, \ldots, q\}$, they take only those colourings of the Cayley tree $T_r$ with colours $1, 2, \ldots, q$ in which adjacent vertices in $T_r$ may receive colours $i$ and $j$ only if $ij$ is an edge or loop of $H$. Second, Brightwell and Winkler replace the uniform distribution of the colours with a "multiplicative" distribution depending on positive reals $\lambda_1, \ldots, \lambda_q$ assigned to the vertices of $H$. If the graph we wish to colour is finite then the probability of a permissible colouring $\phi : V(G) \to \{1, \ldots, q\}$ is proportional to $\prod_{u \in V(G)} \lambda_{\phi(u)}$. If $G$ is infinite then the Gibbs property is used to extend this definition.

By making use of branching walks on certain bipartite graphs, Brightwell and Winkler determine when there is more than one simple semi-invariant Gibbs measure, but leave the reader with several important open problems.

The *random k-assignment problem* is one of the fundamental problems of probabilistic optimization theory. For an $m \times n$ matrix with i.i.d. exponential entries of mean 1, a *k-assignment* is a selection of $k$ elements of the matrix, all in different rows and different columns. The *minimum k-assignment* is the one in which the sum of the elements chosen, i.e., the *cost* of the $k$-assignment, is minimal. Writing $E(k, m, n)$ for the expectation of this minimum, Mézard and Parisi made the surprising conjecture that $E(n, n, n) \to \pi^2/6$ and Aldous gave a remarkable proof of this conjecture. Recently, Aldous made the considerably stronger conjecture that, in fact, $E(n, n, n) = \sum_{i=1}^{n} 1/i^2$.

The article of Coppersmith and Sorkin concerns the even stronger conjecture that $E(k, n, m)$ is precisely the simplest function $F(k, m, n)$, which is the sum of the $k^2$ terms $1/(m-i)(m-j)$, where $i, j \geq 0$ and $i + j \leq k - 1$. Coppersmith and Sorkin prove that, under certain hypotheses, the cost of augmenting a minimum $(m-1)$-assignment in an $(m-1) \times n$ matrix to a minimum $m$-assignment in an $m \times n$ matrix is equal to $F(m, n, m) - F(m-1, m-1, n)$. This seems to be a major step towards justifying the finite conjecture. Unfortunately, the hypotheses are rather strong so this beautiful result does not prove the finite conjecture.

Since the pioneering work of Erdős and Rényi in the 1960's, the theory of random graphs has come a very long way. This is mostly due to the development of combinatorial probability, especially the emergence of a large body of approximation results and concentration of probability inequalities. In his important article on concentration of probability, Janson surveys the methods to obtain sharp concentration results, including the basic inequalities for sums of independent random variables, Janson's own inequality for sums of dependent random variables followed by its variant the Boppana-Spencer inequality, Suen's inequality, the Azuma-Hoeffding inequality for martingales, and Talagrand's inequality. The paper is a wealth of information for everybody interested in probabilistic combinatorics.

It is our hope that the volume would have pleased Paul Erdős, to whom it is dedicated, and that it will inspire much research in combinatorics and related fields.

Thanks are due to Professor Vera T. Sós, who was instrumental in the foundation of the Erdős Institute in Budapest, and to Professor Péter P. Pálffy, the Director of the Institute, who gave all the help he could to the lecturers and the authors of the papers in this volume. The Managing Editor, Professor Dezső Miklós, has done a wonderful job in preparing the papers for publication.

Finally, and most importantly, we thank the lecturers at the Erdős Meeting and the authors of the papers in this volume for their inspiring work.

*Béla Bollobás*
*Memphis, 28 February 2001*

BOLYAI SOCIETY
MATHEMATICAL STUDIES, 10

**Contemporary Combinatorics**
**Budapest, 2002,** pp. 11–28.

# Graph Powers

N. ALON*

The investigation of the asymptotic behaviour of various parameters of powers of a fixed graph leads to many fascinating problems, some of which are motivated by questions in information theory, communication complexity, geometry and Ramsey theory. In this survey we discuss these problems and describe the techniques used in their study which combine combinatorial, geometric, probabilistic and linear-algebra tools.

## 1. Graph Powers

There are several known distinct ways to define the powers of a fixed graph. The $n$-th *AND power* of an undirected graph $G=(V,E)$ is the graph denoted by $G^{\wedge n}$ whose vertex set is $V^n$ in which distinct vertices $(x_1,\dots,x_n)$ and $(x'_1,\dots,x'_n)$ are connected if $\{x_i,x'_i\}\in E$ for all $i\in\{1,2,\dots,n\}$ such that $x_i\neq x'_i$. The $n$-th *OR power* of $G$ is the graph denoted $G^{\vee n}$ whose vertex set is $V^n$ in which distinct vertices $(x_1,\dots,x_n)$ and $(x'_1,\dots,x'_n)$ are connected if distinct $x_i$ and $x'_i$ are connected in $G$ for some $i\in\{1,2,\dots,n\}$.

The study of the asymptotic behaviour of various parameters of these powers of a fixed graph $G$, as well as their behaviour for similarly defined powers of directed and undirected graphs, is motivated by questions in various areas and leads to many intriguing problems. These are discussed in

* Research supported in part by a grant from the the Israel Science Foundation and by the Hermann Minkowski Minerva Center for Geometry at Tel Aviv University.

the following sections, in which we focus our attention mainly to the open problems in the area, and only briefly describe the known results and proof techniques. Proven and disproven conjectures are intermingled throughout the paper with open problems. More detailed proofs can be found in the papers listed in the bibliography.

## 2. Shannon Capacity

The *independence number* $\alpha(G)$ of a graph $G$ is the maximum cardinality of a set of vertices of $G$ no two of which are adjacent.

A *channel* consists of a finite *input set* $X$, a (possibly infinite) *output set* $Y$, and a nonempty *fan-out set* $S_x \subseteq Y$ for every $x \in X$. In each channel *use*, a *sender* transmits an *input* $x \in X$ and a *receiver* receives an arbitrary *output* in $S_x$. Shannon [57] initiated the study of the amount of information a channel can communicate without error.

Associated with a channel $C$ is its *characteristic graph* $G(C)$. Its vertex set is $X$ and two (distinct) vertices are connected if their fan-out sets intersect, namely, both can result in the same output.

Note that every graph $(V, E)$ is the characteristic graph of some channel: its input set is $V$, its output set is $E$, and $S_v$ consists of all edges containing $v$.

The largest number of inputs a channel $C$ can communicate without error in a single use is $\alpha\big(G(C)\big)$, the independence number of its characteristic graph. This is done as follows: the sender and the receiver agree in advance on an independent set $I$. The sender transmits only inputs in $I$. Every received output belongs to the fan-out set of exactly one input in $I$, hence the receiver can correctly determine the transmitted input. Conversely, it is easy to see that a set containing two connected vertices cannot be communicated without error.

When the channel $C$ is used $n > 1$ times the sender transmits a sequence $x_1, \ldots, x_n$ of inputs and the receiver receives a sequence $y_1, \ldots, y_n$ of outputs where each $y_i \in S_{x_i}$. Conceptually, $n$ uses of $C$ can be viewed as a single use of a larger channel $C^n$. Its input set is $X^n$, its output set is $Y^n$, and the fan-out set of $(x_1, \ldots, x_n) \in X^n$ is the Cartesian product $S_{x_1} \times \cdots \times S_{x_n}$.

It is not difficult to check that the characteristic graph of $C^n$ is simply $G^{\wedge n}$, the $n$-th AND product of $G = G(C)$.

It follows that the largest number of messages $C$ can communicate without error in $n$ uses is $\alpha(G^{\wedge n})$. The limit $\lim_{n\mapsto\infty}\left(\alpha(G^{\wedge n})\right)^{1/n}$ is called the *Shannon capacity of* $G$ and is denoted by $c(G)$. It represents the number of distinct messages per use the channel can convey without errors, when used a large number of times. This limit exists, by super-multiplicativity, and is equal to the supremum $\sup_{n\mapsto\infty}\left(\alpha(G^{\wedge n})\right)^{1/n}$. (It is worth noting that sometimes it is customary to call $\log c(G)$ the Shannon capacity of the graph $G$, but we prefer the definition above, following Lovász [50].)

By super-multiplicativity, $c(G) \geq \alpha(G)$ for every $G$. One of the early discoveries of Shannon was that sometimes strict inequality holds. Thus, for example, it is known by the results in [57] and [50] that the Shannon capacity of a cycle of length 5 satisfies $c(C_5) = \sqrt{5}$, whereas, of course, its independence number is 2. It seems interesting to decide how large can the gap between $\alpha(G)$ and $c(G)$ be. In particular, the following conjecture seems plausible.

**Conjecture 2.1.** *For every constant $C > 0$ there is a graph $G$ with independence number 2 satisfying $c(G) > C$.*

As shown implicitly in [21] (and explicitly in [8]) there is a tight connection between the Shannon capacity of graphs and appropriate Ramsey numbers. The Ramsey number $r(3 : k)$ is the maximum number $r$ such that there exists an edge coloring of the complete graph on $r$ vertices by $k$ colors with no monochromatic triangle. It turns out that the conjecture above is equivalent to the conjecture that for every $C > 0$ the Ramsey number $r(3 : k)$ exceeds $C^k$ provided $k$ is sufficiently large as a function of $C$. This is an old question of Erdős (c.f., e.g., [35], [19]). The best known lower bound for $r(3 : k)$ is $\Omega(321^{k/5})$, proved in [22] improving [29], and the best known upper bound is $O(k!)$ ([33]).

A related interesting question, which is more natural from the information theoretic point of view, is the estimation of the largest possible gap between $\alpha(G)$ and $c(G)$ as a function of the number $n$ of vertices of the graph.

**Conjecture 2.2.** *For every $\varepsilon > 0$ and every $n > n_0(\varepsilon)$ there exists a graph $G$ on $n$ vertices satisfying $\alpha(G) < n^{\varepsilon}$ and $c(G) > n^{1-\varepsilon}$.*

Some results about this conjecture appear in [8], where it is shown that there are graphs $G_n$ on $n$ vertices satisfying $\alpha(G_n) \leq O(\log n)$ and $c(G_n) \geq \sqrt{n}$, but the assertion of the conjecture remains wide open and deserves further study.

Another intriguing problem is the computational problem of determining the Shannon capacity $c(G)$ of a given input graph $G$. It is not even known if this problem is decidable, and the value of $c(G)$ is not known even for several extremely simple graphs like the cycle of length 7 (or any odd cycle of length bigger than 5). It is also not known if the capacity of the complement of any odd cycle of length bigger than 5 is 2. The existence of a natural graph invariant which we cannot compute even for such simple finite graphs is highly intriguing, and although the results of Lovász [50], Haemers [38] and those in [3] supply some geometric and algebraic tools for studying this invariant, these methods work only in very limited cases, and it would be desirable to have additional effective techniques.

The following conjecture was raised by Shannon in his original paper in 1956.

**Conjecture 2.3** [57]. *The Shannon capacity of the disjoint union of two graphs is equal to the sum of their capacities.*

He proved that the capacity of a union of two graphs is always at least as large as the sum of the capacities, and equality holds provided the independence number of one of the graphs is equal to the chromatic number of its complement.

In [3] it is proved that this conjecture is false. There are two graphs, each having Shannon capacity at most $k$, so that the capacity of their disjoint union is at least $k^{\Omega(\log k/\log\log k)}$. The proof combines a variant of the beautiful construction of Frankl and Wilson [28], with the ideas in [4] and [8]. These counterexamples have an interesting and somewhat counterintuitive information theoretic interpretation. Indeed, if $G$ and $H$ are graphs of two channels, then their union represents the *sum* of the channels corresponding to the situation where either one of the two channels may be used, a new choice being made for each transmitted letter. Therefore such examples show that by using two channels alternately it is sometimes possible to obtain capacity that exceeds considerably the capacities of each of them. The problem of estimating the largest possible gap between the capacity of a union and the sum of the capacities remains open and deserves further study. A related problem is that of estimating the minimum possible value of $c(G)+c(\overline{G})$, where the minimum is taken over all graphs $G$ with $n$ vertices. The construction in [3] shows that this minimum is at most $e^{O(\sqrt{\log n \log\log n})}$, but it seems plausible that it is, in fact, $\Theta(\log n)$.

The relevance of geometry to the study of the Shannon capacity of graphs was initiated by Lovász in [50]. In this paper he introduced the $\theta$-function of a graph $G = (V, E)$. One of its various equivalent definitions is the maximum possible value of the sum $\sum_{v \in V} (u_v^T \cdot b)^2$, where the maximum is taken over all unit vectors $b$ and $\{u_v, v \in V\}$ in an arbitrary Euclidean space, so that for any two adjacent vertices $w$ and $v$ of $G$ the vectors $u_w$ and $u_v$ are orthogonal. This function has many fascinating properties, including the fact that it is at least the independence number (and even at least the Shannon capacity) of $G$ and at most the chromatic number of its complement, and the fact that it can be computed in polynomial time. See the survey [43] for more details.

A related problem to that of determining the maximum possible value of the Shannon capacity of a graph (on $n$ vertices) with independence number 2 is that of determining or estimating the maximum possible $\theta$-function of such a graph. This problem was raised by Lovász, and is equivalent to the geometric problem of estimating the maximum possible Euclidean norm of a sum of $n$ unit vectors in $R^n$, so that among any three of them some two are orthogonal. Konyagin [45], and Kashin and Konyagin [44] showed that this maximum is at least $\Omega\left(n^{2/3}/(\log n)^{1/2}\right)$ and at most $O(n^{2/3})$ and in [1] it is shown that it is $\Theta(n^{2/3})$. The corresponding graph of [1] has interesting Ramsey theoretic properties as well, and in particular it provides an explicit example of a triangle-free graph on $n$ vertices with independence number $O(n^{2/3})$. Another example of a graph with this property has recently been constructed in [20].

Let $G(n, 1/2)$ denote, as usual, the random graph on $n$ labeled vertices denoted $1, 2, \ldots, n$ , obtained by picking each pair of distinct vertices $i, j$ to be an edge randomly and independently with probability $1/2$. We say that $G(n, 1/2)$ satisfies a property $A$ *almost surely* if the probability that it satisfy $A$ tends to 1 as $n$ tends to infinity.

**Conjecture 2.4.** *There is an absolute constant $b$ so that the Shannon capacity of $G(n, 1/2)$ is at most $b \log_2 n$ almost surely.*

As proved by Juhasz [41], the $\theta$-function of $G(n, 1/2)$ is $\Theta\left(\sqrt{n}\right)$ almost surely. This, together with the well known fact (cf., e.g., [16] or [9]), that the independence number of $G(n, 1/2)$ is $\big(2 + o(1)\big) \log_2 n$ almost surely, implies that the Shannon capacity of $G(n, 1/2)$ is at least $\Omega(\log n)$ and at most $O\big(\sqrt{n}\,\big)$ almost surely.

## 3. Witsenhausen's Rate

The *chromatic number* $\chi(G)$ of a graph $G$ is the minimum number of colors needed to color its vertices so that no two adjacent vertices have the same color.

A *dual source* consists of a finite set $X$, a (possibly infinite) set $Y$, and a *support set* $S \subseteq X \times Y$. In each *dual-source instance,* a *sender* $P_X$ is given an $x \in X$ and a *receiver* $P_Y$ is given a $y \in Y$ such that $(x, y) \in S$. Witsenhausen [61] and Ferguson and Bailey [25] initiated the study of the number of bits that $P_X$ must transmit in the worst case in order for $P_Y$ to learn $x$ without error. (See [56] for the case when $P_X$ and $P_Y$ are allowed to interact.)

The *fan-out* of $x \in X$ is the set $S_x = \big\{y : (x, y) \in S\big\}$ of $y$'s that are *jointly possible* with $x$. Associated with a dual source $S$ is its *characteristic graph* $G = G(S)$. Its vertex set is $X$, and two (distinct) vertices $x, x'$ are connected if their fan-out sets intersect, namely, there is a $y$ that is jointly possible with both. Note that as is the case with the characteristic graphs of channels, here, too, every graph $(V, E)$ is the characteristic graph of some dual source: $X = V$, $Y = E$, and $S = \big\{(x, y) : x \in y\big\}$.

The smallest number of possible messages that enable $P_X$ to transmit one of them for a single instance of $S$ is $\chi(G)$, the chromatic number of $S$'s characteristic graph. To do so, $P_X$ and $P_Y$ agree in advance on a coloring of $G$. Given $x$, $P_X$ transmits its color. $P_Y$, having $y$, can determine $x$ because there is exactly one element of $X$ with this color that is jointly possible with $y$. Conversely, it is easy to see that if two connected vertices are assigned the same message, an error can result.

In $n > 1$ instances of the dual source $S$, $P_X$ knows $x_1, \ldots, x_n$ while $P_Y$ knows $y_1, \ldots, y_n$ such that each $(x_i, y_i) \in S$ and wants to learn $x_1, \ldots, x_n$. Conceptually, $n$ instances can be viewed as a single instance of a larger dual source whose support set is $S^n \subseteq X^n \times Y^n$. The characteristic graph of this larger dual source is $G^{\wedge n}$, the $n$-th AND power of the characteristic graph of $S$. It follows that the smallest number of messages that enable $P_X$ to to convey $n$ instances of $S$ without error is $\chi(G^{\wedge n})$. Let $R(G)$ denote the limit $\lim_{n\mapsto\infty}\left(\chi(G^{\wedge n})\right)^{1/n}$. This quantity, called the *Witsenhaussen's rate* of $G$, measures the number of messages per instance that enable $P_X$ to convey $P_Y$ a large number of instances with no error. By sub-multiplicativity this limit exists and is always at most $\chi(G)$. (Here, too, it is sometimes customary to call $\log R(G)$ the rate of $G$ but we prefer to avoid the logarithm). Witsenhausen [61] showed that for some dual sources, fewer messages suffice for each use of the larger system (than for a single instance). For example, for the cycle of length 5, $\chi(C_5) = 3 > \sqrt{5} = R(C_5)$.

It is interesting to study the maximum possible gap between $\chi(G)$ and $R(G)$. A result of Linial and Vazirani [48] implies that for every constant $C$ there are graphs $G$ for which $R(G) \leq 6$ while $\chi(G) > C$. In [8] it is shown that for every $C$ and every $\varepsilon > 0$ there are graphs $G$ with $R(G) \leq 2 + \varepsilon$ and $\chi(G) > C$. This is done by observing that the chromatic number of $G^{\wedge n}$ does not exceed that of $G^{\vee n}$, and by applying to the Kneser graphs the results of McEliece and Posner [52], and of Berge and Simonovits [13] that relate the chromatic number of $G^{\vee n}$ with the fractional chromatic number of $G$. This, together with the result of Lovász [49] about the chromatic number of Kneser graphs, gives the above mentioned bound.

The study of the maximum possible gap between $\chi(G)$ and $R(G)$ as a function of the number of vertices $n$ of $G$ seems more complicated. As shown in [8] the ratio between these two may be at least $\Omega\left(\sqrt{n}/\log^2 n\right)$, but the following conjecture remains open.

**Conjecture 3.1.** *For every $\varepsilon > 0$ and $n > n_0(\varepsilon)$ there is a graph $G$ on $n$ vertices satisfying $\chi(G) > n^{1-\varepsilon}$ and $R(G) \leq n^{\varepsilon}$.*

As is the case with the Shannon capacity, the computational problem of determining $R(G)$ for a given input graph $G$ seems, in general, far beyond reach of the existing techniques, and even the computation of $R(G)$ for some simple, small graphs appears to be extremely difficult.

## 4. Cayley graphs

The partial results obtained in [8] concerning the study of Conjectures 2.2 and 3.1 combine algebraic and probabilistic techniques. The idea is to prove the existence of self complementary *Cayley graphs* in which the largest independent set is very small compared to the size of the graphs. For a group $H$ and a subset $S \subset H$ satisfying $S = S^{-1}$, the *Cayley graph* of $H$ with respect to $S$ is the graph whose vertices are all elements of $H$ and $g, h \in H$ are adjacent iff $h^{-1}g \in S$. The following Ramsey theoretic conjecture seems plausible.

**Conjecture 4.1.** *There exists an absolute constant $b$ such that for every group $H$ on $n$ elements there is a Cayley graph of this group containing neither a complete subgraph nor an independent set on more than $b \log n$ vertices.*

A weaker version of this conjecture, obtained by replacing the $\log n$ term by a $\log^2 n$ term, is proved in [8], and this is the main tool in obtaining a partial result in the study of Conjecture 3.1. Although the solution of the last conjecture will not lead immediately to progress in the attempts to solve Conjecture 3.1, it is interesting in its own right.

Cayley graphs appear naturally in the study of the maximum possible value of the $\theta$-function of a graph on $n$ vertices with independence number 2. The construction in [1], in which this problem is solved up to a constant factor, uses Cayley graphs of Abelian groups. Their spectral properties, obtained via known bounds on character sums, play a crucial role in proving their properties. These questions are strongly related to problems in Combinatorial Geometry. In fact, by using related techniques together with some of the ideas in [24] and [14] it is proved in [10] that that there is an absolute positive constant $\delta > 0$, so that for all positive integers $k$ and $d$, there are sets of at least $d^{\delta \log_2(k+2)/\log_2 \log_2(k+2)}$ nonzero vectors in $R^d$, in which any $k+1$ members contain an orthogonal pair. This settles a problem of Füredi and Stanley [30]. The problem of estimating more accurately the maximum possible size of such a collection of vectors remains open.

## 5. Sperner Capacity

For a digraph $D = (V, E)$ and for a positive integer $n$, let $w(D^n)$ denote the maximum possible cardinality of a subset $S$ of $V^n$ in which for every ordered pair $(u_1, u_2, \ldots, u_n)$ and $(v_1, v_2, \ldots, v_n)$ of members of $S$ there is some $i$, $1 \leq i \leq n$ such that $(u_i, v_i)$ is a directed edge of $D$. It is easy to see that the function $g(n) = w(D^n)$ is super-multiplicative, and hence the limit

$$\lim_{n \mapsto \infty} \left[ \left( w(D^n) \right)^{1/n} \right]$$

exists and is equal to the supremum of the quantity in the square brackets. This limit, denoted by $C(D)$, is called the *capacity* of the digraph.

The study of the capacity of directed graphs was introduced by Körner and Simonyi and by Gargano, Körner and Vaccaro in [47], [31], where the authors study the quantity $\sum(D) = \log C(D)$, which they call the *Sperner capacity* of $D$, and show that it generalizes the Shannon capacity of an undirected graph. In several subsequent papers [32], [46] they apply some properties of this invariant in the asymptotic solution of various problems in extremal set theory.

The problem of computing $C(D)$ for a given directed graph $D$ is even more difficult in general than that of computing the Shannon capacity $c(G)$ of an undirected graph $G$ , since if $D$ is obtained from the complement of $G$ by directing each edge in both directions then $C(D) = c(G)$. It is thus natural to try and restrict attention to some limited yet interesting classes of directed graphs. One such class is the class of all *tournaments*.

A *tournament* $T$ is a digraph in which for every pair $u, v$ of distinct vertices exactly one of the ordered pairs $(u, v)$, $(v, u)$ is a directed edge. The tournament $T$ is *transitive* if there is a linear order on its vertices such that $(u, v)$ is a directed edge iff $u$ is smaller than $v$ in this order.

It is easy to see that the capacity $C(T_n)$ of the transitive tournament on $n$ vertices is $n$. Therefore, the capacity of any tournament that contains a transitive subtournament on $n$ vertices is at least $n$. Using algebraic techniques, Calderbank, Frankl, Graham, Li and Shepp [18] proved that the capacity of the cyclically directed triangle is 2, namely, the number of vertices in the largest transitive subtournament in it. Blokhuis [15] gave a simpler proof of this result. This inspired Körner and Simonyi ([46]) to conjecture that for every tournament $T$, the capacity $C(T)$ is the maximum number of vertices in a transitive subtournament of $T$.

Using probabilistic arguments it is shown in [2] that this conjecture is false. This is based on some simple properties of *random tournaments* obtained by deciding, for each pair of vertices $u$, $v$ randomly and independently, if $(u, v)$ or $(v, u)$ is a directed edge with equal probability. However, the assertion of the conjecture is true for all tournaments with at most 5 vertices. In order to prove the assertion of the conjecture for small tournaments one can develop an algebraic method, related to the technique of Haemers in [38], which enables one to bound the capacities of other digraphs as well.

The following conjecture seems plausible.

**Conjecture 5.1.** *There is an absolute constant $c$ so that the probability that the capacity of a random tournament on $n$ vertices exceeds $c\log_2 n$ tends to 0 as $n$ tends to infinity.*

Let $t(T)$ denote the maximum number of vertices in a transitive subtournament of $T$. It would be interesting to estimate the maximum possible value of the ratio $C(T)/t(T)$, as $T$ ranges over all tournaments of $n$ vertices. It can be shown that this maximum is at least $\Omega\left(\sqrt{n}/\log n\right)$, but we suspect it might be bigger.

Another interesting problem, suggested by Körner, is to characterize all tournaments $T$ in which for every subtournament $T'$, $C(T') = t(T')$.

Given a directed graph $D = (V, E)$ and a field $F$, a *representation of $D$ of dimension $d$ over $F$* is an assignment of a vector $u_v \in F^d$ to each vertex $v$ of $D$, so that no vector $u_v$ lies in the span of the vectors assigned to the in-neighbors of $v$. This notion seems promising in the study of the capacities $C(D)$ of directed graphs, as it can be shown, using some linear-algebra tools, that if $D$ has such a representation (over any field) then its capacity $C(D)$ is at most $d$. This result, obtained jointly by the author and G. Tardos, together with some simple combinatorial arguments, suffices to determine the capacity of all tournaments with at most 5 vertices. It can be used to show that, as proved in [2], the Sperner capacity of any digraph with maximum outdegree $d$ is at most $d + 1$. More generally, it implies that, as proved in [27], the Sperner capacity of any digraph whose vertex set can be partitioned into $k$ classes, where $d_i$ is the maximum outdegree in the $i$-th class, is at most $\sum_{i=1}^{k}(d_i + 1)$. It does not suffice, however, to solve Conjecture 5.1, whose study seems to require additional techniques. Indeed, the minimum possible value of $\sum_{i=1}^{k}(d_i + 1)$, where the minimum is taken over all tournaments on $n$ vertices and all partitions of their vertices, where the numbers $d_i$ are defined as above, is easily seen to be at least $(n + 1)/2$. Combining probabilistic and combinatorial arguments one can show that the probability that the capacity of a random tournament on $n$ vertices is bigger than $\varepsilon n$ tends to 0 for every fixed positive $\varepsilon$, as $n$ tends to infinity, but the assertion of the conjecture is far stronger.

Sali and Simonyi [58] proved that any undirected vertex transitive self-complementary graph on $n$ vertices has an orientation whose Sperner capacity is $\sqrt{n}$. As mentioned in their paper, it might be that any undirected graph $G$ has an orientation $D$ whose Sperner capacity is equal to the Shannon capacity of the complement of $G$.

## 6. Additional Graph Powers

The $n$-th *(sparse) power* of an undirected graph $G = (V, E)$, denoted simply by $G^n$, is the graph whose vertex set is $V^n$ in which distinct vertices $(x_1, \ldots, x_n)$ and $(x'_1, \ldots, x'_n)$ are connected iff there exists a single index $i$ such that $x_j = x'_j$ for all $j \neq i$ and $x_i$ and $x'_i$ are connected in $G$. The study of the asymptotic behaviour of the independence number of $G^n$, for a fixed graph $G$, can be motivated by the search for appropriately defined error-detecting codes in a certain channel. Indeed, if the vertices of the graph are the possible inputs, and each input may be altered, during the transmission process in a channel, to one of its neighbors, then this independence number represents the maximum number of messages of length $n$ that can be sent in a way that enables us to detect the occurrence of one error. In [36] and some of its references the authors obtained several results about the problem of estimating these independence numbers. In particular, they proved that for every fixed graph $G$, the limit $\lim_{n \mapsto \infty} \frac{\alpha(G^n)}{|V|^n}$ exists, and is at least $1/\chi(G)$ and at most the reciprocal of the fractional chromatic number of $G$. Moreover, for every Cayley graph of an Abelian group this limit is precisely the reciprocal of the fractional chromatic number of $G$. Several examples suggest that this invariant has interesting properties and its study seems to require algebraic and combinatorial tools.

Another interesting topic dealing with sparse powers is the study of isoperimetric inequalities in such powers. For $G = (V, E)$ and $G^n$ as above, and for an integer $d$, let $f(G, n, d)$ denote the maximum possible number of vertices in $V^n$ of distance at least $d$ from a set of half the vertices of $V^n$, divided by $|V|^n$. The study of the asymptotic behaviour of $f(G, n, d)$ attracted a considerable amount of attention, as it appears naturally in the study of many combinatorial and geometric problems. See, for example, [53, 7, 55, 54, 17, 59, 37]. In all these references it is shown that for every fixed $G$ there exists a positive constant $b = b(G)$ such that $f(G,n,d) \le \exp\big[-b(d^2/n)\big]$ for all $d$ and $n$, but it seems interesting to get a more precise estimate, like the tight inequality of Harper [39] for powers of a single edge. This inequality asserts that for the graph of the $n$-cube $G$ where $n$ is, say, $2k+1$, $f(G,n,d)$ is precisely the set of all vertices of the cube whose distance from a Hamming ball of radius $(n-1)/2$ is at least $d$. That is, in this case

$$f(G,n,d) = \sum_{i=0}^{\frac{n+1}{2}-d} \binom{n}{i}.$$

In [5] the authors obtain, for every fixed $G$, an asymptotic formula *with the right constant* for $\log f(G,n,d)$ if $n$ is large and $d$ is not too small as a function of $n$. This is done by a combination of large deviation techniques and ideas borrowed from Game Theory, and provides an asymptotic formula for all relevant $d \gg \sqrt{n}$.

The results for $d = o(n)$ are somewhat simpler and lead to a definition of a constant called in [5] the spread constant of a graph $G$, which appears in the expression for $f(G,n,d)$ for values of $d \gg \sqrt{n}$ that satisfy $d = o(n)$. For a graph $G = (V,E)$, call a function $X$ mapping $V$ to the set of reals *Lipschitz* if $\big|X(u) - X(v)\big| \le 1$ for all $uv \in E$. Any such function can be considered as a random variable on the symmetric probability space $V$ (in which each vertex has the same probability) and as such has a variance $\mathrm{VAR}[X]$. The *spread constant* $c = c(G)$ is the maximum possible value of $\mathrm{VAR}[X]$ over all Lipschitz $X$. In [5] it is proved that for every fixed graph $G$, if $d/\sqrt{n}$ and $n/d$ tend to infinity, then

$$f(G,n,d) = e^{-\frac{d^2}{2cn}(1+o(1))}.$$

For linear distances the situation is more complicated. Let $m = m(G)$ denote the maximum, over all vertices $v$ of $G$, of the average distance of a vertex of $G$ from $v$. It is not too difficult to check that if $d \ge mn + \sqrt{cn}$ then $f(G,n,d) = 0$. For values of $d$ satisfying $\sqrt{n} \ll d$ and $mn - d \gg \sqrt{n}$ the asymptotic value of $f(G,n,d)$ is given as follows. For a real $\lambda$, let $L(\lambda)$ be the maximum possible value of $\ln E[e^{\lambda X}]$, where the maximum is taken over all Lipschitz $X : V \mapsto \mathbf{R}$ with expectation $E[X] = 0$. For a real $t$ define

$$R(t) = \sup_{\lambda \in R} \left[ \lambda t - L(\lambda) \right].$$

Then, for $d$ as above,

$$f(G, n, d) = e^{-R(d/n)n(1+o(1))}.$$

If, for example, $G$ is a triangle, then a simple though tedious computation shows that for every $0 \le t < 2/3$,

$$e^{-R(t)} = 2(2 - 3t)^{t-2/3}(6t + 2)^{-1/3-t},$$

supplying a tight isoperimetric inequality for the space of all vectors of length $n$ over an alphabet of 3 letters with the Hamming metric.

The *tensor product* of two graphs $G = (V, E)$ and $H = (V', E')$ is the graph whose vertex set is the Cartesian product $V \times V'$ of the vertex sets of $G$ and $H$, where $(u, u')$ is joined to $(v, v')$ if either $uv \in E$ or $u'v' \in E'$, but not both. The *$n$-th tensor power of $G$* is the tensor product of $n$ copies of $G$. These powers are studied in [60], where the author shows how to use their properties to construct edge colorings of complete graphs by two colors in which the number of monochromatic copies of $K_4$ is smaller than the expected number of such copies in random colorings. Similar results hold for other graphs including all graphs containing a $K_4$, as shown in [40]. This disproves conjectures of Erdös [23] and of Burr and Rosta [12].

## 7. Some computational aspects

One of the main reasons for the fast development of Combinatorics during the recent years is certainly the widely used application of combinatorial methods in the study and the development of efficient algorithms. It is therefore natural to study the computational aspects of the problems considered here.

The properties of the Lovász $\theta$-function and the fact that it can be computed in polynomial time using semidefinite programming (see [34]) suggest that it can be used for the design of approximation algorithms for the chromatic number of a graph and for its independence number. It is well known that if, as is widely believed, the complexity classes $P$ and $NP$ differ, then there is no polynomial time algorithm that provides a reasonable approximation for these quantities (see [26], [11], [51]). In addition, examples of Feige [24] show that the ratio between the $\theta$-function of a graph on $n$ vertices and its independence number (or chromatic number) may be as large as $n/e^{O(\sqrt{\log n})}$. Despite these facts it is possible to apply the properties of the $\theta$-function and obtain some meaningful approximation of these invariants for graphs with small chromatic number or with large independence number, and indeed there are algorithms along these lines for both problems developed in [42] for coloring and in [6] for the independence number. A better understanding of the possible gaps between the $\theta$-function and these two parameters may lead to improved algorithms.

The computational problem of determining the Shannon capacity of a given input graph seems very difficult. It would be interesting, however, to develop any (possibly exponential or even doubly exponential time) algorithm that provides a reasonable approximation for this quantity . The techniques in [8] or [3] can be extended to show that the independence number of any fixed AND-power of a large graph provides essentially no information on its Shannon capacity. We conclude the paper with the following plausible conjecture

**Conjecture 7.1.** *There exists a function $f(n) \leq O(2^n)$ such that for any graph $G$ on $n$ vertices, $\big(\alpha\big(G^{\wedge f(n)}\big)\big)^{1/f(n)}$ is always at least $0.99c(G)$.*

## REFERENCES


[1] N. Alon, Explicit Ramsey graphs and orthonormal labelings, *The Electronic Journal of Combinatorics,* **1** (1994), R12, 8pp.

[2] N. Alon, On the capacity of digraphs, *European J. Combinatorics,* **19** (1998), 1–5.

[3] N. Alon, The Shannon Capacity of a union, *Combinatorica,* **18** (1998), 301–310.

[4] N. Alon, L. Babai and H. Suzuki, Multilinear polynomials and Frankl–Ray–Chaudhuri–Wilson type intersection theorems, *J. Combinatorial Theory, Ser. A,* **58** (1991), 165–180.

[5] N. Alon, R. Boppana and J. H. Spencer, An asymptotic isoperimetric inequality, *Geometric and Functional Analysis,* **8** (1998), 411–436.

[6] N. Alon and N. Kahale, Approximating the independence number via the $\theta$-function, *Math. Programming,* **80** (1998), 253–264.

[7] N. Alon and V. D. Milman, $\lambda_1$, isoperimetric inequalities for graphs, and superconcentrators, *J. Combinatorial Theory Ser. B,* **38** (1985), 73–88.

[8] N. Alon and A. Orlitsky, Repeated communication and Ramsey graphs, *IEEE Transactions on Information Theory,* **41** (1995), 1276–1289.

[9] N. Alon and J. H. Spencer, **The Probabilistic Method,** Wiley (1992).

[10] N. Alon and M. Szegedy, Large sets of nearly orthogonal vectors, *Graphs and Combinatorics,* **15** (1999), 1–4.

[11] S. Arora, C. Lund, R. Motwani, M. Sudan, and M. Szegedy, Proof verification and intractability of approximation problems, in: *Proceedings of the 33rd IEEE Symposium on Foundations of Computer Science,* IEEE Computer Society Press (1992), 14–23.

[12] S. A. Burr and V. Rosta, On the Ramsey multiplicities of graph problems and recent results, *J. Graph Theory,* 4 (1980), 347–361.

[13] C. Berge and M. Simonovits, The coloring numbers of direct product of two hypergraphs, in: C. Berge and D. Ray-Chaudhuri, editors, *Hypergraph Seminar,* Lecture Notes on Mathematics, # 411. Springer Verlag (1974).

[14] P. Berman and G. Schnitger, On the complexity of approximating the independent set problem, *Information and Computation,* **96** (1992), 77–94.

[15] A. Blokhuis, On the Sperner capacity of the cyclic triangle, *J. Algebraic Combin.,* **2** (1993), 123–124.

[16] B. Bollobás, *Random Graphs,* Academic Press (1985).

[17] B. Bollobás and I. Leader, Compressions and isoperimetric inequalities, *J. Combinatorial Theory Ser. A,* **56** (1991), 47–62.

[18] R. Calderbank, P. Frankl, R. L. Graham, W. Li and L. Shepp, The Sperner capacity of the cyclic triangle for linear and nonlinear codes, *J. Algebraic Combin.,* **2** (1993), 31–48.

[19] F. Chung, Open problems of Paul Erdös in Graph Theory, *J. Graph Theory,* **25** (1997), 3–36.

[20] B. Codenotti, P. Pudlák and G. Resta, Some structural properties of low rank matrices related to computational complexity, *Theoret. Comput. Sci.*, **35** (2000), 89–107.

[21] P. Erdős, R.J. McEliece, and H. Taylor, Ramsey bounds for graph products, *Pacific Journal of Mathematics*, **37(1)** (1971), 45–46.

[22] G. Exoo, A lower bound for Schur numbers and multicolored Ramsey numbers, *Electronic J. Comb.*, **1** (1994), R8.

[23] P. Erdös, On the number of complete subgraphs contained in certain graphs, *Publ. Math. Inst. Hung. Acad. Sci. VII, Ser. A3* (1962), 459–464.

[24] U. Feige, Randomized graph products, chromatic numbers, and the Lovász $\theta$-function, in: *27th Annual ACM Symposium on Theory of Computing,* ACM Press (1995), 635–640.

[25] M. J. Ferguson and D. W. Bailey, Zero-error coding for correlated sources, manuscript (1975).

[26] U. Feige, S. Goldwasser, L. Lovász, S. Safra, and M. Szegedy, Approximating Clique is almost NP-complete, in: *Proceedings of the* 32*nd IEEE Symposium on Foundations of Computer Science,* IEEE Computer Society Press (1991), 2–12.

[27] E. Fachini and J. Körner, Chromatic number, capacity and perfectness of directed graphs, to appear.

[28] P. Frankl and R. Wilson, Intersection theorems with geometric consequences, *Combinatorica,* **1** (1981), 259–286.

[29] H. Fredricksen, Schur numbers and the Ramsey numbers $N(3,3,\ldots,3;2)$, *J. Combinatorial Theory, Ser. A,* **27** (1979), 376–377.

[30] Z. Füredi and R. Stanley, Sets of vectors with many nearly orthogonal pairs (Research Problem), *Graphs and Combinatorics,* **8** (1992), 391–394.

[31] L. Gargano, J. Körner, U. Vaccaro, Sperner theorems on directed graphs and qualitative independence, *J. Comb. Theory, Ser. A,* **61** (1992), 173–192.

[32] L. Gargano, J. Körner, U. Vaccaro, Capacities: from information theory to extremal set theory, *J. Comb. Theory, Ser. A,* **68** (1994), 296–316.

[33] R. E. Greenwood and A. M. Gleason, Combinatorial relations and chromatic graphs, *Canad. J. Math.,* **7** (1955), 9–20.

[34] M. Grötschel, L. Lovász, and A. Schrijver, The ellipsoid method and its consequences in combinatorial optimization, *Combinatorica,* **1** (1981), 169–197.

[35] R. L. Graham, B. L. Rothschild, and J. H. Spencer, *Ramsey Theory* (Second Edition), Wiley Interscience (1990).

[36] G. Hahn, P. Hell and S. Poljak, On the ultimate independence ratio of a graph, *Europ. J. Combinatorics,* **16** (1995), 253–261.

[37] C. Houdré and P. Tetali, Concentration of measure for products of Markov kernels and graph products via functional inequalities, *Combinatorics, Probability and Computing,* to appear.

[38] W. Haemers, On some problems of Lovász concerning the Shannon capacity of a graph, *IEEE Transactions on Information Theory,* **25(2)** (March 1979), 231–232.

[39] L. Harper, Optimal numbering and isoperimetric problems on graphs, *J. Combinatorial Theory,* **1** (1966), 385–394.

[40] C. N. Jagger, P. Šťovíček and A. Thomason, Multiplicities of subgraphs, *Combinatorica,* **16** (1996), 123–141.

[41] F. Juhasz, The asymptotic behaviour of Lovasz' $\theta$ function for random graphs, *Combinatorica,* **2** (1982), 153–155.

[42] D. Karger, R. Motwani, and M. Sudan, Approximate graph coloring by semi-definite programming, in: 35*th Symposium on Foundations of Computer Science,* IEEE Computer Society Press (1994), pp. 2–13.

[43] D. E. Knuth, The sandwich theorem, *The Electronic Journal of Combinatorics,* **A1** (1994), 48pp.

[44] B. S. Kashin and S. V. Konyagin, On systems of vectors in a Hilbert space, *Trudy Mat. Inst. imeni V. A. Steklova,* **157** (1981), 64–67, English translation in: *Proceedings of the Steklov Institute of Mathematics* (AMS 1983), 67–70.

[45] S. V. Konyagin, Systems of vectors in Euclidean space and an extremal problem for polynomials, *Mat. Zametki,* **29** (1981), 63–74. English translation in: *Mathematical Notes of the Academy of the USSR,* **29** (1981), 33–39.

[46] J. Körner, On clique growth in products of directed graphs, *Graphs and Combinatorics,* **14** (1998), 25–36.

[47] J. Körner, G. Simonyi, A Sperner-type theorem and qualitative independence, *J. Comb. Theory,* **59** (1992), 90–103.

[48] N. Linial and U. Vazirani, Graph products and chromatic numbers, in: *Proceedings of the 30th Annual Symposium on Foundations of Computer Science,* (1989), 124–128.

[49] L. Lovász, Kneser's conjecture, chromatic number and homotopy, *Journal of Combinatorial Theory,* **25** (1978), 319–324.

[50] L. Lovász, On the Shannon capacity of a graph, *IEEE Transactions on Information Theory,* **IT-25** (1979), 1–7.

[51] C. Lund and M. Yannakakis, On the hardness of approximating minimization problems, *Proc.* $25^{\text{th}}$ *ACM STOC* (1993), 286–293.

[52] R. J. McEliece and E. C. Posner, Hide and seek, data storage, and entropy, *The Annals of Mathematical Statistics,* **42(5)** (1971), 1706–1716.

[53] B. Maurey, Construction de suites symétriques, *Compt. Rend. Acad. Sci. Paris,* **288** (1979), 679–681 (in French).

[54] C. J. H. McDiarmid, On the method of bounded differences, in: *Surveys in Combinatorics 1989,* London Math. Society Lecture Notes Series 141 (J. Siemons, ed.), Cambridge Univ. Press (1989), 148–188.

[55] V. D. Milman and G. Schechtman, *Asymptotic Theory of Finite Dimensional Normed Spaces,* Lecture Notes in Mathematics 1200, Springer-Verlag (Berlin and New York, 1986).

[56] A. Orlitsky, Worst-case interactive communication I: Two messages are almost optimal, *IEEE Transactions on Information Theory,* **36** (1990), 1111–1126.

[57] C. E. Shannon, The zero-error capacity of a noisy channel, *IRE Transactions on Information Theory,* **2(3)** (1956), 8–19.

[58] A. Sali and G. Simonyi, Orientations of self-complementary graphs and the relation of Shannon and Sperner capacities, *European J. Combin.,* **20** (1999), 93-99.

[59] M. Talagrand, Concentration of measure and isoperimetric inequalities in product spaces, *Inst. Hautes Études Sci. Publ. Math.,* **81** (1995), 73–205.

[60] A. Thomason, Graph products and monochromatic multiplicities, *Combinatorica,* **17** (1997), 125–134.

[61] H. Witsenhausen, The zero-error side information problem and chromatic numbers, *IEEE Transactions on Information Theory,* **22(5)** (September 1976), 592–593.

Noga Alon

*Department of Mathematics*
*Raymond and Beverly Sackler Faculty*
*of Exact Sciences*
*Tel Aviv University*
*Tel Aviv*
*Israel*

e-mail: noga@math.tau.ac.il

BOLYAI SOCIETY
MATHEMATICAL STUDIES, 10

**Contemporary Combinatorics**
**Budapest, 2002,** pp. 29–91.

# On the Amount of Dependence in the Prime Factorization of a Uniform Random Integer

R. ARRATIA*

How much dependence is there in the prime factorization of a random integer distributed uniformly from 1 to $n$? How much dependence is there in the decomposition into cycles of a random permutation of $n$ points? What is the relation between the Poisson–Dirichlet process and the scale invariant Poisson process? These three questions have essentially the same answers, with respect to total variation distance, considering only small components, and with respect to a Wasserstein distance, considering all components. The Wasserstein distance is the expected number of changes — insertions and deletions — needed to change the dependent system into an independent system.

In particular we show that for primes, roughly speaking, $2 + o(1)$ changes are necessary and sufficient to convert a uniformly distributed random integer from 1 to $n$ into a random integer $\prod_{p \le n} p^{Z_p}$ in which the multiplicity $Z_p$ of the factor $p$ is geometrically distributed, with all $Z_p$ independent. The changes are, with probability tending to 1, one deletion, together with a random number of insertions, having expectation $1 + o(1)$.

The crucial tool for showing that $2 + \varepsilon$ suffices is a coupling of the infinite independent model of prime multiplicities, with the scale invariant Poisson process on $(0, \infty)$. A corollary of this construction is the first metric bound on the distance to the Poisson–Dirichlet in Billingsley's 1972 weak convergence result. Our bound takes the form: there are couplings in which

$$\mathbb{E} \sum \left| \log P_i(n) - (\log n) V_i \right| = O(\log \log n),$$

---

* Lectures given June 26,27,29,30, 1998, at the workshop on Probabilistic Combinatorics at the Paul Erdős Summer Research Center of Mathematics.

where $P_i$ denotes the $i^{\text{th}}$ largest prime factor and $V_i$ denotes the $i^{\text{th}}$ component of the Poisson–Dirichlet process. It is reasonable to conjecture that $O(1)$ is achievable.

## Contents

For the reader impatient to get to business, the main results are given by the bound (72) in Theorem 3, and the bound (81) in Theorem 5. A \$500 conjecture, related to Theorem 3, is given by relation (27), and a \$100 conjecture, related to Theorem 5, is given by the bound (80).

## Lecture 1. Growing a random integer

I would like to thank the János Bolyai Mathematical Society and the organizers of this conference for the honor of speaking here, where the spirit of Erdős seems so close. At the conference reception last night Imre Csiszar, student of Rényi, suggested that Erdős, now in heaven, has read the book where all best proofs are given. But I prefer to believe that in heaven, Erdős by choice does not ask to see the book; he only asks of a proof, "Is there an even better one in the book?" And he watches us fellow mathematicians here on earth, as we lecture and write and discover; by not diving into the book, he can continue to compete against us. For the joy of discovery through one's own effort is far more rewarding than even reading the book. I would also like to thank my collaborators, Andrew D. Barbour and Simon Tavaré, as their work and ideas pervade these lectures. I hope that our book on logarithmic combinatorial structures [5], in preparation since 1992, will soon see publication!

The guiding question for today's lecture is to ask, by analogy with the Erdős-Rényi notion of *growing* a random graph, "can we grow a random integer?" For guidance, we compare with the simpler task of "growing" a random permutation.

NOTATION: $n$ is always the parameter, rather than the random object. We consider

— a permutation chosen uniformly from $\mathcal{S}_n$

— an integer $N$ chosen uniformly from 1 to $n$.

We may emphasize the role of the parameter $n$ by writing it explicitly:

$$\mathbb{P}_n(N = i) \equiv \mathbb{P}\big(N(n) = i\big) = 1/n \quad \text{for} \quad i = 1, 2, \ldots, n.$$

For the prime factorization we write

$$N(n) = \prod p^{C_p(n)}, \quad \text{so} \quad \big(C_p(n)\big)_p = \big(C_2(n), C_3(n), C_5(n), \ldots\big)$$

is a dependent process, with $C_p(n)$ identically zero if $p > n$.

The baby fact: as $n \to \infty$

$$\text{(1)} \qquad \big(C_p(n)\big)_p \Rightarrow (Z_p)_p = (Z_2, Z_3, Z_5, \ldots)$$

with independent, geometrically distributed coordinates, with $\mathbb{P}(Z_p \geq k) = p^{-k}$.

**Proof.** Given $j$ distinct primes $p_1, \ldots, p_j$, and integers $i_1, \ldots, i_j \geq 0$, write $d = p_1^{i_i} \ldots p_j^{i_j}$ so that as events,

$$\{C_{p_1} \geq i_1, \ldots, C_{p_j} \geq i_j\} = \{d \,|\, N\}.$$

Thus

$$\text{(2)} \qquad \mathbb{P}_n(d \,|\, N) = \frac{1}{n}\left\lfloor \frac{n}{d} \right\rfloor \to \frac{1}{d} = \mathbb{P}(Z_{p_1} \geq i_1, \ldots, Z_{p_j} \geq i_j),$$

and $j$-dimensional differencing yields

$$\text{(3)} \quad \mathbb{P}_n(C_{p_1} = i_1, \ldots, C_{p_j} = i_j) \to \mathbb{P}(Z_{p_1} = i_1, \ldots, Z_{p_j} = i_j). \qquad \blacksquare$$

In fact, the approximation error in (2) is at most $1/n$, so the error in (3) is at most $2^j/n$ — a crude upper bound which comes from taking the absolute value inside. The sum over all integers $d = \prod_{p \le b} p^{a_p}$ whose largest prime factor is at most $b$, i.e.

$$(4) \qquad \sum_{d:\, P^+(d) \le b} \left| \mathbb{P}\big( C_p(n) = a_p \ \forall p \le b \big) - \mathbb{P}\big( Z_p = a_p \ \forall p \le b \big) \right|,$$

gives the total variation distance $d_{TV}(b, n)$, restricting to primes not exceeding $b$, and surprisingly, the crude upper bound $u(b, n)$, formed by taking absolute values inside, gives pretty good information. But the information is not nearly as good as the result of Kubilius; we will describe these bounds later, in Sections 1.5.2 and 1.5.3.

Since $\mathbb{E} Z_p = \frac{1}{p-1}$ with $\sum \mathbb{E} Z_p = \infty$, it follows from the independence of the $Z_p$ that

$$1 = \mathbb{P}\Big( \sum Z_p = \infty \Big).$$

Thus we call the multiset, having $Z_p$ copies of $p$ for each prime, the *natural random infinite multiset of primes.* The guiding question for today's lecture, can we grow random integers, is now stated more precisely as: can we construct $N(1), N(2), \ldots, N(n), \ldots$ all on a single probability space, together with the natural random infinite multiset of primes, so that the $C_p(n)$ *evolve smoothly,* with $C_p(n) \to Z_p$ as $n \to \infty$. The answer is yes in a sense; the construction culminating with (75) at the end of Lecture 3 has $C_p(n) \to Z_p$ *in probability,* but with probability one, $\liminf C_p(n) = Z_p$ and $\limsup C_p(n) = Z_p + 1$.

## 1.1. Overview: the limits for primes, small and large

This section presents the material from one of the two transparencies which were shown repeatedly throughout the lectures; the material from the other transparency appears in this writeup as Conjecture 1 in Section 2.2 and as Conjecture 2 in Section 4.

| | |
|---|---|
| $N(n)$ is uniform 1 to $n$, with | to focus on— |
| $N(n) = \prod p^{C_p(n)}$ | —*small factors,* |
| $N(n) = P_1(n) P_2(n) \ldots,\ P_1 \ge P_2 \ge \ldots$, prime or 1, | —*large factors* |

**LIMITS** in distribution as $n \to \infty$

$\big(C_p(n)\big)_p \Rightarrow (Z_p)_p$, independent Geometric

$$\left(\frac{\log P_1(n)}{\log n}, \frac{\log P_2(n)}{\log n}, \ldots\right) \Rightarrow (V_1, V_2, \ldots), \quad \text{Poisson–Dirichlet}$$

proved by Billingsley in 1972, where the limit is now known as the Poisson–Dirichlet process, with parameter 1.

**HOW CLOSE?** One coupling has

$$\mathbb{E} \sum_{p \le n} \big| C_p(n) - Z_p \big| \le 2 + O\left(\frac{(\log\log n)^2}{\log n}\right), \tag{5}$$

$$\mathbb{E} \sum \big| \log P_i(n) - (V_i)(\log n) \big| = O(\log\log n). \tag{6}$$

We note that in (5), $2 - \varepsilon$ is not possible, while in (6), $O(1)$ should be possible. The reason that loglog $n$ appears in our bound for (6) is made clear by equation (63) in conjunction with the proof of Theorem 5.

### 1.2. Sketch of the coupling to grow $N(n)$

Take a size biased permutation, say $Q_1, Q_2, \ldots$, of the prime factors in $2^{Z_2} 3^{Z_3} \ldots$.

Let $J(n) =$ the largest partial product $Q_1 Q_2 \ldots Q_j \le n$. Write $L \equiv L(n)$ for the number of factors, so that $J(n) = Q_1 Q_2 \ldots Q_{L(n)}$.

Show: $J(n)$ has approximately the same distribution as $H(n)$, where by definition $H(n)$ has the harmonic distribution on $[n]$,

$$\mathbb{P}\big(H(n) = i\big) = \frac{1/i}{1 + 1/2 + \ldots + 1/n} = \frac{1}{i\, h_n}, \quad i = 1, 2, \ldots, n. \tag{7}$$

Fill in one extra factor, $P_0(n)$, to be prime or one. Use a random uniform $U \in (0, 1]$ to choose uniformly from the $1 + \pi\big(n/J(n)\big)$ possibilities with $J(n) P_0(n) \le n$.

Show that the resulting random integer, $JP_0 \equiv J(n)P_0(n)$ is close to uniform. In fact, from Lemma 3

$$d_{TV}\big(J(n), H(n)\big) = O\left(\frac{1}{\log n}\right), \tag{8}$$

and from Lemma 4

$$d_{TV}\big(J(n)P_0(n), N(n)\big) = O\left(\frac{\log\log n}{\log n}\right). \tag{9}$$

Finally, modify the coupling on the event whose probability is the left side of (9), so that the modified versions of $JP_0$ are exactly uniform, for all $n$.

Noga Alon asked, "Does $\mathbb{P}\big(J(n)P_0(n) = 1\big) \sim 1/n$? The total variation distance bound (9) does not determine the answer. Since $\mathbb{P}\big(P_0(n) = 1 \,|\, J(n) = 1\big) = 1/\big(1 + \pi(n)\big) \sim \log n/n$, Alon's question is equivalent to, "Does $\mathbb{P}\big(J(n) = 1\big) \sim 1/\log n$? Now if $Z_p = 0$ for all $p \le n$ then we must have $J(n) = 1$, and $\mathbb{P}(Z_p = 0 \;\forall p \le n) = \prod_{p \le n}(1 - 1/p) \sim e^{-\gamma}/\log n$ by Mertens' Theorem. Thus Alon's question is equivalent to asking: does $(1 - e^{-\gamma})/\log n$ give the asymptotic probability that *there are* one or more primes less than or equal to $n$ in the infinite multiset *and, in the size biased permutation*, some prime greater than $n$ comes before all of them. The answer, which I did not give at the workshop, but was more or less evident from (48) and (50), is yes; see Lemma 3 for further details. While this affirmative answer might give us hope that $J(n)P_0(n)$ is close to uniform in the sense that for all $i \le n$, $\mathbb{P}\big(J(n)P_0(n) = i\big) \sim 1/n$, it is clearly not so. In fact, (for the coupling in Lecture 3, which is slightly different from the coupling described in this lecture,) for fixed $i$, having $\omega(i)$ distinct prime factors, as $n \to \infty$, $\mathbb{P}\big(J(n)P_0(n) = i\big) \sim \big(1 + \omega(i)\big)/n$; see (59) and (60) for details. Thus $i = 1$ is the only fixed integer having the correct asymptotic probability!

How does $J(n)P_0(n)$ evolve as $n$ grows? Write $p^{\#}$ for the smallest prime larger than $p$, with $1^{\#} = 2$. The following properties hold for all $n > 1$ and for all outcomes.

$$J(n)P_0(n) \in [1, n].$$

Each factor has its own smoothness:

$$J(n)/J(n-1) \quad \text{is one or a prime,}$$

$$P_0(n) = P_0(n-1), P_0(n-1)^{\#}, \quad \text{or} \quad 1,$$

and the jumps in the two factors are linked:

$$J(n) \neq J(n-1) \quad \text{implies} \quad \big(J(n) = n \text{ and } P_0(n) = 1\big),$$

$$P_0(n) < P_0(n-1) \quad \text{implies} \quad \big(J(n) = n \text{ and } P_0(n) = 1\big).$$

### 1.3. Size biased permutations

Several in the audience requested clarification: what is a size biased permutation? Given $k$ objects, with "sizes" or "weights" $r_1, r_2, \ldots, r_k > 0$, we can carry out a size biased permutation using $k$ independent standard exponentially distributed "alarm clocks" $S_1, S_2, \ldots, S_k$ with $\mathbb{P}(S_i > t) = e^{-t}$ for all $t > 0$. The labels $W_i := S_i/r_i$ are exponentially distributed with $\mathbb{P}(W_i > t) = e^{-r_i t}$, (we say "rate $r_i$" or "mean $1/r_i$,") and the labels are independent, with $\mathbb{P}(W_i = W_j) = 0$ for every $i \neq j$. The ranking of the labels induces a size biased permutation of the objects. Observe that $\mathbb{P}(W_i$ is the smallest among $W_1, W_2, \ldots, W_k) = r_i/(r_1 + r_2 + \ldots + r_k)$, which is a familiar fact from the study of finite-state Markov chains in continuous time, where the $r_i$ are jump rates. This calculation shows the distribution of the first item; iterating and using the memoryless property of exponentials gives a product formula for the distribution of the full size biased permutation.

For primes, the size of $p$ is $\log p$, and such a size biased permutation was used in the 1984 Ph.D. thesis of Eric Bach [13] to generate uniformly distributed random integers, factored into primes, and independently by Donnelly and Grimmett in 1993 [22] to give a simple proof of Billingsley's Poisson–Dirichlet limit for prime factors — discussed at (77) below. In our context, the infinite random multiset of primes, there are infinitely many labels, but with probability one, the only limit point is zero, and there is a largest label.. Our size biased permutation starts with the prime having this largest label; listing the labels from largest downward tends to put smaller primes toward the front of the list.

## 1.4. An example of the growth of an integer

Take for example an outcome of the experiment with

$$Z_2 = 3, Z_3 = 1, Z_5 = 0, Z_7 = 1, Z_{11} = 1, \ldots \tag{10}$$

and size biased permutation

$$Q_1, Q_2, \ldots, Q_6, \ldots = 3, 2, 2, 11, 2, 7, \ldots. \tag{11}$$

The first seven partial products, one through $3 \cdot 2 \cdot 2 \cdot 11 \cdot 2 \cdot 7$, shown on a logarithmic scale, are

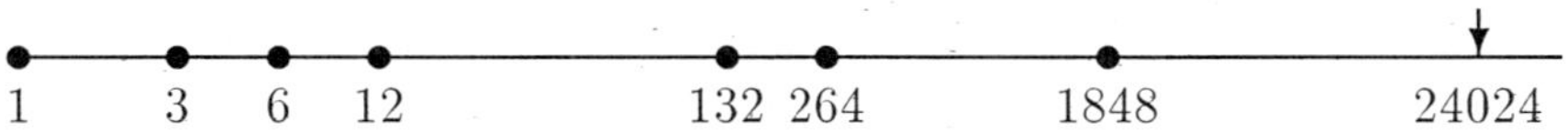


The arrow pointing to $24024 = 1848\ 13$ is to show that the next partial product will lie on or to the right of this location. We know this because the information $6 = \sum_{p<13} Z_p$, together with the good luck that the first six primes in the size biased permutation are less than 13, implies that the seventh prime in the size biased permutation will be 13 or greater. In particular, for $1848 \leq n < 13 \times 1848$, we know that $J(n) = 1848$, even though we don't know the seventh prime in the size biased permutation, accounting for the last line of the table below.

The jumps in $J(n)$ are shown by double horizontal lines in the two tables below.

To continue the above example, we will consider three cases: the first being $U \in (.5, .6]$, the next being $U \in (.6, 2/3]$, and the last being $U > 5/6$.

Recall that $d_{TV}\big(J(n)P_0(n), N(n)\big) \to 0$, but the random integer we grow is *not exactly uniform.* For comparison, random permutations have similar behavior and can be grown exactly. Eric Bach's procedure gets an *exactly uniform* random integer, but not with $n$ *evolving.*

| $n$ | $J(n)$ | $1+\pi(n/J(n))$ | $P_0(n)$ |
|---|---|---|---|
| 1 | 1 | 1 | 1 |
| 2 | 1 | 2 | 1 if $U \le .5$, 2 if $U > .5$ |
| 3,4,5 | 3 | 1 | 1 |
| 6 to 11 | 6 | 1 | 1 |
| 12 to 23 | 12 | 1 | 1 |
| 24 to 35 | 12 | 2 | 1 if $U \le .5, 2$ if $U > .5$ |
| 36 to 59 | 12 | 3 | 1 if $U \le 1/3, \ldots, 3$ if $U > 2/3$ |
| 60 to 83 | 12 | 4 | 1 if $U \le 1/4, \ldots, 5$ if $U > 3/4$ |
| 84 to 131 | 12 | 5 | 1, 2, 3, 5, or 7 |
| 132 to 263 | 132 | 1 | 1 |
| $[264, 2 \cdot 264)$ | 264 | 1 | 1 |
| $[2 \cdot 264, 3 \cdot 264)$ | 264 | 2 | 1 if $U \le .5, 2$ if $U > .5$ |
| $[3 \cdot 264, 5 \cdot 264)$ | 264 | 3 | 1 if $U \le 1/3, \ldots, 3$ if $U > 2/3$ |
| $[1320, 1848)$ | 264 | 4 | 1, 2, 3, or 5 |
| $[1848, 13 \cdot 1848)$ | 1848 | 1, 2, 3, 4, 5, or 6 | 1, 2, 3, 5, 7, or 11 |

## 1.5. Notions of distance

### 1.5.1. The expected number of insertions and deletions needed.

The material below on $d_W$ was delivered at the workshop at the start of Lecture 2; the material on $d_{TV}$ was drawn out in workshop conversations with Joel Spencer, and I prepared a transparency for the lecture but did not deliver it, for lack of time.

We consider the metric $d$ on positive integers which counts the number of changes needed to convert the prime factorization of one integer into that of the other. For example, $d(40, 500) = d(2^3 5^1, 2^2 5^3) = 3$, $d(8,3) = 4$, and $d(i, ip) = 1$ for any integer $i$ and prime $p$. Writing $(i,j)$ for the greatest common divisor, and $\Omega(i)$ for the number of prime factors, including multiplicities, we have in general

$$d(i,j) = \Omega\left(\frac{i}{(i,j)}\right) + \Omega\left(\frac{j}{(i,j)}\right).$$

| | $.5 < U \le .6$ | | $.6 < U \le 2/3$ | | $5/6 < U$ | |
|---|---|---|---|---|---|---|
| $n$ | $P_0(n)$ | $J(n)P_0(n)$ | $P_0(n)$ | $J(n)P_0(n)$ | $P_0(n)$ | $J(n)P_0(n)$ |
| 1 | 1 | 1 | 1 | 1 | 1 | 1 |
| 2 | 2 | 2 | 2 | 2 | 2 | 2 |
| $3, 4, 5$ | 1 | 3 | 1 | 3 | 1 | 3 |
| $[6, 11)$ | 1 | 6 | 1 | 6 | 1 | 6 |
| $[12, 23)$ | 1 | 12 | 1 | 12 | 1 | 12 |
| $[24, 35)$ | 2 | 24 | 2 | 24 | 2 | 24 |
| $[36, 59)$ | 2 | 24 | 2 | 24 | 3 | 36 |
| $[60, 83)$ | 3 | 36 | 3 | 36 | 5 | 60 |
| $[84, 131)$ | 3 | 36 | 5 | 60 | 7 | 84 |
| $[132, 263)$ | 1 | 132 | 1 | 132 | 1 | 132 |
| $[264, 528)$ | 1 | 264 | 1 | 264 | 1 | 264 |
| $[528, 792)$ | 2 | 528 | 2 | 528 | 2 | 528 |
| $[792, 1320)$ | 2 | 528 | 2 | 528 | 3 | 792 |
| $[1320, 1848)$ | 3 | 792 | 3 | 792 | 5 | 1320 |

If we think of converting $j$ to $i$, the first term above is the number of *insertions* needed, and the second term is the number of *deletions*, so we think of $d$ as the insertion/deletion distance, analogous to the string edit distance of Levenstein [36] or Ulam [16]; see the book of Kruskal and Sankoff [33] for more history.

For the sake of comparing the uniform random $N(n)$ with the infinite random multiset of primes, clearly primes $p > n$ should not be considered. Thus, we code up the relevant part of the multiset by *defining*

$$M(n) := \prod_{p \le n} p^{Z_p}, \quad \text{with} \quad Z_2, Z_3, \ldots \quad \text{independent geometric.} \tag{12}$$

Recall that

$$N(n) = \prod_p p^{C_p(n)} = \prod_{p \le n} p^{C_p(n)} \quad \text{is uniform 1 to } n.$$

The insertion-deletion distance between these two random integers is a random, nonnegative integer

$$d(N,M) = \Omega\left(\frac{NM}{(N,M)^2}\right) = \sum_{p\le n} \left| C_p(n) - Z_p \right| .$$

The Wasserstein distance between two random objects $M$ and $N$, for a given metric $d$, is by definition the infimum, over all couplings, of the expected value of $d(N,M)$. Recall that a coupling means a construction of $M$ and $N$ simultaneously on a single probability space; it is understood that the *marginal distributions* for $M$ and for $N$ have been specified in advance, but there is no other constraint on their *joint distribution*. A compactness argument shows that the inf is achieved; see for example [23]. To emphasize the role of the parameter $n$, which determines the marginal distributions of $N(n)$ and $M(n)$, we define

$$d_W(n) := \min_{\text{couplings}} \mathbb{E} d\big(N(n), M(n)\big) . \tag{13}$$

Our result is

$$\lim_{n\to\infty} d_W(n) = 2. \tag{14}$$

We will prove the hard part of this, that $\limsup d_W(n) \le 2$, in Lecture 3, essentially by analyzing the growth of a random integer. The matching lower bound, from [2], is that $\liminf d_W(n) \ge 2$; this is relatively easy, and the reader is challenged to discover a proof, before considering the hint given by the paragraph following (72). A related discussion appears in Section 22 of [3].

For perspective on the content of (14), we note that even the bound $d_W(n) = O(1)$ is very strong. For instance, by comparing $\big(C_p(n)\big)_{p\le n}$ with the independent process $(Z_p)_{p\le n}$, the following consequences can be derived easily: (see [2])

$d_W(n) = o(\log\log n)$ implies the Hardy-Ramanujan Theorem for the normal order of the number of prime divisors.

$d_W(n) = o\big(\sqrt{\log\log n}\,\big)$ implies the Erdős-Kac Central Limit Theorem.

$d_W(n) = O(1)$ gives another proof of the "conjecture of LeVeque," that the error in the Central Limit Theorem is $O\big(1/\sqrt{\log\log n}\,\big)$; the first proof was given by Rényi and Turán in 1957 [40].

$d_W(n) = o(\text{logloglog}\, n)$ yields that the optimal rate in the simplest case of the Brownian motion convergence of Billingsley and Philipp, [18, 19, 38], for the expected sup norm, is order of logloglog $n/\sqrt{\log\log n}$, and no smaller. The underlying idea is that for coupling Brownian motion with the rate one (centered) Poisson process, with both processes run until the variance is $t$, and *without* rescaling by $\sqrt{t}$, the coupling distance grows on the order of $\log t$. For primes this is applied with $t := \sum_{p\le n} 1/p \sim \log\log n$. See Kurtz (1978) [35] for the upper bound, Rio (1994) [41] for the lower bound, and [2] for the connection with primes.

**1.5.2. The total variation distance.** This section gives a "one transparency" overview of the situation involving total variation distance for primes.

NOTATION: $\beta \in [0,1]$, fixed or $\beta = \beta(n) \to 0$; $u \equiv 1/\beta$.

$$d_{TV} := d_{TV}(n^\beta, n) := \min_{\text{couplings}} \mathbb{P}\Big(\big(C_p(n)\big)_{p\le n^\beta} \ne (Z_p)_{p\le n^\beta}\Big)\,. \tag{15}$$

Kubilius [34] in the 1950's showed a) below with an upper bound of the form $\exp(-cu)$, Barban and Vinogradov [14] improved this to the form $\exp(-cu\log u)$, and Elliott [24] gave the particular constants in b) below.

a) If $\beta \to 0$ then $d_{TV} \to 0$, and

b) $d_{TV} = O\big(\exp\big(-\frac{1}{8}u\log u\big) + n^{-1/15}\big)$.

Elliott [24] gives a partial converse: if $\beta \to 1$ then $d_{TV} \not\to 0$, and it is not hard to see the full converse, that $d_{TV} \to 0$ implies $\beta \to 0$. In Spring 1996 I stated a conjecture: that for fixed $\beta$,

c) $H(\beta) := \lim_{n\to\infty} d_{TV}$ exists, and is the same as the limit for permutations. This limit is given explicitly as an integral in [11], and proved to be the limit for permutations in Stark's 1994 PhD Thesis; see [43].

Later in 1996 Tenenbaum [45] improved b) to

$$d_{TV} = O\Big(\exp\big(-u\big(\log u + \log\log u - (1+\log 2+\varepsilon)\big)\big) + n^{\varepsilon-1}\Big)$$

and Arratia and Stark [10] proved c). Soon after Tenenbaum [45] also proved c), with a rate. The limit was further identified [8] as a distance between the restrictions to $[0,\beta]$ of two processes which will be discussed in lectures 3 and 4, the Poisson–Dirichlet process with parameter 1, and the scale invariant Poisson process with intensity $(1/x)\,dx$: for every $\beta \in [0,1]$,

$$H(\beta) = d_{TV}\big(\{V_i : V_i < \beta\}, \{X_i : X_i < \beta\}\big) \tag{16}$$

$$= \min_{\text{couplings}} \mathbb{P}\big(\{V_i : V_i < \beta\} \neq \{X_i : X_i < \beta\}\big)\,.$$

**1.5.3. The bound from taking absolute values inside.** The crude procedure of "taking the absolute values inside" described following (4) shows that the total variation distance $d_{TV}(n^\beta, n)$ is at most $u(n^\beta, n)$ where

$$u(b,n) = \frac{1}{n} \sum_{d \geq 1 \,:\, P^+(d) \leq b} \left\{\frac{n}{d}\right\} 2^{\omega(d)}. \tag{17}$$

The notation here is $\{x\}$ for the fractional part of $x$, $\omega(d)$ for the number of distinct prime factors of $d$, and $P^+(d)$ for the largest prime factor of $d$, with $P^+(1) = 1$. Analysis of $u(b,n)$ (see [2]) shows that when $b, n \to \infty$ together, the threshold for whether $u(b,n)$ tends to zero or infinity is $\log b = (.5 \pm \varepsilon) \log n \operatorname{logloglog} n / \operatorname{loglog} n$. In fact if $\log b \leq \log n \operatorname{logloglog} n/(c \operatorname{loglog} n)$ with $c > 2{+}a > 2$, then $u(b,n) = o\big((\log n)^{-a}\big)$, while if $\log b > \log n \operatorname{logloglog} n/(c \operatorname{loglog} n)$, for $c < 2$, then $u(b,n) \to \infty$. While much weaker than the upper bound of Kubilius, the upper bound $u(b,n)$ on the total variation distance has the virtue that it also serves as an upper bound on the Wasserstein distance: for all $1 \leq b \leq n$,

$$d_{TV}(b,n) \leq d_W(b,n) \leq u(b,n), \tag{18}$$

where $d_W(b,n)$ is the minimum expected number of insertions and deletions needed to convert $\prod_{p \leq b} p^{C_p(n)}$ to $\prod_{p \leq b} p^{Z_p}$. [The $d_W(n)$ in (13) is the special case $b = n$ of this, i.e. $d_W(n) = d_W(n,n)$.] While we know the limit of (16) for $d_{TV}(n^\beta, n)$ as $n \to \infty$ with $\beta \in (0,1)$ fixed, the corresponding behavior of $d_W(b,n)$ remains unknown.

**Open question.** What is the limit of $d_W(n^\beta, n)$ as $n \to \infty$ for fixed $0 < \beta < 1$?

We know that for $\beta = 1$, the limit is 2 — this is (14). For all $\beta \in [0, 1]$, it can be seen from the coupling in Lecture 4 that $\limsup d_W(n^\beta, n) \le 2\beta$ — but only for $\beta = 1$ is there a matching lower bound. The natural candidate for the limit of the distance is the Wasserstein distance for the limit systems,

$$H_W(\beta) := d_W\big(\{V_i : V_i < \beta\}, \{X_i : X_i < \beta\}\big) \tag{19}$$

$$\equiv \min_{\text{couplings}} \mathbb{E}\big|\,\{V_i : V_i < \beta\}\Delta\{X_i : X_i < \beta\}\big|.$$

Thus, the open question involves two tasks. First, prove (or disprove!) that the limit exists and equals $H_W(\beta)$.

**Conjecture 0.** $\forall \beta \in [0, 1]$, $H_W(\beta) = \lim d_W(n^\beta, n)$.

The second task in our open question is to find an explicit formula for $H_W(\beta)$. We know only $H_W(1) = 2$, $H_W(\beta) \le 2\beta$, and trivially, $H_W(0) = 0$ and $H_W$ is monotone.

## Lecture 2. Growing a random permutation

This is a fresh start — the following can be understood easily from scratch, without worrying about the connection with prime factorizations. However, our example for permutations below has been carefully cooked to match the example for primes in Lecture 1.

Write $C_i(n)$ for the number of cycles of length $i$ in our random permutation $\pi \in \mathcal{S}_n$, so that always $C_i(n) = 0$ if $i > n$, and $n = \sum iC_i(n)$. The analog of Theorem 1, provable easily with inclusion-exclusion, is that as $n \to \infty$

$$\big(C_i(n)\big)_i \Rightarrow (Z_i)_i = (Z_1, Z_2, Z_3, Z_4, \ldots) \tag{20}$$

with independent coordinates $Z_i$, Poisson distributed with $\mathbb{E}Z_i = 1/i$. (Note, we are deliberately re-using the same notation that we used for prime factorizations, although the index must be a prime in the latter case. To contrast the two situations, $Z_p$ is either geometrically distributed, with $\mathbb{P}(Z_p \geq k) = p^{-k}$ and $\mathbb{E} Z_p = 1/(p-1)$, or else $Z_p$ is Poisson, with $\mathbb{E} Z_p = 1/p$.) That there is an independent process limit, without rescaling, for both prime factorizations of random integers, and the cycle structure of random permutations, is the most basic ingredient in the analogy between these two structures.

Historical notes: the similarity between primes and permutations appears to have been noted first in 1976 by Knuth and Trabb Pardo [32]. The similarity they note is based on the common limit behavior for the $i^{\text{th}}$ largest prime factor and the $i^{\text{th}}$ longest cycle; further aspects of the similarity, are discussed in [6], which sets forth the role of "conditioning" the independent limit process as the fundamental reason for this similarity. In contrast, in these lectures we will focus on the closeness of primes and permutations to the scale invariant Poisson process, with its relations involving spacings and size biased permutations, as the underlying reason for the similarity. A reasonable attribution for the construction below, based on canonical cycle notation, is Feller 1945 [26], but the explicit connection with cycle *lengths* seems to appear first in the unpublished lecture notes [20]; it was independently discovered in [15] and elaborated on in [4], which attached the name "Feller coupling" to this construction.

Start with the canonical cycle notation for a permutation $\pi$. For example, the permutation $\pi$ with $1 \mapsto 5$, $2 \mapsto 2$, $3 \mapsto 1$, $4 \mapsto 4$, $5 \mapsto 3$, $6 \mapsto 7$, $7 \mapsto 6$ is written as $\pi = (153)(2)(4)(67)$. In writing the canonical cycle notation for a random $\pi \in \mathcal{S}_7$, one always starts with "(1 ", and then makes a seven-way choice, between "(1)(2 ", "(1 2 ", ..., and "(1 7 ". One continues with a six-way choice, a five-way choice, ..., a two-way choice, and finally a one-way choice.

Let $\xi_i$ be defined as the indicator function $\xi_i = 1$(close off a cycle when there is an $i$-way choice). Thus

$$\mathbb{P}(\xi_i = 1) = \frac{1}{i}, \quad \mathbb{P}(\xi_i = 0) = \frac{i-1}{i}, \quad \text{and} \quad \xi_i, \xi_2, \ldots, \xi_n \quad \text{are independent.}$$

An easy way to see the independence of the $\xi_i$ is to take $D_i$ chosen from 1 to $i$ to make the $i$-way choice, so that $\xi_i = 1(D_i = 1)$. Absolutely no computation is needed to verify that the map constructing canonical cycle notation, $(D_1, D_2, \ldots, D_n) \mapsto \pi$, from $[1] \times [2] \times \ldots \times [n]$ to $\mathcal{S}_n$, is a bijection. The "decision" variables $D_1, D_2, \ldots, D_n$ determine the random permutation on $n$ points, while the Bernoulli variables $\xi_1, \xi_2, \ldots, \xi_n$ determine the cycle structure, and something more — a size biased permutation SB $(n)$ of the cycle lengths! The total number of cycles is

$$K_n := \#\text{ cycles} = \xi_1 + \xi_2 + \ldots + \xi_n,$$

$$\text{with} \quad \mathbb{E} K_n = 1 + \frac{1}{2} + \ldots + \frac{1}{n} = \gamma + \log n + o(1).$$

[Incidentally, the relations above give a quick and dirty, but pretty, way to see that the entropy $h\big((C_i(n))_i\big)$ for the cycle structure of a random permutation of $n$ objects is asymptotically $(\log n)^2/2$. Namely, the entropy of a size biased permutation of $k$ objects is at most the entropy of a *uniform* permutation of $k$ objects, which is $\log k! \leq k \log k$. Thus the entropy $h\big(\text{SB}(n)\big)$ of a size biased permutation of the $K_n$ cycle lengths of a random $n$-permutation is at most $\mathbb{E} K_n \log K_n \sim \log n \operatorname{loglog} n$. The entropy of the Bernoulli $(1/i)$ random variable $\xi_i$ is $h(\xi_i) = -\left(\frac{1}{i} \log \frac{1}{i} + \frac{i-1}{i} \log \frac{i-1}{i}\right) = \log i - \log(i-1) + \frac{1}{i} \log(i-1)$ for $i \geq 2$, so

$$\sum_1^n h(\xi_i) = \sum_2^n \left(\log \frac{i}{i-1} + \frac{1}{i} \log(i-1)\right)$$

$$= \log(n+1) + \sum_1^{n-1} \frac{\log i}{i+1} \sim \frac{(\log n)^2}{2}.$$

Since the cycle structure together with the size biased permutation of the cycle lengths determine $\xi_1, \ldots, \xi_n$, we have $(\log n)^2/2 \sim \sum_1^n h(\xi_i) = h\big((C_i(n))_i\big) + h\big(\text{SB}(n)\big) = h\big((C_i(n))_i\big) + O(\log n \operatorname{loglog} n)$, hence $h\big((C_i(n))_i\big) \sim (\log n)^2/2$. As an exercise, the reader can try to verify this directly from Cauchy's formula, and perhaps (open problem?) give an asymptotic expansion!]

The coupling which "grows" a random permutation requires a very simple idea: *for all* values of $n$, use the same $D_1, D_2, \ldots$, and hence the same $\xi_1, \xi_2, \ldots$.

The Feller coupling, as motivated by the process of writing out canonical cycle notation, "reads" $\xi_1 \xi_2 \ldots \xi_n$ from right to left: the length of the first cycle is the waiting time to the first one, the length of the next cycle is the waiting time to the next one, and so on. The multiset of cycle lengths can be determined without regard to right or left: every $i$-spacing in $1\xi_2\xi_3\ldots\xi_n 1$, that is, every pattern of two ones separated by $i-1$ zeros, corresponds to a cycle of length $i$. The spacing from the rightmost one in $1\xi_2\xi_3\ldots\xi_n$ to the "artificial" one at position $n+1$ corresponds to the first cycle in canonical cycle notation, and also to the factor $P_0$ in our construction for growing an almost uniform random integer $J(n)P_0(n)$.

Since we are interested in growing with $n$, we will always read $\xi_1\xi_2\xi_3\ldots$ from left to right. Recall that $\xi_1 = 1$ identically, so $\xi_1\xi_2\xi_3\ldots$ is a random infinite word in the alphabet $\{0,1\}$, starting with a 1. Almost surely, this sequence has infinitely many ones, since the $\xi_i$ are independent with $\sum_{i\geq 1}\mathbb{E}\xi_i = \sum_{i\geq 1} 1/i = \infty$. We *define* the inter-one spacings $B_1, B_2, \ldots \in \mathbb{N}$ by the requirement that

$$\xi_1\xi_2\xi_3\ldots \text{ has ones at } 1, 1+B_1, 1+B_1+B_2, \ldots, \quad \text{and nowhere else.}$$

The example which matches the example (11) from Lecture 1, which was $3, 2, 2, 11, 2, 7, p$ with $p \geq 13$, and partial products $1, 3, 6, 12, 132, 264, 1848, 1848p$, is the sequence

$$\xi_1\xi_2\ldots\xi_{20} = 10111000011000100000, \tag{21}$$

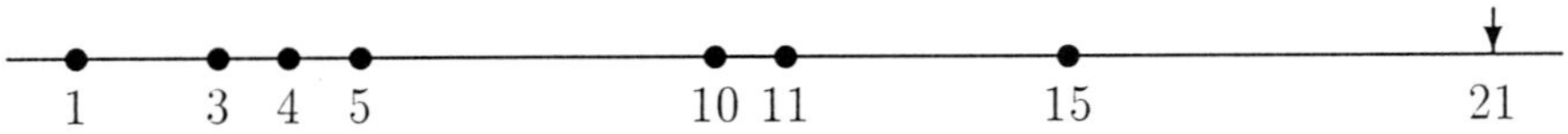


or equivalently,

$$B_1, \ldots, B_6, B_7 = 2, 1, 1, 5, 1, 4, B_7, \quad \text{with} \quad B_7 \geq 6.$$

Do you see the correspondence between $3, 2, 2, 11, 2, 7, p$ with $p \geq 13$, and $10111000011000100000\ldots$? Writing $p_i$ for $i^{\text{th}}$ smallest prime $p_i$, the prime and permutations examples match in that the $k^{\text{th}}$ prime on the list of primes is $p_{B_k}$.

In the Feller coupling, the cycle structure of a random $\pi \in \mathcal{S}_n$ has been realized via $C_i(n) = \#i$-spacings in $1\xi_2\xi_3\ldots\xi_n 1$. For comparison, consider

$$C_i(\infty) := \#i - \text{spacings in } 1\xi_2\xi_3\ldots = \sum_{k\geq 1} 1(B_k = i).$$

An easy calculation, (22) and (23) below, shows $\mathbb{P}\big(C_i(n) \neq C_i(\infty)\big) \leq 2/(n+1)$. Recall that convergence in distribution for infinite sequences, as in (20), is equivalent to having convergence for the restriction to the first $k$ coordinates, for all $k$. Thus

$$\mathbb{P}\big((C_1(n), \dots, C_k(n)) \neq (Z_1, \dots, Z_k)\big) \leq 2k/(n+1) \to 0$$

for all $k$, so that $\big(C_1(n), \dots, C_k(n)\big) \Rightarrow (Z_1, \dots, Z_k)$, and hence $\big(C_1(n), C_2(n), \dots\big) \Rightarrow \big(C_1(\infty), C_2(\infty), \dots\big)$. Comparison with (20) shows that the $C_i(\infty)$ are independent, Poisson, with $\mathbb{E}C_i(\infty) = 1/i$.

### 2.1. A size biased permutation of the multiset having $i$ with multiplicity $Z_i \sim$ Poisson $(1/i)$

The above indirect argument, that the $C_i(\infty)$ are independent, Poisson $(1/i)$ was presented at Oberwolfach one morning in August 1993; Erdős and Svante Janson were in the audience. Svante asked if there were a direct proof; I said I didn't know of one. Before lunch time, Svante Janson found and presented the following direct argument. Start with the $Z_i$, given to be independent, Poisson $(1/i)$. Take a random infinite multiset, having $Z_i$ copies of $s_i := 0^{i-1}1$, the string of length and weight $i$. Take a size biased permutation of this multiset to get a list $R_1, R_2, \dots$, so that by construction, the number of $i$-spacings in the string $1R_1R_2\dots$ is $Z_i$, for each $i$. Calculation ([12], Section 9.1) shows that the random string $1R_1R_2\dots$ of zeros and ones has the same distribution as $\xi_1\xi_2\dots$. And as an historical note: there already existed yet another direct argument, by marking Poisson processes. Jim Pitman describes it this way: "As observed in Diaconis–Pitman [20], the fact that the numbers of $i$-spacings in the Bernoulli $(1/j)$ sequence are independent Poisson $(1/i)$ is an immediate consequence of the structure of records of a sequence of i.i.d. uniform $(0,1)$ variables $U_1, U_2, \dots$. For if $N_1 < N_2 < \dots$ are the successive record indices, then by a well known result of Rényi the indicators $1(N_k = j$ for some $k)$ are independent Bernoulli $(1/j)$, and as shown by Ignatov [29] the numbers of $i$-spacings in the record sequence are independent Poisson $(1/i)$."

### 2.2. Keeping score for the Feller coupling

Starting from the independent Bernoulli $\xi_i$ with $\mathbb{P}(\xi_i = 1) = 1/i$, and defining $Z_i$ as $Z_i := C_i(\infty)$, we have a coupling of the cycle structures $\big(C_i(n)\big)_i$ for $n = 1, 2, \ldots$, together with the independent Poisson $Z_i$ with $\mathbb{E} Z_i = 1/i$. Note that for the event that an $i$-spacing occurs with right end at $k$, the probability is a simple telescoping product:

$$\mathbb{P}(\xi_{k-i} \cdots \xi_k = 1\,0^{i-1}1) = \frac{1}{k-i}\,\frac{k-i}{k-(i-1)} \cdots \frac{k-3}{k-2}\,\frac{k-2}{k-1}\,\frac{1}{k} = \frac{1}{(k-1)k}.$$

We can have $C_i(n) < Z_i$, due to $i$-spacings whose right end occurs after position $n+1$; the expected number of times this occurs is

$$\sum_{k>n+1} \mathbb{P}(\xi_{k-i} \cdots \xi_k = 1\,0^{i-1}1) = \sum_{k>n+1} \frac{1}{(k-1)k} = \frac{1}{n+1} \tag{22}$$

The only way that $C_i(n) > Z_i$ can occur is if the "artificial" one at position $n+1$ in $1\xi_2 \ldots \xi_n 1$ creates an extra $i$-spacing; for each $n$ this can occur for at most one $i$, and it occurs for each $1 \le i \le n$ with the same probability,
(23)

$$\mathbb{P}(\xi_{n-i+1} \cdots \xi_n \xi_{n+1} = 1\,0^{i-1}0) = \frac{1}{n-i+1}\,\frac{n-i+1}{n-i+2} \cdots \frac{n}{n+1} = \frac{1}{n+1}.$$

The length $A(n)$ of the first cycle in canonical cycle notation is precisely the value $i$ for which this "extra" $i$-cycle may occur,

$$A(n) = n + 1 - \max\{j \le n : \xi_j = 1\},$$

so that with

$$J(n) := \max\left\{ \sum_1^l B_l : \sum_1^l B_l \le n-1 \right\} \tag{24}$$

we have $J(n) + A(n) = n$.

We now summarize the Feller coupling in a way which matches the coupling for primes in Subsection 1.2. Start with independent Poisson random variables $Z_i$ with $\mathbb{E} Z_i = 1/i$. Take the infinite multiset having $Z_i$ copies of $i$. Take a size biased permutation $B_1, B_2, \ldots$ of this multiset. Let $J(n) \in [0, n-1]$ be the largest partial sum not exceeding $n-1$. (There is no need to calculate the distribution of $J(n)$, but it happens to be exactly uniform on $0, 1, \ldots, n-1$, which matches the harmonic distribution in (7), in the sense that $\log H(n)$ is approximately uniform on $[0, \log n]$.) Fill in one extra cycle length, $A(n) := n - J(n)$; this corresponds to $P_0(n)$. The resulting cycle structure, with cycles of lengths $B_1, B_2, \ldots, B_L$ and $A(n)$ (where $J(n) = B_1 + \ldots + B_L$), is the cycle structure of a random permutation chosen uniformly from $\mathcal{S}_n$.

Write $\mathbf{e}_i := (0, 0, \ldots, 0, 1, 0, \ldots)$ for the unit vector with all zero coordinates except for a one in position $i$. The Feller coupling shows that the Wasserstein insertion-deletion distance for permutations compared to their independent limit process is at most 2, for all $n$, with a monotonicity relation as a bonus:

$$\mathbb{E} \sum_1^n |C_i(n) - Z_i| \le \frac{2n}{n+1} < 2, \quad \text{and} \tag{25}$$

$$\text{always} \quad \big(C_1(n), C_2(n), \ldots\big) \le (Z_1, Z_2, \ldots) + \mathbf{e}_{A(n)}. \tag{26}$$

The result (26) for random permutations is stronger than the analogous result for prime factorizations, (9), in that there is no exceptionally probability for an event where monotonicity may fail. The result for permutations suggests the following conjecture from [6], for which, in the style of Erdős, I now offer a five hundred dollar prize.

**Conjecture 1** (\$500 prize offered). *For all $n \ge 1$, it is possible to construct $N(n)$ uniformly distributed from 1 to $n$, $M(n)$ defined by (12), and a prime $P(n)$ such that*

$$\textit{always} \quad N(n) \mid M(n) P(n). \tag{27}$$

My reason for believing the conjecture to be true is that permutations "fit together perfectly," as witnessed by the Feller coupling, and that primes do so also. One sense in which "primes fit together perfectly" is that the weights $\log 2, \log 3, \log 5, \log 7, \log 11, \ldots$ are such that 1) all multisets of primes have distinct weights, and 2) the weights of these multisets are evenly spaced: $\log 1, \log 2, \log 3, \log 4, \ldots$.

A restatement of Conjecture 1 in the language of stochastic monotonicity, with respect to the partial order of divisors and multiples, is that for every $n$, *for some randomized choice* of $P_0(n)$ to be 1 or a prime factor of $N(n)$, $N(n)/P_0(n)$ lies below $M(n)$ in distribution. A more specific version of the conjecture, from [2], is that $P_0(n)$ can be chosen as the first prime factor of $N(n)$ under a size biased permutation.

In terms of the usual combinatorial language of matchings, Conjecture 1 may be stated as follows. For any set $D$ of positive integers, define

$$l(D) := \{i \,:\, \exists m \in D,\ p \text{ prime}, \quad i \,|\, mp\},$$

so that for example $l(\{1\})$ is the set of primes, together with 1. Write $[n]$ for $\{1, 2, \ldots, n\}$. The conjecture is that $\forall n \geq 1$, $\forall D \subset [n]$,

$$\frac{1}{n}\big|l(D) \cap [n]\big| \geq \sum_{m \in D} \mathbb{P}\big(M(n) = m\big) = \prod_{p \leq n}\left(1 - \frac{1}{p}\right) \sum_{m \in D} \frac{1}{m}.$$

## 2.3. At least $2 - \varepsilon$ indels are needed

The following extensions to (25) aren't obvious, but help give a full picture of the qualitative behavior of the Feller coupling. Namely, the positive and negative parts of the quantity inside the absolute value in the left side of (25) have

$$\sum_1^n \big(C_i(n) - Z_i\big)^+ \to 1 \quad \text{in probability and expectation, and}$$

$$\text{for} \quad k = 0, 1, 2, \ldots, \mathbb{P}\left(\sum_1^n \big(Z_i - C_i(n)\big)^+ = k\right) \to p_k > 0,$$

$$\text{with} \quad \sum k p_k = 1.$$

In words, the coupling converts $\big(C_1(n),\ldots,C_n(n)\big)$ to $(Z_1,\ldots,Z_n)$ with, in the limit, one deletion and a random, mean 1 number of insertions, using $k$ insertions with probability $p_k$.

A separate argument (see [2]) shows that, for *any* coupling, Feller or otherwise, with probability approaching one, at least one deletion is necessary, i.e. $1 = \lim \mathbb{P}\big(\sum_1^n \big(C_i(n) - Z_i\big)^+ \geq 1\big)$. In particular, $1 \leq \liminf \mathbb{E} \sum_1^n \big(C_i(n) - Z_i\big)^+$. Then, since $\mathbb{E}C_i(n) = 1/n = \mathbb{E}Z_i$ for $i = 1$ to $n$, the number of insertions must, on average, equal the number of deletions. Hence $1 \leq \liminf \mathbb{E} \sum_1^n \big(Z_i - C_i(n)\big)^+$, so that $2 \leq \liminf \mathbb{E} \sum_1^n \big|C_i(n) - Z_i\big|$. This shows that the Feller coupling, from the point of view of (25), is asymptotically optimal.

# Lecture 3. Rescaling space — to get a scale invariant Poisson process

## 3.1. Review: couplings for primes and permutations

To review, we have two couplings for growing a system with parameter $n$. The coupling for primes grows a random integer $J(n)P_0(n)$ which is almost uniformly distributed from 1 to $n$. The coupling for permutations grows a random permutation which is distributed exactly uniformly in $\mathcal{S}_n$.

The coupling for primes takes a list $Q_1, Q_2, \ldots$ of primes, forms their partial products, and pays attention to $J(n) = Q_1 Q_2 \ldots Q_{L(n)}$, the largest partial product not exceeding $n$. On a logarithmic scale, this means that we plot the points $0, \log Q_1, \log Q_1 + \log Q_2, \ldots$, and consider the largest point not exceeding $\log n$. Think of the values $\log Q_i$ as spacings. We finish by filling in one extra spacing, $\log P_0(n)$, to get to the point $\log\big(J(n)P_0(n)\big)$ close to, but not to the right of, $\log n$. The resulting integer is $J(n)P_0(n) = Q_1 Q_2 \ldots Q_L P_0$, with either $L$ or $L+1$ prime factors, depending on whether $P_0 = 1$ or not.

The coupling for permutations takes a list $B_1, B_2, \ldots$ of positive integers, forms their partial sums, and pays attention to $J(n)$, the largest partial sum not exceeding $n-1$. We then fill in one extra spacing, of size $A(n)$, to get to the point $n = J(n) + A(n) = B_1 + B_2 + \ldots + B_L + A(n)$. The resulting cycle structure has $L(n) + 1$ cycles.

We review once again, looking only at the sequence of points plotted, and their spacings. For primes, the points plotted are 0, $\log Q_1$, $\log Q_1 + \log Q_2, \ldots$, with spacings $\log Q_1, \log Q_2, \log Q_3, \ldots$. Conditional on the multiset of spacings, the order in which they are taken is given by a size biased permutation. For each prime $p$, the number of $k$ such that $\log Q_k = \log p$ is $Z_p$, and the $Z_p$ are independent, geometrically distributed. For permutations, the points plotted are 0, $B_1$, $B_1 + B_2, \ldots$, with spacings $B_1, B_2, B_3, \ldots$. Conditional on the multiset of spacings, the order in which they are taken is given by a size biased permutation. For each positive integer $i$, the number of $k$ such that $B_k = i$ is $Z_i$, and the $Z_i$ are independent, Poisson $(1/i)$.

The underlying reason that primes and permutations have similar behavior is that for both systems, the *spacings* have the same *logarithmic* property: the expected total number of spacings of size at most $x$ grows like $\log x$. For primes, this is the property that

$$\mathbb{E} \sum_{\log p \leq x} Z_p = \sum_{p \leq e^x} \frac{1}{p-1} \sim \log x$$

and for permutations,

$$\mathbb{E} \sum_{i \leq x} Z_i = \sum_{i \leq x} \frac{1}{i} \sim \log x.$$

We pursue this further; even the expression $\gamma + \log x + o(1)$ is common to the two systems — see (41).

### 3.2. Limits after scaling space

What does the sequence $\xi_1 \xi_2 \ldots$ look like when viewed from a distance? Encode $\xi_1 \xi_2 \ldots \in \{0,1\}^\infty$ as the point process, i.e. random measure,

$$\sum_{i \geq 1} \xi_i \, \delta(i),$$

where $\delta(i)$ is the measure placing unit mass at $i$. Our example was $\xi_1\xi_2\ldots =$

$$101110000110001000000\ldots$$

$$\longleftrightarrow \delta(1)+\delta(3)+\delta(4)+\delta(5)+\delta(10)+\delta(11)+\delta(15)+\ldots.$$

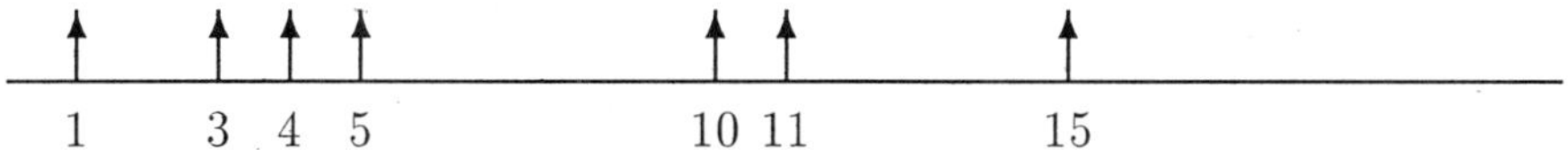


We rescale space: divide the locations by $x$, to get $\sum_i \xi_i\,\delta(i/x)$ and take $x\to\infty$. The limit is the "scale invariant" Poisson process $\mathcal{X}$ on $(0,\infty)$ with intensity $\frac{1}{x}\,dx$. The intensity gives the expected number of points in any interval $(a,b)$ with $0<a<b<\infty$, which is

$$\mathbb{E}\mathcal{X}(a,b) = \int_a^b \frac{1}{x}\,dx = \log(b/a).$$

The Poisson property is that for any disjoint $I_1, I_2, \ldots \subset (0,\infty)$, the counts of points in these sets, $\mathcal{X}(I_1), \mathcal{X}(I_2), \ldots$ are independent, and Poisson distributed. In particular, for $0<a<b<\infty$

$$\mathbb{P}\big(\mathcal{X}(a,b)=0\big) = \exp\big(-\mathbb{E}\mathcal{X}(a,b)\big) = \frac{a}{b}.$$

Likewise, consider the point process $\sum Z_i\,\delta_i$. This process encodes the multiset of spacings used in the Feller coupling, without keeping track of the order in which the spacings are used. In our example, $\xi_1\xi_2\ldots = 101110000110001000000\ldots$, which implies $Z_1\geq 3$, $Z_2\geq 1$, $Z_4\geq 1$, $Z_5\geq 1$. We now reveal more about the outcome in this example, and declaring that $Z_1=3, Z_2=1, Z_3=0, Z_4=1, Z_5=1$, which corresponds to the example for primes, (10). And to have a nice picture, we also declare that $Z_i=0$ for $i=6$ to 13, $Z_{14}=2$, $Z_{15}=Z_{16}=0$, $Z_{17}=1$, and $Z_i=0$ for $i=18$ to 25. Our random measure is

$$\sum_{i\geq 1} Z_i\,\delta(i) = 3\delta(1)+\delta(2)+\delta(4)+\delta(5)+\delta(14)+\delta(17)+\ldots.$$

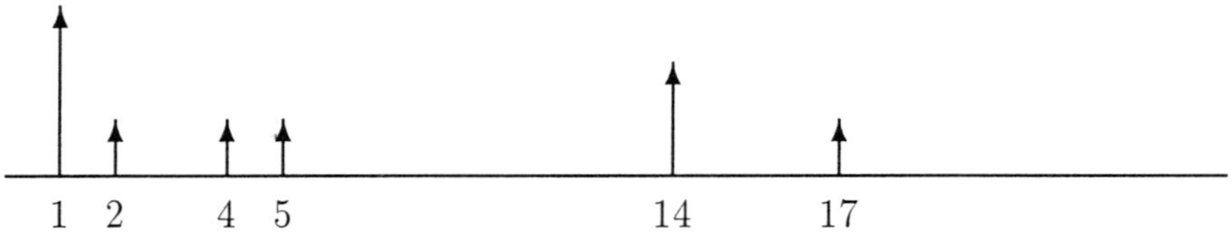


This process rescaled, $\sum Z_i \, \delta(i/x)$, converges in distribution to $\mathcal{X}$:

$$\sum_{i \geq 1} Z_i \, \delta(i/x) \Rightarrow \mathcal{X} \quad \text{as} \quad x \to \infty. \tag{28}$$

Recall that $Z_i$ is the number of $i$-spacings of $\xi_1 \xi_2 \ldots$. We have a process $\xi$ whose rescaled limit is $\mathcal{X}$, and the spacings of this process $\xi$ also have rescaled limit $\mathcal{X}$. Hence it is plausible to guess that $\mathcal{X}$ is equal in distribution to its own process of spacings.

### 3.3. The scale invariant spacing lemma

Indeed, the process $\mathcal{X}$ *is equal in distribution* to its own spacings. Since $\mathcal{X}$ has no multiple points (unlike $\sum Z_i \delta(i)$, which can have multiple points), $\mathcal{X}$ can be encoded as a random set $\mathcal{X} = \{X_i : i \in \mathbb{Z}\} \subset (0, \infty)$, with the points indexed so that for all $i \in \mathbb{Z}$, $X_i < X_{i+1}$. With such indexing, the spacings are the points

$$Y_i := X_{i+1} - X_i,$$

and the statement that the spacings of $\mathcal{X}$ have the same distribution as $\mathcal{X}$ is

$$\{Y_i : i \in \mathbb{Z}\} \stackrel{\mathrm{d}}{=} \mathcal{X}, \quad \text{and} \quad 1 = \mathbb{P}(Y_i \neq Y_j \; \forall i \neq j \in \mathbb{Z}).$$

The proof of this, from [2], is inspired by Janson's proof in Section 2.1 that if the multiset with $Z_i$ copies of $i$, to be used as spacings, is placed in a size biased permutation, then the set of points is encoded by the Bernoulli $(1/i)$ sequence $1\xi_2\xi_3 \ldots$. Viewing this in terms of the limits under spatial rescaling, it suggests that if the points of $\mathcal{X}$, to be used as spacings, are placed in a size biased permutation, then the set of partial sums also is distributed as the scale invariant Poisson process.

The size biased permutation of $\mathcal{X}$ has a very pleasing, symmetric expression. Recall from Section 1.3 that a size biased permutation can be generated by using, for an object of weight $x$, an exponentially distributed label $W$, with density $f_W(w) = xe^{-wx}$. To emphasize that the weights are also random, and we have conditioned on the values of these weights, we write this with the notation for a conditional density given $x$, that is $f_{W\,|\,X}(w\,|\,x)\,dw = xe^{-wx}\,dw$. Starting with the scale invariant Poisson process $\mathcal{X} = \{X_i\}$, with intensity $f_X(x)\,dx = (1/x)\,dx$ on $(0,\infty)$, and attaching label $W_i$ to $X_i$, the process $\{(W_i, X_i) : i \in \mathbb{Z}\}$ of points with labels is a Poisson process with intensity $f_X(x)\ dx f_{W\,|\,X}(w\,|\,x)\,dw =$

$$f_{W,X}(w,x)\,dw\,dx = e^{-wx}\,dw\,dx, \quad (w,x) \in (0,\infty)^2.$$

Note the beautiful and perfect symmetry between the points and labels: the distribution of the process is invariant under $(w,x) \mapsto (x,w)$. No such symmetry is possible for our other size biased permutations, since the points have discrete support, such as $\mathbb{N}$ or $\{\log p : p \text{ is prime}\}$, while the labels are continuously distributed in $(0,\infty)$. The distribution of the process is also invariant under rescaling of the form $(w,x) \mapsto (cw, x/c)$; we apply this with $c = 2$ in the picture below, and with $c = \log n$ in the proof of Lemma 2.

The picture shows the Poisson process with intensity $e^{-wx}$ restricted to the region $(0,b]^2$ with $b = 5$. The number of points in this region is Poisson with mean $\int_0^b (1 - e^{-bx})\,dx$; this mean is 3.796 for $b = 5$, and would be 8.401 for $b = 50$.

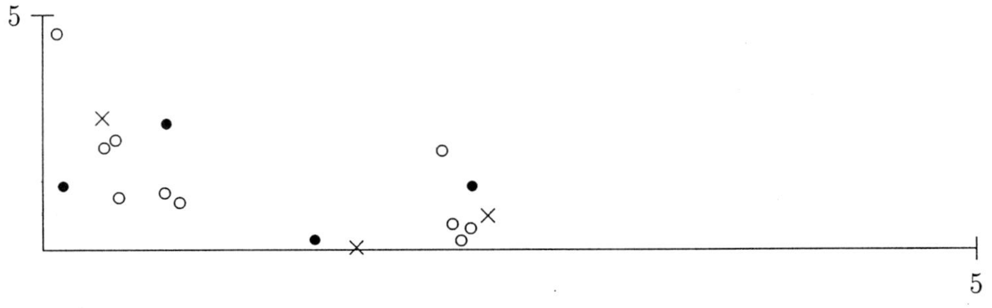

We show three realizations; the first experiment, shown with solid circles, had 4 points in this region, the second experiment, shown with open circles had 10 points, and the third experiment, shown with "×" had 3 points. This is honest simulation; I did only three runs, even though they hardly look typical to my naive eye. Try to visualize each of the three runs by itself.

The proof of the scale invariant spacing lemma goes as follows. Start with the point process on $(0,\infty)^2$ with intensity $e^{-wy}\,dw\,dy$, and index the points as $(W_i, Y_i)$ with $W_i > W_{i+1}$ for all $i \in \mathbb{Z}$. (The labels $W_i$ are all distinct, with probability one, and we remove from the probability space the complementary event.) Notice the reverse direction for the deterministic inequality; small indices $i \longleftrightarrow$ large labels $W_i$, which tend to go with small weights $Y_i$. This gives us a set of points $\{Y_i\}$ having the distribution of the scale invariant Poisson process, indexed in order of a size biased permutation, tending from small to large. Define the points $X_j$ to be the partial sums of the $Y_i$ in this order:

$$\text{for} \quad i \in \mathbb{Z}, \quad X_j := \sum_{-\infty<i<j} Y_i$$

This gives a set of points $0 < \ldots < X_i < X_{i+1} < \ldots < \infty$, and further calculation shows that the distribution of $\{X_i : i \in \mathbb{Z}\}$ is that of a scale invariant Poisson process with intensity $(1/x)\,dx$. By construction, the spacings of the $X_i$ are the points $Y_i$, with the desired distribution.

### 3.4. Primes and the scale invariant Poisson

How does the process of points in the coupling for primes appear, viewed from a distance? Recall, the points are $0, \log Q_1, \log Q_1 + \log Q_2, \ldots,$ and their spacings, $\log Q_1, \log Q_2, \ldots$ are such that $\log p$ occurs $Z_p$ times, where the $Z_p$ are independent and geometrically distributed. A process which encodes the multiset of spacings, in a form suitable for spatial rescaling, is the random measure

$$\sum_{i\geq 1} \delta(\log Q_i) = \sum_p Z_p\ \delta(\log p). \tag{29}$$

The direct analog of (28) would be the statement that

$$\sum_p Z_p\,\delta(\log p/x) \Rightarrow \mathcal{X} \quad \text{as} \quad x \to \infty. \tag{30}$$

Standard probability theory reduces this to showing that for $0 < a < b < \infty$, the expected mass that the rescaled measure gives to $(a, b)$ converges to $\log(b/a)$ as $x \to \infty$. The mass is $\sum_{ax<\log p<bx} \mathbb{E}\, Z_p = \sum_{e^{ax}<p<e^{bx}} 1/(p-1)$, so that sufficient number-theoretic knowledge needed to prove (30) is that as $y \to \infty$, $\sum_{p\le y} 1/p = B + \log\log y + o(1)$, for some constant $B$, a statement weaker than the prime number theorem.

But the random measure $\sum_p Z_p\,\delta(\log p) = \sum_{i\ge 1} \delta(\log Q_i)$ is much closer to $\mathcal{X}$ than the rescaling relation (30) shows. Namely, we can match up points so that the total amount of displacement needed to convert one sequence to the other is finite (in expectation, and therefore almost surely.) This coupling comes from [2], where it is combined with the total variation distance approximation of the small prime factors of a uniform integer, to give a metrized version of the Poisson process approximation for the "intermediate" prime divisors, from De Koninck and Galambos [25]. The points $X_i$ of $\mathcal{X}$ are indexed by $i \in \mathbb{Z}$, while the points $Q_i$ are indexed by $i \in \mathbb{N}$, so we define $Q_i = 1$ for $i \le 0$. The claim is that we can construct the $Q_i$ and $X_i$ on a single probability space, so that

$$\mathbb{E} \sum_{i\in\mathbb{Z}} |X_i - \log Q_i| < \infty. \tag{31}$$

**3.4.1. Ignoring the difference between geometric and Poisson.** In (31) the chief obstacle is that $\mathcal{X}$ is intrinsically a Poisson process, while the prescribed multiplicity of $p$ in the sequence $Q_i$ is not Poisson, but rather a geometric, $Z_p$. We make our task easier if we change the prescription to the following: for every $p$, $A_p = \sum_i 1(Q_i = p)$, where the $A_p$ are independent, Poisson $(1/p)$. The two versions of the prescribed counts, $Z_p$ and $A_p$ are close, and can be coupled with $\mathbb{E}|Z_p - A_p| = 1/\big(p(p-1)\big)$. To convert the coupling with Poisson multiplicities into a coupling with geometric multiplicities, we can move $|Z_p - A_p|$ copies of $\log p$, each at most through distance $\log p$, because there is an infinite supply of points $\log Q_j = 0$ to swap with. Since

$$\sum \log p / \big(p(p-1)\big) < \infty, \tag{32}$$

this perturbation is absorbed by the right side of (31).

A simple way to handle the scale invariant Poisson process on $(0,\infty)$ is to start with the translation invariant Poisson $\mathcal{L}$ with intensity $1\,dx$ on $\mathbb{R}$ — a process for which the number of points in an interval of length $x$ is Poisson distributed, with mean $x$. If the points of this process are $L_i$ for $i \in \mathbb{Z}$, then setting

$$X_i := e^{L_i} \tag{33}$$

gives the points of the scale invariant Poisson process $\mathcal{X}$. To get $A_p$ to be Poisson with mean $1/p$, all we need to do is assign some interval of length $1/p$, and let $A_p$ count how many points of $\mathcal{L}$ land in that interval. Using disjoint intervals for different $p$ makes the $A_p$ mutually independent.

The error estimate for the prime number theorem (see e.g. [42], or [44] Section 4.1) implies the well known estimate

$$\sum_{p \le x} \frac{1}{p} = B + \log\log x + O\Big(\exp\big(-c\sqrt{\log x}\,\big)\Big) \tag{34}$$

as $x \to \infty$, for some $c > 0$, with constant

$$B := \gamma - \sum_{k \ge 2} \sum_{p} \frac{1}{kp^k} \doteq .261497,$$

where $\gamma$ is Euler's constant, $\gamma \doteq .5772$. We use this in the form

$$f(x) := -B + \sum_{p \le e^x} \frac{1}{p} = \log x + O\Big(\exp\big(-c\sqrt{x}\,\big)\Big). \tag{35}$$

The best upper bound for (34) is $O\Big(\exp\big(-c(\log x)^{3/5}(\mathrm{loglog}\, x)^{-1/5}\big)\Big)$, but more easily stated estimate leads to the same order error bounds in our work.

We define a function $g : \mathbb{R} \to \{0, \log 2, \log 3, \ldots\}$ to be essentially the inverse of $f$. Specifically, $g$ has $(-\infty, -B] \mapsto 0$, $(-B, -B+1/2] \mapsto \log 2$, $(-B+1/2, -B+1/2+1/3] \mapsto \log 3, \ldots$. Since $f$ is close to the log function, $g$ is close to the exponential function. Now let $h : (0,\infty) \to \{0, \log 2, \log 3, \ldots\}$ be defined by $h(x) := g(\log x)$, so that $h$ is close to the identity function on $(0,\infty)$, and can be applied directly to the points $X_i$ of the scale invariant Poisson process on $(0,\infty)$. We have, under $h$, $(0, e^{-B}] \mapsto 0$, $(e^{-B}, e^{-B+1/2}] \mapsto \log 2$, $(e^{-B+1/2}, e^{-B+1/2+1/3}] \mapsto \log 3, \ldots$. This map $h$ is such that the number of $X_i$ with $h(X_i) = \log p$ is Poisson $(1/p)$, independently for all primes $p$, and $h$ is so close to the identity function that $\mathbb{E} \sum_{i \in \mathbb{Z}} \left| X_i - h(X_i) \right| < \infty$.

The illustration below show the "bins" for the translation invariant Poisson process on $\mathbb{R}$, with all arrivals to the left of $-B$ representing ones, arrivals between $-B$ and $-B+1/2$ representing twos, and so on. The length of the bin corresponding to $p$ is $1/p$. We have faked arrivals to correspond to the example (10) from Lecture 1, with $Z_2 = 3, Z_3 = 1, Z_5 = 0, Z_7 = 1, Z_{11} = 1$. The reader should *imagine* an arrow labeled "exponential map," pointing to a second picture, labeled "Bins for Poisson $(1/x\, dx$ on $(0,\infty))$", with dividing markers at $e^{-B}, e^{-B+1/2}, e^{-B+1/2+1/3}$, and so on.

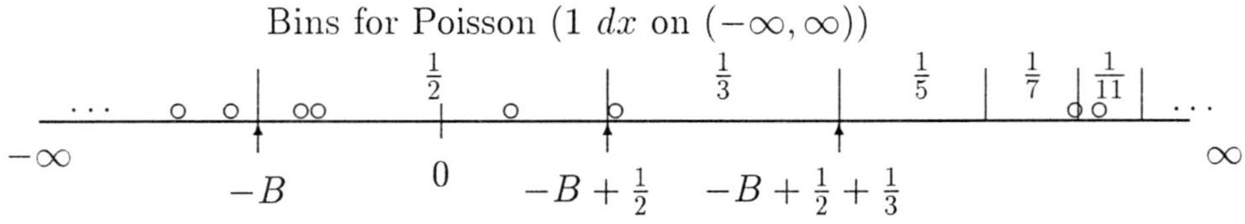


**Theorem 1.** *Let $\mathcal{X}$ be the scale invariant Poisson process on $(0,\infty)$, with intensity $(1/x)\, dx$, and points $X_i$, $i \in \mathbb{Z}$. Define $Q_i$ by $\log Q_i = h(X_i)$. Then every $Q_i$ is either one or a prime, and for every prime $p$, $A_p := \sum_i 1(Q_i = p)$ is Poisson distributed, with $\mathbb{E} A_p = 1/p$. The $A_p$ are mutually independent, and with $f$ defined by (35),*

$$\mathbb{E} \sum_{i \in \mathbb{Z}} |X_i - \log Q_i| = \int_0^\infty \left| f(x) - \log x \right| dx < \infty. \tag{36}$$

**Proof.** Constructing $\mathcal{X}$ from $\mathcal{L}$ as in the discussion above, so that $X_i = \exp(L_i)$, we have $\log Q_i = h(X_i) = g(L_i)$. This construction makes it obvious that the $A_p$ are independent, Poisson, with $\mathbb{E} A_p = 1/p$. To prove (36), we argue that

$$\mathbb{E} \sum_{i \in \mathbb{Z}} |\log Q_i - X_i| = \mathbb{E} \sum \left| g(L_i) - \exp(L_i) \right| \tag{37}$$

$$= \int_{-\infty}^{\infty} \left| g(l) - e^l \right| dl =: a_0,$$

where we have used Fubini to express an expectation of a sum over points of $\mathcal{L}$ in terms of the intensity of $\mathcal{L}$. Now

$$a_0 = \int_{-\infty}^{\infty} \left| g(l) - e^l \right| dl = \int_0^{\infty} \left| f(x) - \log x \right| dx < \infty, \tag{38}$$

where the equality holds because $a_0$ may be interpreted as the area between the graphs of $g$ and the exponential function, and equals the area between the graphs of $f$ and the logarithm (with vertical segments added to span the jumps of $f$ and of $g$.) The convergence of this last integral at infinity follows from the bound (35), and the convergence of the integral at 0 follows simply as $\int_0^{\log 2} \left| f(x) - \log x \right| dx = \int_0^{\log 2} (-B - \log x)\, dx < \infty$. ■

**3.4.2. Exploiting the relation between geometric and Poisson** The argument at (32), for changing the coupling with $A_p \sim$ Poisson $(1/p)$ copies of $p$ to one with $Z_p \sim$ Geometric, is not pretty, but more importantly, it is a non-explicit procedure and makes it hard to control the size biased permutation of the $\log Q_i$ and their partial products. Carrying out the following explicit coupling turned out to be the key to being able to analyze the coupling for primes described in Lecture 1. A simple observation, that the geometric distribution is compound Poisson, makes everything work! Every baby *should know* the following standard lemma, which we will apply with $a = 1/p$ and corresponding random variables $Z_p$ and $A_p^{(k)} \equiv A_{p^k}$. The added entropy involved in the integer partition occurring in (39) is discussed later, at (46).

**Lemma 1.** *Let $Z$ be geometric with parameter $a \in (0,1)$, i.e. $\mathbb{P}(Z \geq k) = a^k$. Let $A \equiv A^{(1)}$ be Poisson with $\mathbb{E}A = a$, and more generally, for $k = 1,2,3,\ldots$, let $A^{(k)}$ be Poisson with $\mathbb{E}A^{(k)} = a^k/k$, with the $A^{(k)}$ independent. Then*

$$Z \stackrel{\mathrm{d}}{=} \sum_{k\geq 1} kA^{(k)} = A + \sum_{k\geq 2} kA^{(k)}. \tag{39}$$

**Proof.** Recall that the probability generating functions of geometric and Poisson are given by

$$\mathbb{E}s^Z = \sum_{j\geq 0} s^j(1-a)a^j = \frac{1-a}{1-as}, \quad \mathbb{E}s^A = \sum_{j\geq 0} s^j e^{-a}a^j/j! = e^{a(s-1)},$$

with $|s| < 1$, so that $\mathbb{E}s^{kA} = e^{a(s^k-1)}$ and $\mathbb{E}s^{kA^{(k)}} = e^{((as)^k - a^k)/k}$. Writing

$$\log\frac{1-a}{1-as} = \sum_{k\geq 1}\frac{(as)^k - a^k}{k}$$

proves (39). ∎

Notice also that the last expression in (39) demonstrates stochastic domination $Z_p \geq_d A_p$, so that in the expressions $|Z_p - A_p|$ in the argument before (32), the absolute value signs weren't needed!

For us,

$$Z_p = \sum_{k\geq 1} kA_{p^k}$$

with $p$ prime, $k \geq 1$, and

$$q = p^k, \quad A_q \sim \text{Poisson}, \quad \mathbb{E}A_q = \frac{1}{kq}.$$

The coupling for primes described in Lecture 1 had a multiset with $Z_p$ copies of $p$, taken in order of a size biased permutation. A far more tractable coupling has a multiset of primes and prime powers, with $A_q$ copies of the prime power $q = p^k$, and the objects $Q_i$ (re-using the same notation $Q_i$ as before) to be taken in size biased permutation are these prime powers. In the remainder of this, Section 3.4.2, we will study the coupling of this multiset of prime powers to the scale invariant Poisson. Then in Section 3.5 we will take a size biased permutation, to give our second coupling for growing a random integer. We will analyze this second coupling in detail; the first coupling for growing a random integer, from Lecture 1, can be analyzed as a perturbation of the second coupling — but we won't present the details of the comparison, which are similar in spirit to proofs of Lemmas 5 and 6.

The modified version of (34) is

$$\sum_{q=p^k \le x} \frac{1}{kq} = \gamma + \log\log x + O\Big(\exp\big(-c\sqrt{\log x}\,\big)\Big) \tag{40}$$

and our modified version of (35) is

$$f(x) := -\gamma + \sum_{q=p^k \le e^x} \frac{1}{qk} = \log x + O\Big(\exp\big(-c\sqrt{x}\,\big)\Big). \tag{41}$$

Likewise, we modify $g$ to be essentially the inverse of this $f$, with $g(t) = 0$ if $t \in (-\infty, -\gamma]$, and if $q > 1$ is a prime power, $l = \log q$ and $\log x \in \big(f(l-), f(l)\big]$ then $g(x) = l$. We define $h = g \circ \log$, so that our modified version of the map $h$, which can be applied to the points of the scale invariant Poisson process, has

$$(0, e^{-\gamma}] \mapsto 0, \quad (e^{-\gamma}, e^{-\gamma+1/2}] \mapsto \log 2, \quad (e^{-\gamma+1/2}, e^{-\gamma+1/2+1/3}] \mapsto \log 3,$$

$$\big(\exp(-\gamma + 1/2 + 1/3), \exp(-\gamma + 1/2 + 1/3 + 1/8)\big] \mapsto \log 4, \tag{42}$$

and so on. Notice that in the last endpoint given above we have $1/8 = 1/(kq)$ for $q = p^k = 4$. To summarize formally, $h$ is given by the recipe: $h(x) = 0$ if $0 < x \le e^{-\gamma}$, and if $q > 1$ is a prime power, $l = \log q$ and $\log x \in \big(f(l-), f(l)\big]$, then $h(x) = l$, with $f$ as in (41).

**Theorem 2.** *Let $\mathcal{X}$ be the scale invariant Poisson process on $(0, \infty)$, with intensity $(1/x)\,dx$, and points $X_i$, $i \in \mathbb{Z}$. Define points $Q_i$ by $\log Q_i = h(X_i)$,* *for the function $h$ described at (42). Then every $Q_i$ is either one or a prime power, and for every $q = p^k$, with $p$ prime and $k \geq 1$, $A_q := \sum_i 1(Q_i = q)$ is Poisson distributed, with $\mathbb{E} A_q = 1/(kq)$. The $A_q$ are mutually independent, and with $f$ defined by (41),*

$$\mathbb{E} \sum_{i \in \mathbb{Z}} |X_i - \log Q_i| = \int_0^\infty \left| f(x) - \log x \right| dx < \infty. \tag{43}$$

**Proof.** Constructing $\mathcal{X}$ from $\mathcal{L}$ as in the discussion above, so that $X_i = \exp(L_i)$, we have $\log Q_i = h(X_i) = g(L_i)$. This construction makes it obvious that the $A_q$ are independent, Poisson, with $\mathbb{E} A_{p^k} = 1/(kq)$. To prove (43), we argue that

$$\mathbb{E} \sum_{i \in \mathbb{Z}} |\log Q_i - X_i| = \mathbb{E} \sum \left| g(L_i) - \exp(L_i) \right| = \int_{-\infty}^\infty \left| g(l) - e^l \right| dl =: b_0, \tag{44}$$

where we have used Fubini to express an expectation of a sum over points of $\mathcal{L}$ in terms of the intensity of $\mathcal{L}$. Now

$$b_0 = \int_{-\infty}^\infty \left| g(l) - e^l \right| dl = \int_0^\infty \left| f(x) - \log x \right| dx < \infty, \tag{45}$$

where the equality holds because $b_0$ may be interpreted as the area between the graphs of $g$ and the exponential function, and equals the area between the graphs of $f$ and the logarithm. The convergence of this last integral at infinity follows from the bound (41), and the convergence of the integral at 0 follows simply as $\int_0^{\log 2} \left| f(x) - \log x \right| dx = \int_0^{\log 2} (-\gamma - \log x)\, dx < \infty$. ∎

### 3.5. The size biased permutation of the multiset having $\log p^k$ with multiplicity $A_{p^k} \sim \text{Poisson}\left(1/(kp^k)\right)$

Just as the multiset with $Z_p$ copies of each prime $p$, for independent $Z_p \sim$ geometric$(1/p)$, is the "natural random infinite multiset of primes," the multiset with $A_q$ copies of each prime power $q = p^k > 1$, for independent $A_q \sim \text{Poisson}\left(1/(kq)\right)$, is the *natural random infinite multiset of prime powers.* The latter multiset may be viewed as the former, with auxiliary randomization, picking a partition of the integer $Z_p$, independently for each $p$.

[Incidentally, the amount of additional information in our partitioning of the $Z_p$ is small; it is approximately .612433379 bits, computed as follows. Recall that for a discrete random variable $X$ with $\mathbb{P}(X = X_i) = p_i > 0$, $\sum p_i = 1$, the *entropy* is $h(X) := -\sum p_i \log p_i$. For the geometrically distributed $Z$ in (39), $h(Z) = -\log(1-a) - a/(1-a)\log a = \sum_{k\geq 1}(a^k/k)\big(1 + \log(1/a^k)\big)$, and for $A \sim$ Poisson $(x)$ we have $h(A) = x + x\log(1/x) + \sum_{j\geq 2}\mathbb{P}(A \geq j)\log j$, which we apply with $x = a^k/k$ for $k = 1, 2, 3, \ldots$. Recall that the entropy of an independent process, such as $\big(A^{(1)}, A^{(2)}, \ldots\big)$, is the sum of the entropies of the coordinates. Thus in (39), the additional information needed to partition $Z$ into $\sum_{k\geq 1} kA^{(k)}$ is $d(a) := \sum_{k\geq 1} h\big(A^{(k)}\big) - h(Z) =$

$$\sum_{k\geq 1}\left(\frac{a^k}{k} + \frac{a^k}{k}\log\frac{k}{a^k} + \sum_{j\geq 2}\mathbb{P}\big(A^{(k)} \geq j\big)\log j\right) - \sum_{k\geq 1}\frac{a^k}{k}\left(1 + \log\frac{1}{a^k}\right) \tag{46}$$

$= a^2\log 2 + O(a^3\log a)$ as $a \to 0+$. Numerical evaluation, with logs taken base 2, gives $d(1/2) \doteq .375076$ as the additional information needed to partition $Z_2$, approximately .13879 for $Z_3$, and approximately .612433379 for the sum over all primes.]

In Lecture 1 we described a procedure for "growing" a random integer, based on a size biased permutation of the multiset having each prime $p$, taken as an object of weight $\log p$, with multiplicity $Z_p \sim$ Geometric $(1/p)$. While relatively simple to describe, this coupling is rather hard to analyze directly. The reason that it is hard to analyze directly is that the size biased permutation involves using exponentially distributed labels $W_i$, and the resulting two-dimensional process, with points of the form $(W_i, \log Q_i)$, does not have a simple structure. By changing the objects to be prime powers $q = p^k$, with weight $\log q$ and multiplicity $A_q \sim$ Poisson $1/(kq)$ the total number of factors of $p$ is still distributed as $Z_p \sim$ Geometric $(1/p)$, because $\sum_{k\geq 1} kA_{p^k} \stackrel{\mathrm{d}}{=} Z_p$, but now the size biased permutation is tractable.

The size biased permutation is tractable because attaching conditionally independent labels $W_i$ to the Poisson process with points $\log Q_i$ yields a two-dimensional Poisson process with points $(W_i, \log Q_i)$ — this is an instance of the "labeling" theorem; see for example [31].

This process $\{(W_i, \log Q_i) : i \in \mathbb{N}\}$ is very close to the Poisson process of Section 3.3 on $(0,\infty)^2$ with intensity $e^{-wy}\,dw\,dy$, which was used to give a size biased permutation of the scale invariant Poisson process. However, our two-dimensional process now is neither discrete nor continuous. It is supported on a one dimensional subset of the positive quadrant, formed by the half lines $w > 0, y = \log q$ for some prime power $q = p^k > 1$. Its intensity on the line $y = \log q$ is the product of the discrete intensity, $f_Y(y) = 1/(kq)$ for the weight $y$, times the continuous density for exponentially distributed label $w$, $f_{W\,|\,Y}(w\,|\,y)\,dw = ye^{-wy}\,dw$. That is, for $y = \log q$, $q = p^k$,

$$f_{W,Y}(w,y)\,dw = f_Y(y)\; f_{W\,|\,Y}(w\,|\,y)\,dw = \frac{\log q}{kq}\; e^{-wy}\,dw. \tag{47}$$

The next lemma gives an exact expression for the density of $J(n)$, and will let us see in Lemma 3 that the distribution of $J(n)$ is close to the harmonic distribution (7) on $1, 2, \ldots, n$: in (50), $d(\log\log n^\beta)$ is close to $d\beta/\beta$, and the zeta function has a simple pole at one, so the expression in (50) is close to

$$\frac{1}{i}\int_{\beta>1-\alpha}\frac{d\beta}{\beta}\int_{c>0}\beta e^{-\beta c}e^{-\alpha c}\frac{c}{\log n}\,dc = \frac{1}{i\log n}. \tag{48}$$

**Lemma 2.** *For the $Q_i$ taken in order of decreasing labels $W_i$, let $J(n)$ be the largest partial product with $J(n) \le n$. Then for $i = n^\alpha = 1$ to $n$, with $q = p^k = n^\beta$ ranging over prime powers,*

$$\mathbb{P}(\,J(n) = i\,) = \sum_{q\,:\,\beta>1-\alpha}\frac{1}{kq}\int_{c>0}\beta\; e^{-\beta c}\frac{i^{-1-c/\log n}}{\zeta(1+c/\log n)}\,dc \tag{49}$$

$$= \frac{1}{i}\sum_{q\,:\,\beta>1-\alpha}\frac{1}{kq}\int_{c>0}\beta\; e^{-\beta c}\frac{e^{-\alpha c}}{\zeta(1+c/\log n)}\,dc. \tag{50}$$

**Proof.** Write the labels $W_i$ as $W_i = S_i/\log Q_i$ as in Section 1.3, with $S_1, S_2, \ldots$ being iid standard exponentials, independent of $Q_1, Q_2, \ldots$. For any prime power $q > 1$ and constant $t \ge 0$, the number $A_q(t) := \sum_i \mathbf{1}(Q_i = q,\ S_i/\log q > t)$ of occurrences of $q$ with label greater than $t$ is Poisson distributed with $\mathbb{E} A_q(t) = \mathbb{E} A_q \ \mathbb{P}(S_i/\log q > t) = 1/(kq)\ q^{-t}$. The $A_q(t)$ jointly for all $q$ are independent. For each prime $p$ let

$$Z_p(t) := \sum_{k \geq 1} k A_{p^k}(t). \tag{51}$$

Note that by (39) the distribution of $Z_p(t)$ is geometric with $\mathbb{P}(Z_p(t) \geq j) = (1/p^{1+t})^j$ and

$$\mathbb{P}(Z_p(t) = j) = (1 - p^{-1-t}) \left( \frac{1}{p^{1+t}} \right)^j. \tag{52}$$

For any $t > 0$, consider the product $I_t$ of all primes and prime powers having label strictly greater than $t$, i.e.

$$I_t := \prod_{q = p^k} q^{A_q(t)} = \prod p^{Z_p(t)}.$$

Using (52) and the independence of the $Z_p(t)$ over all primes $p$, for any $i \geq 1$ we have

$$\mathbb{P}(I_t = i) = \prod (1 - p^{-1-t}) i^{-1-t} = \frac{i^{-1-t}}{\zeta(1+t)}. \tag{53}$$

With probability one, all labels $W_i$ are distinct. For any $t > 0$ there are, with probability one, only finitely many labels greater than $t$ — this follows from (53), which summed over $i = 1, 2, \ldots$ yields 1. There are infinitely many labels, with probability one, because the total intensity of our Poisson process is infinite — it is $\sum_{q=p^k} \mathbb{E} A_q = \sum_{q=p^k} 1/(kq) > \sum_p (1/p) = \infty$. For these three reason combined, with probability one, as $t$ decreases from infinity to zero, the partial products $I_t$ increase from 1 to infinity, and each increase corresponds to factoring in one new factor $Q_i$, taken in decreasing order of their labels $L_i = S_i/\log Q_i$.

Let $Q^*(n)$ be the new factor that first takes the partial product beyond $n$, and let $T(n)$ be its label. Our product $J(n)$ will equal $I_t$ for $t = T(n)$. We consider the joint distribution of $(Q^*(n), T(n), J(n))$. Write $i = n^\alpha$ for the test value for $J(n)$, $q = p^k = n^\beta$ for the test value for $Q^*(n)$, and $c$ for

$T(n)\log n$, which, as $\log n$ times the label for $q$, is exponentially distributed with rate $\log q/\log n = \beta$. Summing over $q$ and integrating over $c$ yields (49) as the marginal distribution of $J(n)$, and simplifying yields (50). ∎

**Lemma 3.** *Let $H(n)$ have the harmonic distribution (7) on 1 to $n$, so that for $i \le n$, $\mathbb{P}\big(H(n) = i\big) = 1/(ih_n)$. For the $J(n)$ in Lemma 2, based on a size biased permutation of the natural infinite Poisson multiset of prime powers,*

(54)
$$d_{TV}\big(J(n), H(n)\big) := \sum_{i\le n} \big|\mathbb{P}\big(J(n) = i\big) - \mathbb{P}\big(H(n) = i\big)\big| = O\left(\frac{1}{\log n}\right).$$

*Furthermore, the relative error in approximating the density of $J(n)$ by the harmonic density is $O(1/\log n)$, uniformly:*

$$\max_{1\le i\le n}\left|\frac{\mathbb{P}\big(J(n) = i\big)}{\mathbb{P}\big(H(n) = i\big)} - 1\right| = O\left(\frac{1}{\log n}\right). \tag{55}$$

**Proof.** Define $d_n(i) := (i\log n)\mathbb{P}\big(J(n) = i\big)$ so that (50) can be rewritten as

$$d_n(i) = \sum_{q\,:\,\beta>1-\alpha} \frac{1}{kq}\int_{c>0} \beta\, e^{-\beta c}\frac{e^{-\alpha c}\log n}{\zeta(1 + c/\log n)}\, dc. \tag{56}$$

Since $\mathbb{P}\big(H(n) = i\big) = 1/(ih_n) = 1/(i\log n)\big(1 + O(1/\log n)\big)$, showing (55) is equivalent to showing $\max_{1\le i\le n}\big|d_n(i) - 1\big| = O(1/\log n)$.

In order to simplify (56), we apply the following with $\delta = c/\log n$. From the well known $\zeta(1+\delta) = (1/\delta) + \gamma + O(\delta)$ as $\delta \to 0+$, we get $1/\zeta(1+\delta) = \delta\big(1 - \gamma\delta + O(\delta^2)\big) = \delta - O(\delta^2)$ as $\delta \to 0+$. It follows that $\exists C_1, \big|1/\zeta(1+\delta) - \delta\big| \le C_1\delta^2$ for all $\delta > 0$. This motivates us to consider a first order approximation to the right side of (56), defined by

$$e_n(i) := \sum_{q\,:\,\beta>1-\alpha} \frac{1}{kq}\int_{c>0} \beta\, e^{-\beta c} c\, e^{-\alpha c}\, dc.$$

Our goal is to show that uniformly in $1 \le i \le n$, $e_n(i) = 1 + O(1/\log n)$; having this, the error estimate for $d_n(i)$ versus $e_n(i)$ will be virtually the same computation.

Note that since $\int_{c>0} c\, e^{-(\alpha+\beta)c}\, dc = (\alpha+\beta)^{-2}$, and $\beta = \log q/\log n$, we have

$$e_n(i) = \frac{1}{\log n} \sum_{q>n^{1-\alpha}} \frac{\log q}{kq} (\alpha+\beta)^{-2}.$$

Instead of (41) we only need a crude bound, due to Chebyshev, that

$$R(x) := \sum_{q=p^k \le x} \frac{\log q}{kq} - \log x = O(1).$$

Fix $n$ and $i = n^\alpha$, $1 \le i \le n$, and define

$$S_t := \sum_{n^{1-\alpha}<q\le n^t} \frac{\log q}{kq} = \big(t - (1-\alpha)\big) \log n - R(n/i) + R(n^t),$$

so that by Abel summation

$$e_n(t) = \frac{1}{\log n} \int_{t\in(1-\alpha,\infty)} dS_t (\alpha+t)^{-2} = \frac{1}{\log n} \int_{(1-\alpha,\infty)} S_t\ 2(\alpha+t)^{-3}\, dt.$$

The contribution at infinity to the Abel summation is zero since $S_t \sim t \log n$ as $t \to \infty$. From $\sup_{x>0} R(x) < \infty$, and $\int_{(1-\alpha,\infty)} 2(t+\alpha-1)\,(\alpha+t)^{-3}\, dt = 1$ it follows that $\max_{1\le i\le n} \big| e_n(i) - 1 \big| = O(1/\log n)$.

Finally, using $\big|1/\zeta(1+\delta) - \delta\big| \le C_1\delta^2$, we have

$$\begin{aligned} \big| d_n(i) - e_n(i) \big| &\le \frac{C_1^2}{\log n} \sum_{q\,:\,\beta>1-\alpha} \frac{1}{kq} \int_{c>0} \beta\, e^{-\beta c} c^2\, e^{-\alpha c}\, dc, \\ &= \frac{C_1^2}{(\log n)^2} \int_{(1-\alpha,\infty)} dS_t\ 2(\alpha+t)^{-3}\, dt = O\left(\frac{1}{\log n}\right). \end{aligned}$$

■

## 3.6. Filling in the extra prime factor

As in Lecture 1, we take $P_0(n)$ to be one or prime (and not a prime power!), such that $J(n)P_0(n) \le n$, picking uniformly over the $1+\pi\big(n/J(n)\big)$ possibilities.

With the notation $f_n(i) := \mathbb{P}\big(J(n)P_0(n) = i\big)$, the total variation distance in the next lemma is $d_{TV}(J\,P_0, N) =$

$$(57)\quad \sum_{1\le i\le n}\left(f_n(i)-\frac{1}{n}\right)^+ = \sum_{1\le i\le n}\left(f_n(i)-\frac{1}{n}\right)^- = \frac{1}{2}\sum_{1\le i\le n}\left|f_n(i)-\frac{1}{n}\right|.$$

**Lemma 4.** *The total variation distance between the distribution of $J(n)P_0(n)$ and the uniform distribution satisfies*

$$(58)\qquad d_{TV}\big(N(n), J(n)\,P_0(n)\big) = O\left(\frac{\operatorname{loglog} n}{\log n}\right).$$

**Proof.** Since $P_0$ is one or prime, for $1 \le m \le n$,

$$(59)\quad f_n(m) := \mathbb{P}\big(J(n)\,P_0(n) = m\big) = \sum_{p\,|\,m}{}^{*}\,\mathbb{P}\left(J(n) = \frac{m}{p}\right)\frac{1}{1+\pi(np/m)},$$

where $\sum^*$ indicates that the index $p$ ranges over prime divisors of $m$ *also allowing* $p = 1$. Using $1/(i\log n)$ as an approximation for $\mathbb{P}\big(J(n) = i\big)$, we consider the simpler expression

$$g_n(m) := \frac{1}{n}\sum_{p\,|\,m}{}^{*}\,\frac{1}{\log n}\,\frac{np}{m}\,\frac{1}{1+\pi(np/m)}.$$

We have

$$(60)\qquad \sup_{1\le m\le n}\big|1 - f_n(m)/g_m(n)\big| = O(1/\log n)$$

thanks to (55), and hence also $\sum_{m\le n}\big|f_n(m) - g_n(m)\big| = O(1/\log n)$. Thus (58) is equivalent to $\sum_{m\le n}\big|g_n(m) - \frac{1}{n}\big| = O(\operatorname{loglog} n/\log n)$.

¿From the well known error bound for the prime number theorem, $\pi(x) = \mathrm{li}(x) + O\big(xe^{-c\sqrt{\log x}}\big)$, together with the approximation $\mathrm{li}(x) = (x/\log x)\big[1 + 1/\log x + O\big((\log x)^{-2}\big)\big]$ we have

$$r(x) := \left|\frac{x}{1+\pi(x)} - (\log x - 1)\right| = O(1/\log x).$$

Thus a good approximation to $g_n(m)$ will be given by

$$h_n(m) := \frac{1}{n\,\log n}\sum_{p\,|\,m}^{*}\left(\log\left(\frac{np}{m}\right) - 1\right).$$

Writing $s(i)$ for the largest squarefree divisor of $i$, and $\omega(i)$ for the number of distinct prime divisors of $i$, we have

$$h_n(m) = \frac{\log\big(s(m)\big) + \big(1+\omega(m)\big)\big(\log(n/m) - 1\big)}{n\,\log n}. \tag{61}$$

With $N \equiv N(n)$ to represent the uniform distribution on 1 to $n$, the total variation distance in (58) is approximately

$$\frac{1}{2}\sum_{m\le n}\left|h_n(m) - \frac{1}{n}\right| \tag{62}$$

$$= \frac{1}{2}\,\mathbb{E}\left|\frac{\log\big(s(N)\big) - \log n}{\log n} + \frac{1+\omega(N)}{\log n}\big(\log(n/N) - 1\big)\right|.$$

Before completing the proof of (58), we outline the analysis to focus on the source of the loglog $n$ factor in (58), and the reason that it cannot be decreased. The net contribution from the first term inside the expectation in (62) is $O(1/\log n)$, and for the second term, the two factors are approximately uncorrelated, with $\mathbb{E}\big(1+\omega(N)\big)/\log n \sim \mathrm{loglog}\, n/\log n$ for the first factor. The second factor has $\mathbb{E}\big|\log(n/N) - 1\big| \to \mathbb{E}|S-1| = 2/e$, where

$S$ has the standard exponential distribution, with $\mathbb{P}(S > t) = e^{-t}$. Thus it *should* be possible to show that

$$d_{TV}\big(N(n), J(n)\, P_0(n)\big) \sim \frac{1}{e}\left(\frac{\operatorname{loglog} n}{\log n}\right). \tag{63}$$

The estimates for the simpler task (58) in place of (63) are as follows. We need only $K := \sup_{x \geq 1} r(x) < \infty$ to conclude that $\big|g_n(m) - h_m(n)\big| \leq 1/(n \log n) \sum^*_{p\,|\,m} r(np/m) \leq K\big(1 + \omega(m)\big)/(n \log n)$, and hence $\sum_{1 \leq m \leq n} \big|g_n(m) - h_n(m)\big| \leq K\mathbb{E}\big(1 + \omega\big(N(n)\big)\big)/\log n = O(\operatorname{loglog} n/\log n)$. In (62), for the first term with $\mathbb{E} \log\big(s(N)\big) - \log n$ we have $\mathbb{E} \log N(n) - \log n \to 1$ by Stirling's formula, and

$$\begin{aligned} \mathbb{E} \log\big(N(n)/s\big(N(n)\big)\big) &\leq \sum_{p \leq n} \log p\, \mathbb{E}\big(C_p(n) - 1\big)^+ \\ &\leq \sum_p \log p \sum_{k \geq 2} p^{-k} < \infty. \end{aligned}$$

For the second term, we apply Cauchy-Schwarz, with $\mathbb{E}\big(1 + \omega\big(N(n)\big)\big)^2 \sim (\operatorname{loglog} n)^2$ and $\mathbb{E}\big(\log\big(n/N(n)\big)\big)^2 \to 1$. These bounds combine to show that $\sum_{m \leq n} \big|h_n(m) - \frac{1}{n}\big| = O(\operatorname{loglog} n/\log n)$. Together with our previous bounds comparing $f, g$, and $h$, we have proved (58). ■

### 3.7. Keeping score: 1 insertion and on average $1 + O\big((\operatorname{loglog} n)^2/\log n\big)$ deletions suffice for primes

We need to control the expected number of deletions used to convert $M(n)$ to $N(n)$, which correspond to the number of primes not exceeding $n$ but occurring in the size biased permutation *after* the partial product $J(n)$. A priori it seems reasonable to believe that one would have to calculate something along the lines of (22) for permutations, crossed with (49) for primes. Happily, the idea of "matching intensity" can be used to finesse the calculation.

¿From (58) in Lemma 4 it follows (see Section 3.8 iff you want to know the details) that on a single probability space we can construct independent $Z_p$, together with $J(n)$ and $P_0(n)$, and an exactly uniform $N(n)$, so that always $J(n) \mid M(n)$, and the good event

$$E_n = \{ J(n)P_0(n) = N(n) \} \tag{64}$$

has

$$\mathbb{P}(E_n^c) = d_{TV}(J(n)\, P_0(n), N(n)) = O\left(\frac{\log\log n}{\log n}\right). \tag{65}$$

On the uncoupled event, $E_n^c$, how many prime factors, that might contribute to $d_W(n)$, can we expect to see? Recall our notation from Section 1.5.1. Lemma 6 in [2] states that for events $E$ of small, but not too small probability, the expected number of prime factors is $O((\log\log n)\mathbb{P}(E))$. The precise statement is: uniformly in $\delta \in [0, 1]$,

$$\sup_{E : \mathbb{P}(E) \le \delta} \mathbb{E}(1(E)\ \Omega(N(n))) = O\left(\max\left(\delta \log\log n, \frac{1}{\log n}\right)\right). \tag{66}$$

With $E = E_n^c$ and $\delta = \mathbb{P}(E_n^c)$, the combination of (65) with (66) shows that

$$\mathbb{E}\left(1(E_n^c)\Omega\left(\frac{N}{(N, M)}\right)\right) \le \mathbb{E}(1(E_n^c)\Omega(N)) = O\left(\frac{(\log\log n)^2}{\log n}\right). \tag{67}$$

Now $J(n) \mid M(n)$ always, and on $E_n$ we have $N(n) = J(n)P_0(n)$ so that $N/(N, M) \mid P_0$, so that

$$\mathbb{E}\left(1(E_n)\Omega\left(\frac{N}{(N, M)}\right)\right) \le 1. \tag{68}$$

Adding gives

$$\mathbb{E}\Omega\left(\frac{N(n)}{(N(n), M(n))}\right) \le 1 + O\left(\frac{(\log\log n)^2}{\log n}\right). \tag{69}$$

Now *any coupling* has

$$\mathbb{E}\Omega\left(\frac{N(n)}{(N(n),M(n))}\right) - \mathbb{E}\Omega\left(\frac{M(n)}{(N(n),M(n))}\right) = O(1/\log n),$$

because (see e.g. [44] p. 41)

$$\mathbb{E}\Omega(N(n)) - \mathbb{E}\Omega(M(n)) = O(1/\log n). \tag{70}$$

Combining this with (69) yields

$$\mathbb{E}\Omega\left(\frac{M(n)}{(N(n),M(n))}\right) \le 1 + O\left(\frac{(\mathrm{loglog}\, n)^2}{\log n}\right). \tag{71}$$

Adding (69) and (71) proves

**Theorem 3.** *The coupling of Section* 3.5, *based on a size biased permutation of the natural Poisson multiset of prime powers, and extended to include a uniform random integer* $N(n)$, *has*

$$\mathbb{E}\sum_{p\le n}\left|C_p(n) - Z_p\right| \le 2 + O\left(\frac{(\log\log n)^2}{\log n}\right), \tag{72}$$

*and hence* $d_W(n) \le 2 + O\big((\log\log n)^2/\log n\big)$.

A separate argument (see [2]) shows that, for *any coupling*, with probability approaching one, at least one insertion is necessary to convert $M(n)$ to $N(n)$, i.e. $1 = \lim \mathbb{P}\big(\sum_{p\le n}\left(C_p(n) - Z_p\right)^+ \ge 1$. In particular, $1 \le \liminf \mathbb{E}\sum_{p\le n}\left(C_p(n) - Z_p\right)^+$. Using (70), the average number of deletions is within $O(1/\log n)$ of the number of insertions. Hence $1 \le \liminf \mathbb{E}\sum_{p\le n}\left(Z_p - C_p(n)\right)^+$, so that $2 \le \liminf \mathbb{E}\sum_{p\le n}\left|C_p(n) - Z_p\right|$. This shows that that $\lim d_W(n) = 2$, and that our coupling, from the point of view of the insertion-deletion metric in Section 1.5.1, is asymptotically optimal.

### 3.8. Extending the coupling to $N(n)$, constructively

(The reader is invited to skip past this section, which defends the claim at (64).) Our coupling as described so far is fairly natural and explicit. It starts with independent Poisson $A_q$ for $q = p^k$. These determine the prime powers $Q_i$ and the $Z_p$ such that $M(n) := \prod_{p \le n} p^{Z_p} = \prod_{i:\, Q_i = p^k,\ p \le n} Q_i$, where the $Z_p$ for primes $p$ are independent geometric. Use independent exponentially distributed $S_1, S_2, \ldots$ to give a size biased permutation of these prime powers $q$; this determines $J(n)$, a divisor of $M(n)$. A single uniformly distributed random variable $U$, independent of the $Q_i$ and $S_i$ can then be used to determine $P_0$, via the recipe: with $K(n) := 1 + \pi\bigl(n/J(n)\bigr)$, let $P_0 = 1$ if $K(n)U \le 1$, and let

$$P_0 = p_i \quad \text{if} \quad K(n)U \in (i, i+1], \tag{73}$$

where $p_i$ denotes the $i^{\text{th}}$ smallest prime. Equation (58) gives an upper bound on the total variation distance between the distributions of $J(n)P_0(n)$ and of $N(n)$, and (57) emphasizes that this is just about the *distribution* of $J(n)P_0(n)$.

We have constructed $J(n)P_0(n)$ so far, and we have *not yet* constructed $N(n)$. It is a standard coupling argument that there exist couplings in which the good event $E_n = \{J(n)P_0(n) = N(n)\}$ has the maximum possible probability, with $\mathbb{P}(E_n^c) = d_{TV}(JP_0, N)$. It is also true, but less obvious, that there exist such couplings which extend our already given construction of $\bigl((Q_i, S_i)_{i \ge 1}, P_0\bigr)$. We note further that joint distribution of $\bigl((Q_i, S_i)_{i \ge 1}, P_0, N\bigr)$ for such an extension is *not uniquely determined.* The next paragraph offers a constructive choice of joint distribution.

We present a recipe for constructing $N(n)$ as a function of the random variables used above, together with some auxiliary randomization. Take two additional uniform random variables $U_1, U_2$ with $U, U_1, U_2, Q_1, Q_2, \ldots, S_1, S_2, \ldots$ independent. We define a deterministic function $r_n(u, u_1, u_2, q_1, q_2, \ldots, s_1, s_2, \ldots)$ such that for all $\omega \in \Omega$, $N(n) := r_n(U, U_1, U_2, Q_1, Q_2, \ldots, S_1, S_2, \ldots)$ and $\mathbb{P}\bigl(N(n) \ne J(n)P_0(n)\bigr) = d_{TV}(JP_0, N)$. The recipe $r_n$ is determined by two requirements. First, let $b_n(i) := \min\bigl(f_n(i), \frac{1}{n}\bigr)/f_n(i)$, and let $E_n$ be the event

$$E_n := \bigl\{ U_1 \le b_n\bigl(J(n)P_0(n)\bigr) \bigr\}. \tag{74}$$

Note that $\mathbb{P}(E_n) = \sum_{1\le i\le n} f_n(i)b_n(i) = \sum \min\big(f_n(i), \frac{1}{n}\big) = 1 - d_{TV}(JP_0, N)$. On the event $E_n$, we *define* $N(n)$ by $N(n) = J(n)P_0(n)$. Second, let $G_n(j) := \sum_{i\le j}\big(f_n(i) - \frac{1}{n}\big)^- / d_{TV}(JP_0, N)$, so that by (57), $G_n(n) = 1$. On the event $E_n^c$, we *define* $N(n)$, to have one of the values $i$ for which $f_n(i) < 1/n$, by setting:

$$(75) \qquad \text{on} \quad E_n^c, \quad N(n) = j \quad \text{if and only if} \quad G_n(j-1) < U_2 \le G_n(j).$$

It follows that for $i = 1$ to $n$, $\mathbb{P}\big(N(n) = i\big) = 1/n$, and that $E_n$ satisfies (64) and (65).

The above construction yields $N(n)$, uniformly distributed from 1 to $n$, together with random integers $J(n)$ and $P_0(n)$ that evolve smoothly with $n$ growing, such that the event $E_n = \big\{N(n) = J(n)P_0(n)\big\}$ has the maximal possible probability, namely $1 - d_{TV}\big(J(n)\,P_0(n), N(n)\big)$. Is it the case that with probability one, for all sufficiently large $n$, we have $N(n) = J(n)P_0(n)$? This is not a trivial question, as the events $E_n$ are not nested, and the sum of their probabilities is infinite.

**Theorem 4.** *For the above construction,*

$$1 = \mathbb{P}(E_n \textit{ eventually}).$$

**Proof.** First note that as events, $\big\{J(n) \to \infty\big\} = \big\{\sum Z_p = \infty\big\}$, and hence $1 = \mathbb{P}\big(J(n) \to \infty\big)$. Recall our use of two fixed uniform random variables, $U$ in (73) and $U_1$ in (74). For $\delta > 0$ we will show that

$$(76) \qquad \big\{U_1 < 1-\delta,\ U > \delta, \text{ and } J(n) \to \infty\big\} \subset \{E_n \text{ eventually}\}$$

and hence $\mathbb{P}(E_n \text{ eventually}) \ge (1-\delta)^2$.

Assume we are given an outcome in the event on the left side of (76). Since the $J(n)$ are partial products, $J(n) \to \infty$ ensures that $\omega\big(J(n)\big) \to \infty$ as $n \to \infty$. To have $E_n$ fail, we must have $b_n\big(J(n)P_0(n)\big) < 1-\delta$, and hence $nf_n\big(J(n)P_0(n)\big) > 1/(1-\delta) > 1+\delta$. For $n$ and $\omega(m)$ both large, arguing as in the proof of Theorem 4, $f_n(m)/h_n(m) \to 1$ so that $f_n(m) > 1+\delta$ implies that $nh_n(m) > 1 + \delta/2$. [The hypothesis that $\omega(m)$ is large is needed to ensure that terms of $g_n(m)$ having $x = np/m$ small, where we cannot guarantee that $x/\big(1+\pi(x)\big)$ is close to $(\log x - 1)$, make a relatively negligible

contribution.] Using only the bound $\log s(m) \le \log n$ in (61), this implies that for sufficiently large $n$ and $\omega(m)$, $\big(1+\omega(m)\big)\log(n/m) > (\delta/2)\log n$. Pick $x_0 > 1/(2\delta)$ and large enough that $x > x_0$ implies $\pi(2\delta x) > (\delta)\big(1+\pi(x)\big)$. Since $\sup_{1\le i\le n}\omega(i) = o(\log n)$, $\big(1+\omega(m)\big)\log(n/m) > (\delta/2)\log n$ implies that for sufficiently large $n$, $n/m > x_0$. Thus for sufficiently large $n$, if $E_n$ fails then $x = n/J(n) > x_0$, But $U > \delta$ now implies $P_0(n) \ge 2\delta n/J(n)$, which would contradict $n/m > x_0$ with $m = J(n)P_0(n)$. This shows that for the given outcome, there is an $N_0$, (depending on the outcome through the values of $J(1), J(2), \ldots$ and $U, U_1$,) such that for all $n > N_0$, $E_n$ occurs.
■

## Lecture 4: The distance to the Poisson–Dirichlet

For an integer $N(n)$ distributed uniformly from 1 to $n$, write

$$N(n) = P_1(n)P_2(n)\ldots P_{K_n}(n) = P_1(n)P_2(n)\ldots, \qquad P_1 \ge P_2 \ge \ldots,$$

where $K_n = \Omega\big(N(n)\big)$ is the number of prime factors of $N$, and every $P_i(n)$ is either one or prime. Billingsley [17] in 1972 proved that the Poisson–Dirichlet process gives the limit in distribution for the sizes of the large prime factors,

$$\left(\frac{\log P_1(n)}{\log n}, \frac{\log P_2(n)}{\log n}, \ldots\right) \Rightarrow (V_1, V_2, \ldots), \tag{77}$$

where $(V_1, V_2, \ldots)$ has the Poisson–Dirichlet distribution with parameter 1. The marginal distribution of the largest component is given by Dickman's [21] function $\rho$, in the form $\mathbb{P}(V_1 \le 1/u) = \rho(u)$; see [44] Chapter III.5. The characterization of the limit which is useful for us is

$$(V_1, V_2, \ldots) \stackrel{\mathrm{d}}{=} \mathrm{RANK}\,(1 - X_1, X_1 - X_2, X_2 - X_3, \ldots) \tag{78}$$

where RANK is the function which sorts the coordinates in nonincreasing order, and $X_1, X_2, \ldots$ are those points of the scale invariant Poisson process which fall in $(0,1)$, indexed with $1 > X_1 > X_2 > \ldots > 0$. For other characterizations of the Poisson–Dirichlet, see for example [8, 39]. Donnelly and Grimmett [22] gave a very nice proof of (77) by showing that a size biased permutation of the left side of (77) converges in distribution to $(1 - X_1, X_1 - X_2, X_2 - X_3, \ldots)$, and then using the continuity of RANK on the simplex.

We asked: how close are the right and left sides of (77)? One notion of approximation, from Knuth and Trabb Pardo [32], is that for fixed $i$ and $t \in (0,1)$, as $n \to \infty$,

$$\mathbb{P}\left(\frac{\log P_i(n)}{\log n} \leq t\right) = \mathbb{P}(V_i \leq t) + O\left(\frac{1}{\log n}\right).$$

They also give a version with a $O(1/\log n)$ correction term, of the form: for fixed $i \geq i$, and fixed $t \in (0,1)$, $\mathbb{P}\big(\log P_i(n)/(\log n) \leq t\big) = \mathbb{P}(V_i \leq t) + r_i(t)/\log n + O\big(1/(\log n)^2\big)$. That a similar result holds for the *joint* finite dimensional distributions, together with an expansion in negative powers of the logarithm, has recently been shown by Tenenbaum [46]. To state this, for $k \geq 1$ write $F_n(\alpha_1, \ldots, \alpha_k) = \mathbb{P}\big(\log P_i(n)/(\log n) \geq \alpha_i$ for $i = 1$ to $k\big)$, and $\phi_0(\alpha_1, \ldots, \alpha_k) = \mathbb{P}(V_1 \geq \alpha_1, \ldots, V_k \geq \alpha_k)$, so that Billingsley's result (77) is equivalent to: for all $k \geq 1$ and $\alpha_1, \ldots, \alpha_k \in (0,1)$, $F_n(\alpha_1, \ldots, \alpha_k) = \phi_0(\alpha_1, \ldots, \alpha_k) + o(1)$. Tenenbaum's result is the following: for every $k \geq 1$ there exist functions $\phi_1, \phi_2, \ldots$, continuous except on finitely many hyperplanes, such that,

$$(79) \qquad F_n(\alpha_1, \ldots, \alpha_k) = \sum_{0 \leq h \leq H} \frac{\phi_h(\alpha_1, \ldots, \alpha_k)}{(\log n)^h} + O\left(\frac{1}{(\alpha_k \log n)^{H+1}}\right)$$

holds for all fixed $(\alpha_1, \ldots, \alpha_k)$, and uniformly outside an exceptional set of $(\alpha_1, \ldots, \alpha_k)$ with measure $O(\log \log n / \log n)$, specified in [46]

A very natural and useful choice of metric is the $l_1$ distance, since this controls the approximation of the set of logarithms of all divisors, see [27, 37], by its limit, see [3], Section 22. Approximations in this metric are not comparable to results such as (79); an analogous situation is that, from knowledge that the difference between the two sides of (3) is at most $1/n$, Kubilius' fundamental lemma *does not* follow as a consequence.

A very natural and useful choice of metric is the $l_1$ distance, since this controls the approximation of the set of logarithms of all divisors, see

[27, 37], by its limit, see [3], Section 22. For the metrized approximation question, it is not necessary to divide by $\log n$. For a proof or disproof of the following conjecture about the $l_1$ distance, from [6], I now offer a one hundred dollar prize.

**Conjecture 2** (\$100 prize offered). *For all $n \geq 1$, it is possible to construct $N(n)$ uniformly distributed from 1 to $n$, and the Poisson–Dirichlet process $(V_1, V_2, \ldots)$, on one probability space, so that*

$$\mathbb{E} \sum \left| \log P_i(n) - (\log n)V_i \right| = O(1). \tag{80}$$

Note that the liminf of the left side of (80) is at least 1, because $\mathbb{E} \sum \log P_i(n) = \mathbb{E} \log N(n) = \log n - 1 + o(1)$, from Stirling's $n! \sim (n/e)^n \sqrt{2\pi n}$, while $\mathbb{E} \sum (\log n) V_i = \log n$. We finished our workshop lectures by describing a coupling that achieves $O(\text{loglog}\, n)$ in place of $O(1)$ in (80); here in the writeup we also present the proof that this coupling works as claimed.

**Theorem 5.** *The coupling of the Poisson–Dirichlet with a random integer $N(n)$ uniformly distributed from 1 to $n$, described in Section 4.1, achieves*

$$\mathbb{E} \sum \left| \log P_i(n) - (\log n)V_i \right| = O(\log \log n). \tag{81}$$

Historical notes: *all* logarithmic combinatorial structures have a Poisson–Dirichlet limit for the fractions of system size in the largest, second largest, third largest, ..., components. This was shown in 1977 for permutations, by Kingman [30], and independently by Vershik and Schmidt [47, 48]. It was shown for random mappings — where the Poisson–Dirichlet limit has parameter 1/2 — by Aldous [1] in 1983. It was shown for a wide class of combinatorial structures by Hansen [28] in 1994, and with local limit bounds, for a very general scheme, in [7].

The analog of Conjecture 1 is *false* for permutations, for the simple reason that $nV_i$ has a distribution with a continuous density, while the size $L_i(n)$ of the $i^{\text{th}}$ largest cycle has integer support, so that under *any conceivable* coupling, for every $i$, $\liminf_n \mathbb{E} \left| L_i(n) - nV_i \right| \geq 1/4$. For $i > (1+\varepsilon)\log n$, one can match $nV_i$ with $L_i(n) = 0$, and the net result is that $\liminf_n (1/\log n) \mathbb{E} \sum_i \left| L_i(n) - nV_i \right| \geq 1/4$. It is shown in [9] that the coupling for permutations which is analogous to our coupling in Section 4.1 achieves this lower bound, with

$$\mathbb{E} \sum \big| \log L_i(n) - nV_i \big| \sim \frac{1}{4} \log n. \tag{82}$$

## 4.1. Growing a random integer from the Poisson–Dirichlet

This section gives a third coupling for growing a random integer $J(n)P_0(n)$ which is close in distribution to the uniform random integer $N(n)$. Our first coupling, in Lecture 1, determined $J(n)$ from a size biased permutation of the multiset with $Z_p$ spacings of size $\log p$, with $Z_p$ geometrically distributed. Our second coupling, in Lecture 3, used instead a size biased permutation of the Poisson multiset with $A_{p^k}$ spacings of size $\log p^k$, with $A_{p^k} \sim$ Poisson $\big(1/(kp^k)\big)$. This Poisson multiset can be constructed by applying the deterministic function $h$ at (42) to the points of the scale invariant Poisson process, so the second coupling may be viewed as constructing $J(n)$ from a deterministic function of the scale invariant Poisson, together with the auxiliary randomization of a size biased permutation.

For our third coupling, this Poisson multiset, *and its permutation*, will be given by a deterministic function applied to the *spacings* of the Poisson–Dirichlet process. These spacings are the points of the scale invariant Poisson process, in order of a size biased permutation. The sizes $\log p^k$ are slightly different from the sizes of the Poisson–Dirichlet spacings, so the ordering of spacings in our third coupling is a perturbation of that in our second coupling.

Start with the Poisson process on $(0,\infty)^2$, with intensity $e^{-wy}\,dw\,dy$, from Section 3.3, with points $\big\{(W_i, Y_i), i \in \mathbb{Z}\big\}$, but reverse the direction of the indexing, so that $W_i < W_{i+1}$ for all $i \in \mathbb{Z}$. Defining $X_i := \sum_{j \geq i} Y_j$ gives the scale invariant Poisson process $\mathcal{X} = \{X_i \,:\, i \in \mathbb{Z}\}$, indexed in decreasing order, with $X_{i+1} < X_i$. Now *shift* the indexing of $\big\{(W_i, Y_i), i \in \mathbb{Z}\big\}$, so that $X_1$ appears as the first point to the left of $\log n$. To summarize, we have the points of the scale invariant Poisson process $\mathcal{X}$, indexed so that

$$0 < \ldots < X_3 < X_2 < X_1 < \log n \leq X_0 < X_{-1} < \ldots. \tag{83}$$

The Poisson–Dirichlet from (78), scaled up by $\log n$, is realized as

$$((\log n)V_1, (\log n)V_2, \ldots) = \text{RANK}\,(\log n - X_1, X_1 - X_2, \ldots). \tag{84}$$

The points $Y_i = X_i - X_{i+1}$ for $i \in \mathbb{Z}$ are the points of the scale invariant Poisson process, and using $h$ from (42), we construct random $Q_i$ by

$$\log Q_i = h(Y_i), \quad \text{with} \quad Y_i = X_i - X_{i+1}, \quad i \in \mathbb{Z}. \tag{85}$$

An outcome where $e^{-\gamma} \geq Y_i$ for $i = 2, 6, 7, \ldots$, so that $Q_1, Q_3, Q_4, Q_5$ are prime powers; but $1 = Q_2 = Q_6 = Q_7 = \cdots$. $\log Q_i = h(Y_i)$ is close to $Y_i$, and $\log P_0(n)$ is close to $\log n - X_1$.

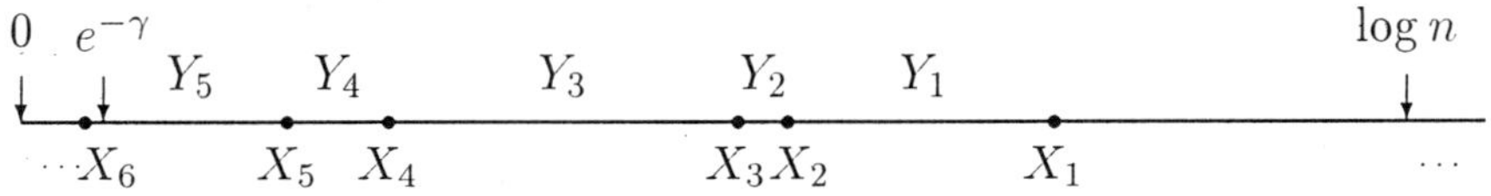


[The following information is motivational, and not part of our proof. Like the $Q_1, Q_2, \ldots$ in Section 3.5, for every $q = p^k$ we have that $A_q := \sum_{i \in \mathbb{Z}} 1(Q_i = k)$ is Poisson $\big(1/(kq)\big)$, with the $A_q$ mutually independent, but there are several differences in the indexing scheme: here the $Q_i$ are indexed from right to left, there may be instances of $Q_i = 1$ in between occurrence of proper prime powers, and most significantly, the sizes used for the size biased permutation are the $X_i - X_{i+1}$ and not the $\log Q_i$. We will need to show that since $h$ is close to the identity function, the second and third couplings are close, in that with high probability, they produce the same $J(n)$.]

Define $J^*(n)$ by

$$J^*(n) = \prod_{i \geq 1} Q_i. \tag{86}$$

It is conceivable that $J^*(n) > n$ if $X_1$ is close enough to $\log n$ and $h$ gets applied in places with $h(y) > y$; in this case we will prescribe $P_0^*(n) = 1$.

When $J^*(n) \leq n$, take $P_0^*(n)$ to be one or prime, such that $J^*(n)P_0^*(n) \leq n$, picking uniformly over the $1 + \pi\big(n/J^*(n)\big)$ possibilities.

We have two tasks. The first, carried out in Lemma 5, is to show that the prime factors $P_1^*(n), P_2^*(n), \ldots$ of this random integer $J^*(n)P_0^*(n)$, listed in nonincreasing order, give a vector of logarithms close to the Poisson–Dirichlet. The second task, carried out in Lemma 6, is to show, like Lemma 4, that the random integer we have constructed is close to uniform. ¿From these two lemma, Theorem 5 follows easily. We will state the lemmas, then give the proof of Theorem 5 to finish this section. The next section provides the proofs of the two lemmas.

**Lemma 5.**

$$\mathbb{E} \sum_{i \geq 1} \big| \log P_i^*(n) - (\log n) V_i \big| = O(1). \tag{87}$$

**Lemma 6.**

$$d_{TV}\big(J^*(n)\, P_0^*(n), N(n)\big) = O\left(\frac{\text{loglog}\, n}{\log n}\right). \tag{88}$$

**Proof of Theorem 5.** As in Section 3.8, the coupling of $J^*(n)P_0(n)$ with the Poisson–Dirichlet process can be extended to include $N(n)$, distributed uniformly on 1 to $n$, in such a way that $N(n) = J^*(n)P_0^*(n)$ except on a "bad" event of probability equal to the total variation distance $d_{TV}\big(J^*(n)\, P_0^*(n), N(n)\big)$. On the bad event we have no control over the $l_1$ distance, apart from the trivial bound: it is at most $2\log n$. Multiplying by an upper bound on the probability of the bad event, from Lemma 6, gives us the main contribution to the error, of size $O(\log\log n)$. On the complement of the bad event, the contribution is $O(1)$ by Lemma 5. Adding these errors gives (81). ■

### 4.2. Proofs of the lemmas for Theorem 5

**Proof of Lemma 2.** There are three contributions to the $l_1$ distance: first $Y_i$ versus $\log Q_i$, second a contribution from $Q_i$ which are prime powers but not prime, and third $\log P_0^*(n)$ versus $(\log n - X_1)$. Observe that for any vectors $\mathbf{x} = (x_1, x_2, \ldots)$ and $\mathbf{y} = (y_1, y_2, \ldots)$ in $[0,\infty)^{\mathbb{N}} \cap l_1$, the $l_1$ distance is not increased if the coordinates of both vectors are sorted: $\|\,\mathrm{RANK}\,(\mathbf{x}) - \mathrm{RANK}\,(\mathbf{y})\|_1 \leq \|\mathbf{x} - \mathbf{y}\|_1$.

Consider the random variable

$$D = \sum_{i \in \mathbb{Z}} \left| h(Y_i) - Y_i \right|. \tag{89}$$

It has

$$\mathbb{E}\, D = \mathbb{E} \sum_{i \in \mathbb{Z}} \left| h(Y_i) - Y_i \right| = b_0 < \infty, \tag{90}$$

using (44) and the scale invariant spacing lemma.

The first contribution to our expected $l_1$ distance is handled by: for all $n$,

$$\mathbb{E} \sum_{i \geq 1} |\log Q_i - Y_i| = \mathbb{E} \sum_{i \geq 1} \left| h(Y_i) - Y_i \right| \leq \mathbb{E}\, D = b_0 < \infty. \tag{91}$$

To handle the second contribution, we "split up" any prime powers $p^k$ with $k > 1$ which may occur among the $Q_i$, defining $Q_1^*, Q_2^*, \ldots$ to be one or prime, so that $J(n) = \prod_{i \geq 1} Q_i = \prod_{i \geq 1} Q_i^*$. To do this, start with $Q_j^* = 1$ whenever $Q_j = 1$; some of these will be changed. Always take $Q_i^* = p$ when $Q_i = p^k$. For any $Q_i = p^k$ with $k > 1$, take $k-1$ indices $j$ for which $Q_j^* = 1$ and change these to $Q_j^* = p$. With $\mathbf{x} = (\log Q_1, \log Q_2, \ldots)$ and $\mathbf{y} = (\log Q_1^*, \log Q_2^*, \ldots)$ we have $\|\mathbf{x} - \mathbf{y}\| = \sum_{p,k,i} 2(k-1)(\log p) 1(Q_i = p^k) \leq \sum_{q = p^k, k > 1} 2(\log q) A_q$. Note that $\mathbb{E} \sum_{q=p^k, k>1} (\log q) A_q = \sum_{q = p^k, k>1} (\log q)/(kq) < \infty$.

To handle the third contribution: using $c = \left(\log n - \sum h(Y_i)\right)^+$, our recipe for $P_0^*(n)$ is to choose uniformly over the $1 + \pi(e^c)$ numbers which are one, or a prime at most $e^c$. Writing $\mathbb{P}_c$ and $\mathbb{E}_c$ for such a choice of $P_0$, we have $\sup_{c\geq 0}\mathbb{E}_c(c-\log P_0)<\infty$, as a simple consequence of the prime number theorem, that $\pi(e^x)\sim e^x/x$ as $x\to\infty$. Using $X_1=\sum_{i\geq 1}Y_i$, we have $\big|(\log n - X_1)-\big(\log n-\sum_{i\geq 1}h(Y_i)\big|=\big|\sum_{i\geq 1}(Y_i-h(Y_i)\big|$, with expectation bounded by $b_0$, using (90). Combining yields $\sup_n \mathbb{E}\big|\log P_0^*(n)-(\log n - X_1)\big|<\infty$.

Combining these three contributions, and using the $l_1$ contraction property of the function RANK, proves (87). ■

**Proof of Lemma 6.** In contrast to (83) and (85), we now re-index the $\{Y_i\,:\,i\in\mathbb{Z}\}$ so that

$$\ldots<Y_{-2}<Y_{-1}<Y_0\leq e^{-\gamma}<Y_1<Y_2<\ldots.$$

The choice of location of $e^{-\gamma}$ has the effect that $Q_1,Q_2,\ldots$ are the primes and prime powers, while $1=Q_i$ for $i\leq 0$. The indexing of the $Y_i$ in the order of their own values has the effect that, with

$$S_i:=W_i/Y_i,$$

the $S_i$ for $i\in\mathbb{Z}$ are, conditional on the values of $Y_i, i\in\mathbb{Z}$, mutually independent, standard exponentials — and this would not have been true under the previous indexing, where $W_i<W_{i+1}$.

We construct the second coupling, of Section 3.5, from this multiset $\{Q_1,Q_2,\ldots\}$. For the size biased permutation of the $\log Q_i$, we use the exponentially distributed labels

$$\hat{W}_i:=S_i/\log Q_i=W_i\frac{Y_i}{h(Y_i)}, \tag{92}$$

so that $J(n)$ is the largest partial product of the $Q_i$ not exceeding $n$, with the $Q_i$ taken in order of decreasing labels $\hat{W}_i$. In contrast, $J^*(n)$ is a partial product of $Q_1,Q_2,\ldots$ taken in order of decreasing $W_i$. Because $h$ is close to the identity, eventually the permutation induced by the $W_i$ agrees with the permutation induced by the $\hat{W}_i$. We will show that

$$\mathbb{P}\big(J^*(n) \neq J(n)\big) = O\left(\frac{\log\log n}{\log n}\right), \tag{93}$$

and thus $d_{TV}\big(J^*(n)P_0^*(n), J(n)P_0(n)\big) = O(\log\log n/\log n)$. Combined with Lemma 4, this yields $d_{TV}\big(J^*(n)P_0^*(n), N(n)\big) = O(\log\log/\log n)$. As a remark, we believe that the quantity in (93) is actually $O(1/\log n)$, but since this improved bound would not improve the overall result, we settle for the looser bound.

First, we show that the effect in (86) of stopping at the largest $X_i < \log n$, rather than stopping with the largest partial product not exceeding $n$, is negligible. Consider the "good" event

$$G = \big\{ D \geq \log\log n \text{ or } \mathcal{X} \cap (\log n - \log\log n, \log n + \log\log n) \neq \emptyset \big\}, \tag{94}$$

with $D$ given by (89). We will show that

$$\mathbb{P}(G^c) = O\left(\frac{\log\log n}{\log n}\right), \tag{95}$$

To show this, first observe that $\mathbb{P}\big(\mathcal{X} \cap (\log n - \log\log n, \log n + \log\log n) \neq \emptyset\big) = O(\log\log n/\log n)$, since the intensity of $\mathcal{X}$ is $1/x\,dx$. Second, observe that $\mathbb{P}(D \geq \log\log n) = O(1/\log n)$, which follows from showing $\mathbb{E}e^{\beta D} < \infty$ with some $\beta > 1$. In fact, $\mathbb{E}e^{\beta D} < \infty$ for all $\beta$; with $g$ as defined following (41), and using (33), we have $D = \sum_{i\in\mathbb{Z}} \big|g(L_i) - \exp(L_i)\big|$ so that

$$\mathbb{E}e^{\beta D} = \exp\left(\int_{-\infty}^{\infty} \left(e^{\beta(g(l)-e^l)} - 1\right) dl\right).$$

The contribution to the integral from the neighborhood of $-\infty$ is finite, using $g(l) = 0$ there, and contribution to the integral from the neighborhood of $\infty$ is finite, using $g(l) - e^l = O\big(e^l \exp(-ce^{l/2})\big)$ as $l \to \infty$, which follows from (41).

Next, we consider a bad event on which the two permutations, one induced by the sizes $W_i$, and the other induced by the sizes $\hat{W}_i := W_i\, Y_i/h(Y_i)$ as in (92), might disagree in a nontrivial way, i.e. giving the opposite ordering to $Q_i, Q_j > 1$, out beyond the place where the partial sum of the $Y_i$ exceeds $(\log n)/2$. Let

$$B = \Big\{ \exists i \neq j,\ Y_i, Y_j > e^{-\gamma}, W_i < W_j, \hat{W}_i \geq \hat{W}_j, \qquad (96)$$
$$\sum Y_k 1(W_k \geq W_i) > (\log n)/2 \Big\}.$$

For $n$ so large that $\log\log n < (\log n)/2$, we have

$$\{ J^*(n) \neq J(n) \} \subset G^c \cup B,$$

so that it only remains to show $\mathbb{P}(B) = O(\log\log n / \log n)$.

Let

$$T_w = \sum_{i \in \mathbb{Z}} Y_i 1(W_i > w).$$

The distribution of $T_w$ is exponential, with $\mathbb{E}T_w = 1/w$ — see for example [31], under the "Moran process", or [2], where this is used as an ingredient in the proof of the scale invariant spacing lemma. We say that $\big((w,y),(w',y')\big)$ is a "potential witness" to the bad event $B$ if $y, y' > e^{-\gamma}$, $w' < w$, the Poisson process $\{(W_i, Y_i)\}$ has points at $(w,y)$ and $(w',y')$, and no points $(W_k, Y_k)$ with $w' < W_k < w$, and $T_w + y + y' > (\log n)/2$, and

$$\log\left(\frac{w}{w'}\right) \leq \big|\log\left(y/h(y)\right)\big| + \big|\log\left(y'/h(y')\right)\big|. \qquad (97)$$

Let $N_B$ denote the number of potential witnesses, and observe that $B \subset \{N_B > 0\}$. Thus, it only remains to calculate that $\mathbb{E}N_B = O(\log\log n / \log n)$; and in fact we will show that it is $O(1/\log n)$.

Now $\mathbb{E}N_B$ is merely a four-fold integral, so the reader is invited to take our claim at face value; but for those declining our invitation, here are the details. Conditional on having points $(W_i, Y_i)$ and $(W_j, Y_j)$ with

$W_i = w, W_j = w'$, the joint distribution of $Y_i, Y_j, T_w$ is that of three independent exponentials, with means $1/w, 1/w'$, and $1/w$ respectively. Let $c_0 := \exp\left(2\sup\left\{\left|\log\left(y/h(y)\right)\right| : y > e^{-\gamma}\right\}\right)$, so that for a potential witness, $w/w' \le c_0$, and $Y_j$ lies below an exponential with mean $c_0/w$. This gives us, for the conditional probability, that $\mathbb{P}_{w,w'}(Y_i+Y_j+T_w > \log n/2) \le 3\mathbb{P}\big((c_0/w)S_1 > \log n/6\big) = 3\exp\big(-w\log n/(6c_0)\big)$.

We need some monotonicity for the next simplification. from (41) we have that

$$h(x) = x + O\big(xe^{-c\sqrt{x}}\big) \quad \text{as} \quad x \to \infty,$$

so that for some constants $c_1, c_2 > 0$, for all $y > e^{-\gamma}$, $\left|\log\left(h(y)/y\right)\right| < c_1 e^{-c_2\sqrt{y}}$. Thus we can relax the notion of "potential witness," replacing (97) with the condition

$$\log\left(\frac{w}{w'}\right) \le r(y) + r(y'), \quad \text{where} \quad r(y) = c_1 e^{-c_2\sqrt{y}}. \tag{98}$$

Write $N_R$ for the number of potential witnesses in this relaxed sense, so that $N_B \le N_R$. Now the indicator of the inequality (98) is a *decreasing* function of $(y, y')$, while the indicator $1(y+y'+t > \log n/2)$ is an *increasing* function, so that we have negative correlations (with respect to $Y_i$, $Y_j$, and $T_w$, which are conditionally independent given $w$, $w'$):

$$\mathbb{P}_{w,w'}\left(Y_i + Y_j + T_w > \frac{\log n}{2}, \log\left(\frac{w}{w'}\right) \le r(Y_i) + r(Y_j)\right)$$

$$\le \mathbb{P}_{w,w'}\left(Y_i + Y_j + T_w > \frac{\log n}{2}\right)\mathbb{P}_{w,w'}\left(\log\left(\frac{w}{w'}\right) \le r(Y_i) + r(Y_j)\right).$$

We use the monotonicity of $r(\cdot)$ again, to justify

$$\mathbb{P}_{w,w'}\left(\log\left(\frac{w}{w'}\right) \le r(Y_i) + r(Y_j)\right) \le \mathbb{P}\left(\log\left(\frac{w}{w'}\right) \le r\left(\frac{S_i}{w}\right) + r\left(\frac{S_j}{w}\right)\right),$$

where $S_i$, $S_j$ represent independent standard exponentials.

Recalling that $\{W_k : k \in \mathbb{Z}\}$ form a copy of the scale invariant Poisson process $\mathcal{X}$, and $\mathbb{P}(\mathcal{X} \cap (w', w) = \emptyset) = w'/w$,

$$\mathbb{E} N_R \le \iint_{w'<w} \frac{dw'}{w'} \frac{dw}{w} \frac{w'}{w}\, 3 \exp\left( - w \log n/(6c_0)\right)$$

$$\times \mathbb{P}\left( \log\left(\frac{w}{w'}\right) \le r\left(\frac{S_i}{w}\right) + r\left(\frac{S_j}{w}\right)\right).$$

Recall further that for two consecutive points $W_j < W_i$ of the scale invariant Poisson process, conditional on $W_i = w$ the distribution of $W_j$ is that of $Uw$, where $U$ uniformly distributed in $(0, 1)$. Thus the right hand side above is equal to

$$\int_{w>0} \frac{dw}{w}\, 3 \exp\left( - w \log n/(6c_0)\right) \mathbb{P}\left( \log\left(\frac{1}{U}\right) \le r\left(\frac{S_i}{w}\right) + r\left(\frac{S_j}{w}\right)\right).$$

Since $-\log U$ is exponentially distributed, with density bounded above by one, we have

$$\mathbb{E} N_R \le \int_{w>0} \frac{dw}{w}\, 3 \exp\left(-w \log n/(6c_0)\right) \mathbb{E}\, 2r\left(\frac{S_i}{w}\right)$$

$$= 6c_1 \int_{w>0} \exp\left( - w \log n/(6c_0)\right) dw \int_{y>0} e^{-wy} e^{-c_2\sqrt{y}}\, dy$$

$$< 6c_1 \int_{y>0} e^{-c_2\sqrt{y}}\, dy \int_{w>0} \exp\left( - w \log n/(6c_0)\right) dw = O\left(\frac{1}{\log n}\right).$$

This completes the proof of (88). ■

## REFERENCES


[1] Aldous, D. J. (1983) Exchangeability and related topics. Springer, Lecture Notes in Mathematics, vol. 1117.

[2] Arratia, R. (1996) Independence of prime factors: total variation and Wasserstein metrics, insertions and deletions, and the Poisson–Dirichlet process. Draft, 70 pages.

[3] Arratia, R. (1998) On the central role of scale invariant Poisson processes on $(0, \infty)$. Microsurveys in Discrete Probability (Princeton, NJ, 1997), 21–41, (edited by D. Aldous and J. Propp) DIMACS Ser. Discrete Math. Theoret. Comut. Sci., 41 (1998), Amer. Math. Soc., Providence, RI.

[4] Arratia, R., Barbour, A. D., and Tavaré, S. (1992) Poisson process approximation for the Ewens Sampling Formula. Ann. Appl. Probab. **2**, 519–535.

[5] Arratia, R., Barbour, A. D., and Tavaré, S. (2001) Logarithmic combinatorial structures. Monograph, in preparation. Draft version available at `http://www-hto.usc.edu/books/tavare/ABT/index.html`

[6] Arratia, R., Barbour, A. D., and Tavaré, S. (1997) Random combinatorial structures and prime factorizations. AMS Notices **44**, 903–910.

[7] Arratia, R., Barbour, A. D., and Tavaré, S. (1999) On Poisson–Dirichlet limits for random decomposable combinatorial structures. Combin., Probab., Comput. **8** 193–208

[8] Arratia, R., Barbour, A. D., and Tavaré, S. (1999) The Poisson–Dirichlet distribution and the scale invariant Poisson process. Combin., Probab., Comput. **8**, 407–416.

[9] Arratia, R., Barbour, A. D., and Tavaré, S. (1999) Expected $l_1$ distance in Poisson–Dirichlet approximations for random permutations: a tale of four couplings. Preprint

[10] Arratia, R., and Stark, D. (1999) A total variation distance invariance principle for primes, permutations and Poisson–Dirichlet. Preprint.

[11] Arratia, R., and Tavaré, S. (1992) The cycle structure of random permutations. Ann. Probab. **20**, 1567–1591.

[12] Arratia, R., and Tavaré, S. (1996) Random partitions, permutations, and primes. Course lecture notes for Math 533. University of Southern California, Department of Mathematics.

[13] Bach, E. (1985) Analytic Methods in the Analysis and Design of Number-theoretic Algorithms. The MIT Press.

[14] Barban, M. B., and Vinogradov, A. I. (1964) On the number theoretic basis of the probabilistic theory of numbers. Dokl. Akad. Nauk SSSR **154**, 495–496.

[15] Barbour, A. D. (1990) Comment on "Poisson approximation and the Chen–Stein method". Statistical Science **5**, 425–427.

[16] Beyer, W., Stein, M., Smith, T., and Ulam, S. (1972) Metrics in Biology, an Introduction. Los Alamos Scientific Laboratory report LA-4973. Reprinted in Analogies between Analogies; the mathematical reports of S. M. Ulam and his Los Alamos collaborators, 1990 University of California Press.

[17] Billingsley, P. (1972) On the distribution of large prime factors. Period. Math. Hungar. **2**, 283–289.

[18] Billingsley, P. (1973) Prime numbers and Brownian motion. Amer. Math. Monthly **80**, 1099–1115.

[19] Billingsley, P. (1974) The 1973 Wald Memorial Lecture: The probability theory of additive arithmetic functions. Ann. Probab. **2**, 749–791.

[20] Diaconis, P., and Pitman, J. (1986) Permutations, record values and random measures. Unpublished lecture notes, Dept. Statistics, U. C. Berkeley.

[21] Dickman, K. (1930) On the frequency of numbers containing prime factors of a certain relative magnitude. Ark. Math. Astr. Fys. **22**, 1–14.

[22] Donnelly, P., and Grimmett, G. (1993) On the asymptotic distribution of large prime factors. J. London Math. Soc. (2) **47**, 395–404.

[23] Dudley, R. M. (1989) Real Analysis and Probability. Wadsworth and Brooks/Cole.

[24] Elliott, P. D. T. A. (1979) Probabilistic Number Theory I, Springer Grundlehren der math. Wissenschaften 239.

[25] De Koninck, J.-M., and Galambos, J. (1987) The intermediate prime divisors of integers. Proc. Amer. Math. Soc. **101**, 213–216.

[26] Feller, W. (1945) The fundamental limit theorems in probability. Bull. Amer. Math. Soc., **51**, 800–832.

[27] Hall, R.R., and Tenenbaum, G. (1988) Divisors. Cambridge Tracts in Mathematics **90**, Cambridge.

[28] Hansen, J. C. (1994) Order statistics for decomposable combinatorial structures. Random Structures and Algorithms **5**, 517–533.

[29] Ignatov, Z. (1981) Point processes generated by order statistics and their applications. In Point processes and queuing problems (Colloq., Keszthely, 1978), 109–116, North-Holland, Amsterdam–New York

[30] Kingman, J. F. C. (1977) The population structure associated with the Ewens sampling formula. Theor. Pop. Biol. **11**, 274–283.

[31] Kingman, J. F. C. (1993) Poisson Processes. Oxford Science Publications

[32] Knuth, D., and Trabb Pardo, L. (1976) Analysis of a simple factorization algorithm. J. Theoret. Comput. Sci. **3**, 321–348.

[33] Kruskal, J., and Sankoff, D. (1983) Time warps, string edits, and macromolecules: the theory and practice of sequence comparison. Addison–Wesley.

[34] Kubilius, J. (1962, translated 1964) Probabilistic Methods in the Theory of Numbers. AMS Translations of Mathematical Monographs, **11**.

[35] Kurtz, T. (1978) Strong approximation theorems for density dependent Markov chains. Stochastic Procs. Appls. **6**, 223–240.

[36] Levenstein, V. (1965) Binary codes capable of correction deletions, insertions, and reversals. Cybernetics and Control Theory **10** 707–710 (1996); Russian Doklady Akademii Nauk SSR **163** 845–848.

[37] Mendès France, M., and Tenenbaum, G. (1993) Systèmes de points, diviseurs, et structure fractale. Bull. Soc. Math. de France **121**, 197–225.

[38] Philipp, W. (1973) Arithmetic functions and Brownian motion. Proc. Sympos. Pure Math **24** 233–246.

[39] Pitman, J., and Yor, M. (1997) The two-parameter Poisson–Dirichlet distribution derived from a stable subordinator. Ann. Probab. **25**, 855–900.

[40] Rényi, A., and Turán, P. (1957) On a theorem of Erdős–Kac. Acta Arith. **4**, 71–84.

[41] Rio, E. (1994) Local invariance principles and their application to density estimation. Probab. Th. Related Fields **98**, 21–45.

[42] Rosser, J. B., and Schoenfeld, L. (1962) Approximate formulas for some functions of prime numbers. Illinois J. Math. **6**, 64–94.

[43] Stark, D. (1997) Explicit limits of total variation distance in approximations of random logarithmic combinatorial assemblies by related Poisson processes. Combin., Probab., Comput. **6**, 87–105.

[44] Tenenbaum, G. (1995) Introduction to analytic and probabilistic number theory. Cambridge studies in advanced mathematics, **46**. Cambridge University Press.

[45] Tenenbaum, G. (1999) Crible d'Ératosthène et modèle de Kubilius, Number theory in progress, Vol. 2 (Zakopane–Kościelisko, 1997), 1099–1129, de Gruyter, Berlin, 1999.

[46] Tenenbaum, G. (2000) A rate estimate in Billingsley's theorem for the size distribution of large prime factors. Quart. J. Math. **51**, 387–405.

[47] Vershik, A. M., and Shmidt, A. A. (1977) Limit measures arising in the theory of groups, I, Theory Probab. Appl. **22**, 79–85.

[48] Vershik, A. M., and Shmidt, A. A. (1978) Limit measures arising in the theory of groups, II, Theory Probab. Appl. **23**, 36–49.

Richard Arratia

*Univ. of Southern California*
*Department of Mathematics*
*Los Angeles CA 90089-1113*
*U.S.A.*

e-mail: rarratia@math.usc.edu

BOLYAI SOCIETY
MATHEMATICAL STUDIES, 10

**Contemporary Combinatorics**
**Budapest, 2002,** pp. 93–138.

# Tic-Tac-Toe

J. Beck

## §1. Positional games

Tic-Tac-Toe and its variants are board games which are won by the first player to complete some kind of winning pattern.

Tic-Tac-Toe (or Noughts and Crosses): Played as follows, the game board is a big square which is partitioned into $3 \times 3 = 9$ congruent small squares. Whoever moves first puts a nought in one of the nine small squares. The opponent puts a cross into any other small square, and then they alternate nought and cross in the remaining empty squares until one player wins by getting three of his own squares in a line. There are precisely eight winning triplets.

Note that the Tic-Tac-Toe is a "3-in-a-row" game on a $3 \times 3$ board. A straightforward generalization is the "$n$-in-a-row" game on an $n \times n$ board that we simply call the $n^2$-game. The multi-dimensional generalization is the Hales–Jewett $n^d$-game, or simply the

### $n^d$-game

We define the $n^d$-game in the same fashion as Tic-Tac-Toe is the $3^2$-game. The two players alternately put their marks in the cells of a $d$-dimensional cube of size

$$n \times \cdots \times n = n^d.$$

The winner is the player to have $n$ of his marks in a line *first.* More precisely, the board of the $n^d$-game is the set of $d$-tuples

$$X = \{\mathbf{a} = (a_1, a_2, \ldots, a_d) : 1 \le a_j \le n \text{ for each } 1 \le j \le d\}.$$

The winning sets of the $n^d$-game are the $n$-in-a-line sets, that is, the $n$-element sequences

$$\left(\mathbf{a}^{(1)}, \mathbf{a}^{(2)}, \ldots, \mathbf{a}^{(n)}\right)$$

of the board $X$ such that, for each $j$, the sequence $a_j^{(1)}, a_j^{(2)}, \ldots, a_j^{(n)}$ composed of the $j$th coordinates is either strictly increasing from 1 to $n$, or strictly decreasing from $n$ to 1, or constant (apart from constant sequences).

In other words, this is the $n$-in-a-line game on the $n^d$ board. Note that in higher dimensions most of the lines are some kind of diagonals.

The special case $n = 3$, $d = 2$ gives the usual Tic-Tac-Toe.

## Strong positional games

An extremely far-reaching generalization of Tic-Tac-Toe is the class of *strong positional games.* Let $(X, \mathcal{F})$ be an arbitrary finite hypergraph. Here $X$ is a finite set, the *board* of the game, and $\mathcal{F}$ is an *arbitrary* family of subsets of $X$, the family of *winning sets.* The two players, First-Player (or **I**) and Second-Player (or **II**), alternately occupy previously unoccupied elements of the board $X$. The player who occupies all the elements of some winning set $A \in \mathcal{F}$ *first* is the winner — otherwise the play ends in a draw.

Sometimes we just give the family $\mathcal{F}$ of winning sets: then the board $X$ is simply the union $\bigcup_{A \in \mathcal{F}} A$ of all winning sets.

## Strategy stealing argument

Since an extra move in a strong game does not harm a player, by the well-known *strategy stealing* argument, whoever plays first can force at least a draw ("drawing strategy"). Heuristically this is obvious because strong positional games are symmetric and the First-Player has the advantage of the first move. The precise proof of this remarkable result is just a little bit more complicated.

First note that a *strategy* for the First-Player (Second-Player) formally means a function Str with domain "the set of all subsequences of even (odd) length of the board $X$" and with range $X$ such that, if the First-Player's points are $x_1, x_2, x_3, \ldots$, and the Second-Player's points are $y_1, y_2, y_3, \ldots$, then the $i^{\text{th}}$ move $x_i$ ($y_i$) is determined from the "past" by

$$x_i = \mathrm{Str}\,(x_1, y_1, x_2, y_2, \ldots, y_{i-1}) \in X \setminus \{x_1, y_1, x_2, y_2, \ldots, y_{i-1}\}$$

$$\left(y_i = \mathrm{Str}\,(x_1, y_1, x_2, y_2, \ldots, y_{i-1}, x_i) \in X \setminus \{x_1, y_1, x_2, y_2, \ldots, y_{i-1}, x_i\}\right).$$

**Theorem 1** ("Strategy Stealing Theorem"). *Let $(X, \mathcal{F})$ be an arbitrary finite hypergraph. Then playing the strong positional game on $(X, \mathcal{F})$, the First-Player has a strategy that guarantees at least a draw, i.e. a win or a draw.*

**Proof.** The following is a purely existence argument. The negation of the statement of Theorem 1 is that the Second-Player (**II**) has a winning strategy. Indeed, by De Morgan's law, the negation of a statement like $\exists x_1 \forall y_1 \exists x_2 \forall y_2 \cdots$ is $\forall x_1 \exists y_1 \forall x_2 \exists y_2 \cdots$.

Let Str be a winning strategy for **II**, and the idea is to see what happens if the First-Player (**I**) steals it. Note that a winning strategy for a player is a very long list of instructions telling the player that if the opponent does this, then he does that, such that if the player follows the instructions, he will always win. Now **I** can use **II**'s winning strategy Str to win as follows. **I** takes an arbitrary first move, and then *pretends* to be the second player, that is, he ignores his first move. After **II**'s each move, **I**, as a fake second player, reads the instruction in Str to take action. If **I** is told to take a move that is still available, he takes it. If this move was taken by him before as his ignored "arbitrary" first move, then he takes another "arbitrary move".

The crucial point here is that an extra move, namely the last "arbitrary move", only *benefits* **I** in a strong game. The conclusion is that if **II** had a winning strategy, so would **I**. But it is impossible that both players win in the same play, so the supposed winning strategy for **II** cannot exist. This implies that **I** can always force at least a draw. ■

Let us make some important remarks. Our main interest is the outcome of the play between *perfect* players. To be a perfect player assumes that the player knows at least one of his *optimal* strategies, even if this knowledge requires "supernatural powers" (like being able to perform a case-study of size $10^{100}$). If First-Player is *not* perfect and makes a "mistake" in playing a strong game, then the opponent might be able to take advantage of the mistake, and wins the play. Of course, this doesn't contradict Theorem 1.

When we say "perfect player", this doesn't mean that the concept of winning (or drawing) strategy *presupposes* "rational" behavior of the opponent. Not at all. A winning (drawing) strategy for player $P$ simply means that $P$ can always force a win (a win or a draw), no matter what his opponent does. If player $P$ has a winning strategy and follows the instructions of his winning strategy, it does not make any difference whether his opponent is clever or foolish, $P$ will win in any case. On the other hand, if player $P$ has a drawing strategy, then of course the opponent can make a difference: a foolish opponent can lose a potential draw-game.

## When draw is impossible — Ramsey Theory

When can the First-Player win in a strong positional game? A simple but useful sufficient condition is: when a draw is impossible.

Of course this condition is not necessary. For example, the full branches of a binary tree with $n$ levels form an $n$-uniform family of $2^{n-1}$ winning sets such that the first player has an easy win. But this strong game has plenty of draw end-positions.

A draw end-position in a strong game $(X, \mathcal{F})$ means a 50–50% two-coloring of the board $X$ such that no winning set $A \in \mathcal{F}$ is monochromatic. A slightly more general concept is when we allow *arbitrary* two-colorings, not just 50-50% colorings.

The *chromatic number* $\chi(\mathcal{F})$ of a hypergraph $\mathcal{F}$ is the least integer $r \geq 2$ such that the elements of $X$ can be colored with $r$ colors yielding no monochromatic $A \in \mathcal{F}$. If the chromatic number of $\mathcal{F}$ is bigger than two, then draw is impossible. In this case First-Player's (at least) drawing strategy in Theorem 1 is actually a winning strategy.

**Theorem 2** ("Ramsey Criterion"). *Suppose that the board $X$ is finite, and the family $\mathcal{F}$ of winning sets has chromatic number at least three. Then First-Player has a winning strategy in the strong positional game $(X, \mathcal{F})$.* ■

Theorem 2 describes a subclass of strong positional games with the remarkable property that one can easily diagnose the winner without being able to say how one wins.

A fundamental result in Ramsey Theory is the Hales–Jewett theorem (see [18–19]). This is basically the combinatorial content of Van der Waerden's famous theorem on arithmetic progressions. It says that there is a finite threshold number $h(n)$ such that the family of winning sets in the $n^d$-game has chromatic number $\geq 3$ if $d \geq h(n)$. Therefore, applying the "Ramsey Criterion" (Theorem 2), First-Player has a winning strategy.

Unfortunately this result is rather poor because our present knowledge on the Hales–Jewett threshold number $h(n)$ is rather poor. The best known upper bound on $h(n)$, a breakthrough result of Shelah [23], is primitive recursive, which is much, much better than the original Van der Waerden–Hales–Jewett threshold (see e.g. in [18]). The original argument could not even provide a simple recursive upper bound. Unfortunately Shelah's bound is still enormous for "pedestrian combinatorics". It is an open problem whether $h(n) < \text{tower}_n(n)$ holds or not, where $\text{tower}_k$ denotes the $k$-fold iteration of the exponential function: $\text{tower}_1(x) = 2^x$ and for $k \geq 2$, $\text{tower}_k(x) = 2^{\text{tower}_{k-1}(x)}$ (so e.g. $\text{tower}_2(x) = 2^{2^x}$). To get a good idea about the order of magnitude of Shelah's bound, the reader is recommended to study the following $n$-fold iteration: let $\text{tower}\,(k;x) = \text{tower}_k(x)$, and

$$\text{supertower}_2(n) = \text{tower}\Big(\text{tower}\,\big(\cdots \text{tower}\,\big(\,\text{tower}\,(2;2);2\big)\,\ldots;2\big)\,;2\Big).$$

In the other direction, the best known *lower* bound is nearly quadratic: $h(n) > cn^2/\log n$. The discrepancy between the quadratic lower bound and Shelah's supertower upper bound is very shocking.

It seems to be highly unlikely that the breaking points for the $n^d$-games are anywhere close to the Hales–Jewett threshold numbers $h(n)$, but no method is known for handling this problem. Note that even the existence of a breaking point is questionable.

## Weak positional games

As we have seen in Theorem 1, Second-Player has no chance to win a strong game against a perfect First-Player. Then, why does he not just concentrate on preventing First-Player from building a winning set, and simply ignores his own desire for building? We thus can name him the Breaker, and the other one the Maker. We can even take one small step further, and consider the Maker-Breaker version even if the Breaker moves first.

In general, on the same hypergraph $(X, \mathcal{F})$, one can play the symmetric strong version, and also the asymmetric *Maker-Breaker version*, where Maker's aim is to pick every element of a winning set $A \in \mathcal{F}$, and Breaker's aim is to prevent Maker from doing so. The winner is the one who achives his goal – so draw is impossible by definition.

If First-Player can force a win in the strong game $(X, \mathcal{F})$, then the *same* play gives him, as Maker, a win in the Maker-Breaker version on the same hypergraph. But the converse is not true. It is possible that Maker as the first player (Maker = **I**) has a winning strategy in the Maker-Breaker version while Second-Player can force a draw in the strong version. This happens for example in the standard $3^2$ Tic-Tac-Toe: the strong game is a draw, but the Maker-Breaker version is a win for Maker = **I**. This is why one can call a Maker's winning strategy in the Maker-Breaker version a "weak win", and refer to the Maker-Breaker version as a *weak positional game*.

Similarly, if Breaker = **II** has a winning strategy in the weak game, then the same strategy is a drawing strategy in the strong game. This is just the negation of the statement above. Naturally Tic-Tac-Toe is again a counter-example for the converse: Second-Player's drawing strategy in the strong game cannot prevent Maker = **I** from occupying a winning triplet.

While playing the strong version on a hypergraph, both players have their own threats, and either of them, fending off the other's, may build his own winning set. Therefore, a play is a delicate balancing between threats and counter-threats and can be of very intricate structure even if the hypergraph itself is simple.

The *weak* version (i.e. the Maker-Breaker version) is usually somewhat simpler. Maker does not have to waste valueable moves fending off his opponent's threats. Maker can simply concentrate on his own goal.

This does not mean, however, that weak games are easy. For example, the well-known *Hex* is equivalent to a weak positional game. We recall that the board of Hex is a rhombus of hexagons of size $n \times n$. The two players, White (first player) and Black take the two pairs of opposite sides of the board. The players alternately put their pieces on unoccupied hexagons. White wins if his pieces connect his opposite sides of the board, and Black wins if his pieces connect the other pair. Note that draw is impossible by topological reasons: in order to win in Hex it is enough to prevent the opponent from making a connecting chain, i.e. to achieve Breaker's goal in the weak game. But the fact that Hex is "weak" does not help at all to find an explicit winning strategy. On a general $n \times n$ board we do not know any explicit winning strategy for White in Hex, not even for small values like $n = 8$ (though we know its existence).

In contrast to the strong version, in weak positional games there is an *explicit* version of the strategy stealing argument: the *copycat strategy.*

**Theorem 3** ("Copycat Theorem"). *Let $(X, \mathcal{F})$ be a finite hypergraph of chromatic number bigger than two, and let $(X', \mathcal{F}')$ be a disjoint copy of $(X, \mathcal{F})$. Let $Y = X \cup X'$ and $\mathcal{G} = \mathcal{F} \cup \mathcal{F}'$. Then* Maker = **II** *has an explicit winning strategy in the weak positional game $(Y, \mathcal{G})$.*

**Proof.** Let $f : X \to X'$ be the isomorphism between $(X, \mathcal{F})$ and $(X', \mathcal{F}')$. Maker's winning strategy is the following copycat strategy: if Breaker = **I**'s last move was $x \in X$ or $x' \in X'$, then Maker's next move is $f(x) \in X'$ or $f^{-1}(x') \in X$. Since the chromatic number of $(X, \mathcal{F})$ is larger than two, one of the two players will completely occupy a winning set. If this player is Maker, we are done. If Breaker occupies some $A \in \mathcal{F}$, then Maker occupies $f(A) \in \mathcal{F}'$, and we are done again. ∎

Consider the weak $n^d$ game where $d \geq h(n) + 1$. This condition guarantees that the board contains at least two disjoint copies of $n^{h(n)}$. The copycat strategy of Theorem 3 provides an *explicit* winning strategy for Maker, and it does not matter whether Maker is the first or the second player.

The copycat strategy of Theorem 3 is a *pairing strategy.* It is a very common technique to guarantee a win or a draw. *A pairing strategy* means a decomposition of the board into disjoint pairs of points, and when your opponent takes one point from a pair, you take the other one.

Observe that the family of winning sets in the $n^d$-game has an important additional feature: the winning sets are lines, and any two lines have at most one point in common. Set-systems with the property that any two sets have at most one point in common are called *almost disjoint.*

If the family of winning sets is almost disjoint, then the question "does the pairing strategy work?" is a standard perfect matching problem. Indeed, in an almost disjoint family two distinct winning sets cannot contain the same pair of points. So the pairing strategy works if and only if one can find a family of disjoint 2-element representatives. But to find a family of disjoint 2-element representatives is a standard problem in Matching Theory.

The following general result, which is a direct application of Matching Theory, gives a *necessary* condition for the existence of a pairing strategy for an arbitrary family of winning sets.

**Theorem 4** ("Matching Criterion"). *Consider the weak game on $(X, \mathcal{F})$, and assume that for any subfamily $\mathcal{G} \subseteq \mathcal{F}$,*

$$\Big| \bigcup_{A \in \mathcal{G}} A \Big| \geq 2|\mathcal{G}|.$$

*Then Breaker can force a win by a pairing strategy. Breaker can be the first or the second player, it does not matter.*

**Proof.** By the well-known König-Hall theorem, we can find disjoint 2-element representatives: $h(A) \subset A$ $(A \in \mathcal{F})$, $|h(A)| = 2$, $h(A) \cap h(B) = \emptyset$ whenever $A$ and $B$ are different winning sets from $\mathcal{F}$. There is a little technical twist here: we have to apply the König-Hall theorem to the "duplicate" of $\mathcal{F}$, i.e. every $A \in \mathcal{F}$ is taken in two copies. Now whenever the first player occupies one point from a 2-element representative, then the second player takes the other one. ■

**Remark.** Note that Matching Theory provides several efficient $O(N^{2.5})$ running time algorithms to actually *find* a family of disjoint 2-elements representatives. So the Matching Criterion is efficient. If the family $\mathcal{F}$ of winning sets is almost disjoint, then Theorem 4 is a *necessary and sufficient* condition. Indeed, the same pair of points cannot block two different winning sets if the family is almost disjoint. In other words, each winning set must have its own pair of points, which is exactly the concept of disjoint 2-elements representatives.

**Theorem 5** ("Matching Degree Criterion"). *Let $\mathcal{F}$ be an $n$-uniform system, i.e., $|A| = n$ for each $A \in \mathcal{F}$. Further assume that each $x \in X$ is contained in at most $n/2$ elements of $\mathcal{F}$. Then Breaker can force a win in the weak game on $\mathcal{F}$ by a pairing strategy.*

**Proof.** Observe that Theorem 4 applies here. Indeed, for any subfamily $\mathcal{G} \subseteq \mathcal{F}$, by a standard double-counting argument,

$$n|\mathcal{G}| = \sum_{A\in\mathcal{G}} |A| = \sum_{x\in\bigcup_{A\in\mathcal{G}} A} \; \sum_{A\in\mathcal{G}:x\in A} 1 \le \left|\bigcup_{A\in\mathcal{G}} A\right| \cdot \frac{n}{2},$$

and dividing by $n/2$,

$$\left|\bigcup_{A\in\mathcal{G}} A\right| \ge 2|\mathcal{G}|.$$

The proof of Theorem 5 is complete. ■

Note that the Matching Criteria (Theorems 4-5) are very general. They are *local* conditions only: they do not give any restriction on the global size of $\mathcal{F}$. Both of them hold for infinite boards as well.

Hales and Jewett [19] proved, by a pioneering application of Theorem 5, that if

(1) $$n \ge 3^d - 1 \quad (n \text{ odd}) \quad \text{and} \quad n \ge 2^{d+1} - 2 \quad (n \text{ even})$$

then the $n^d$-game is a draw. Indeed, if $n$ is odd, there are at most $(3^d - 1)/2$ winning lines through any point and this is achieved only at the center point. (This suggests that the center point is probably the optimum first move for the first player if $n$ is odd.) If $n$ is even, the maximum degree of the family is $2^d - 1$ instead of $(3^d - 1)/2$.

If a pairing strategy can force a draw in an almost disjoint family, then there are at least twice as many points as winning sets. The converse is obviously not true.

Hales and Jewett conjectured [19] that the $n^d$-game is always a draw if there are at least twice as many points (i.e. cells) as winning lines.

How many lines are there? It is easy to see that every winning line of the $n^d$-game is determined by the two cells which extend the line into the surrounding

$$(n+2) \times (n+2) \times \cdots \times (n+2) = (n+2)^d$$

cube. So the total number of lines is

$$\frac{(n+2)^d - n^d}{2}. \tag{2}$$

Therefore, the Hales–Jewett conjecture is that the $n^d$-game is a draw when

$$n^d \geq (n+2)^d - n^d, \tag{3}$$

that is, when $2 \geq \left((n+2)/n\right)^d$. Since $\left((n+2)/n\right)^d = \left(1+\frac{2}{n}\right)^d \approx e^{2d/n}$, (3) is asymptotically equivalent to

$$n \geq \frac{2d}{\log 2}. \tag{4}$$

Observe that the Hales–Jewett conjecture (i.e. (4)) is "exponentially stronger" than (1).

The last statement of the first section does not really deserve the name "theorem": we simply collect two trivial implications.

**Trivial Theorem.**

(i) *If* Breaker ($=$ **II**) *has a winning strategy in the weak game on* $(X, \mathcal{F})$, *then the same strategy guarantees a draw for Second-Player in the strong game on the same hypergraph* $(X, \mathcal{F})$.

(ii) *If First-Player has a winning strategy in the strong game on* $(X, \mathcal{F})$, *then the same strategy guarantees a win for* Maker ($=$ **I**) *in the weak game on the same hypergraph* $(X, \mathcal{F})$. ■

## §2. Combinatorial games in general

### Game-graph and game-tree

Let us return to Theorems 1–2. Since *strategy stealing* is a purely existence argument, it does not give the slightest clue as to actually *find* the existing winning or drawing strategies in strong games. On the other hand, for weak games the copycat strategy in Theorem 3 and the pairing strategies in Theorems 4–5 are explicit.

In order to *find* an explicit winning or drawing strategy we often have no choice but to make an exhaustive search through the usually enormous *game-graph.* (To try out all possible strategies is much worse: the total number of strategies is usually a *double* exponential function of the size of the board.)

Any positional game (strong or weak) can be represented as a labeled directed graph: the vertices (points) are the positions, and the directed edges (arrows) are the options. The terminal positions are labeled by **I**, **II**, and **D** according to the outcome of the concrete play ending at that terminal position: **I** if it is a First-Player's win, **II** if it is a Second-Player's win, and **D** if it is a draw. This labeled digraph is what we call the *game-graph.*

The game-tree gives a somewhat more transparent picture. A play is a directed path (dipath) from the start to any terminal position. We can distinguish between different plays by *duplicating* the points (i.e positions) as necessary. If we duplicate each point (i.e. position) $v$ in as many copies as the number of dipaths from the start to $v$, then we get the rooted *game-tree* in which there is a 1-to-1 correspondence between the full branches of the tree and the plays of the game. The difference between the concepts of game-graph and game-tree is rather minor. Though the game-tree gives a better picture, it is substantially bigger than the game-graph.

The ultimate questions of game theory are "which player has a winning (or drawing) strategy" and "how to find one".

A simple linear time algorithm, the so-called "backward labeling", or "pruning", settles both ultimate questions for any positional game. This algorithm is linear in size of the game-graph.

What is more, the "backward labeling" provides the "complete analysis" of a positional game.

For an arbitrary *combinatorial game* (to be defined below) the computing time of the "backward labeling" is linear in size of the (usually bigger) game-tree. Note that the game-graph and the game-tree are usually at least exponentially large. This is where the "catch" is: this is why the "backward labeling algorithm" is impractical.

Now we define the class of *combinatorial games* in general.

## Combinatorial games

Here is a list of natural requirements.

1. There are two players: First-Player (or first player, or **I**, or Alpha, or White, or Maker or Breaker, etc.), who starts the game, and Second-Player (or second player, or **II**, or Beta, or Black, or Breaker or Maker, etc.).
2. There are finitely many *positions*, and a particular *starting position.*
3. There are clearly defined rules that specify the *moves* that either player can make from a given position to its *options.*
4. The players move alternately.
5. The rules are such that a sequence of alternating moves will always come to an end in a *finite* number of moves. The "ends" are called *terminal positions* (or end-positions).
6. A complete sequence of alternating moves ending with a terminal position is called a *play.*
7. The *outcome* of a play is specified by the rules. It may be a win for the First-Player, or a win for the Second-Player, or a draw.
8. Both players know what is going on (perfect information).
9. There are no chance moves.

Note that the 5th requirement prohibits perpetual draws, that is, we exclude "loopy games".

With each positional game we can associate its game-tree. The standard way to define an "abstract" game is to reverse this process: an "abstract" combinatorial game is a "coin-pushing game" on a labeled tree.

In technical terms, a *combinatorial game* $\mathbf{G} = (T, F)$ is a labeled finite rooted tree directed from the root (i.e. the root has no in-degree, and any other vertex has in-degree exactly one). Here $T = (V, E)$ is the tree, $V$ is a multi-set of the positions (that is, each position has many "copies"), and $u \to v \in E$ if and only if $v$ is an option from $u$ (i.e. there is a move from $u$ to $v$). At the beginning of a play a coin is placed in the starting position (root). At each move the players alternately push the coin along a directed edge of the tree. $F$ denotes the outcome-function ("labeling") which associates with each "leaf" of the tree a label. Note that a "leaf" represents a terminal position, which corresponds to a particular play. The three possible labels are **I**, **II**, and **D**, according to the outcome of the particular play ending at that "leaf" (terminal position). **I** if it is a First-Player's win, **II** if it is a Second-Player's win, and **D** if it is a draw.

We have to insert an important remark here: positional games are special in the sense that each position is available for exactly one player only. Indeed, if the *number* of First-Player's marks, denoted by $\#\mathbf{I}$, equals the number of Second-Player's marks, denoted by $\#\mathbf{II}$, then that position is available for the First-Player only. If the difference is 1, that is, if $\#\mathbf{I} - \#\mathbf{II} = 1$, then that position is available for the Second-Player only. It follows that, for every position, the set of First-Player's options is always disjoint from the set of Second-Player's options — in fact, one of the two sets is always the empty set.

A positional game can be represented as a coin-pushing game on its own game-tree. A more "economic" representation is a coin-pushing game on its own *game-graph.* This representation belongs to the class of *transparent games* (see below).

The game-graph of a positional game (without the labeling) is a *rooted linear digraphs.* A *digraph* (i.e. *directed graph*) is a *linear digraph* if its vertex set has a partition $V_0, V_1, V_2, \ldots, V_k$ ($k \geq 1$) into non-empty sets such that

(a) for each $v \in V_i$ with $1 \leq i \leq k$ there is a $u \in V_{i-1}$ such that the directed edge $u \to v$ which goes from vertex $u$ to vertex $v$ belongs to the linear digraph,

(b) if $u \to v$ is a directed edge of the linear digraph, then there is an $i$ with $1 \leq i \leq k$ such that $u \in V_{i-1}$ and $v \in V_i$.

A *rooted* linear digraph means that the first vertex-class $V_0$ has a single vertex called the root. Note that a rooted linear digraph is always *connected* as an ordinary graph, i.e. when we ignore the directions. In a *rooted linear digraph* every vertex has a well-defined *distance* from the root: if $u \in V_i$ then the distance of vertex $u$ from the root is $i$.

In technical terms, a *transparent game* $\mathbf{G} = (G, F)$ means a coin-pushing game on a finite rooted linear digraph $G$. At the beginning of a play a coin is placed in the root ("starting position"). At each move the players alternately push the coin along a directed edge of digraph $G$. $F$ denotes the outcome-function ("labeling") which associates with each leaf (i.e. vertex with zero out-degree) of digraph $G$ a label. The three possible labels are **I**, **II**, and **D**.

Note that a vertex of the digraph represents a "position", a leaf represents an "end-position", label **I** "means" that the end-position is a first player's win, **II** "means" a second player's win, and **D** "means" a draw.

Since a rooted directed tree is a rooted linear digraph, the class of *combinatorial games* is a subclass of the class of *transparent games.*

A (winning, drawing) *strategy* for player $P$ is a function which associates with every possible play of the opponent a (winning, winning or drawing) counter-play of player $P$. This procedure must be non-anticipatory. This means that at a certain stage of a play a strategy assigns the "next move" in such a way that it can depend on the previously selected moves only. Formally, a winning (drawing) strategy Str for the First-Player means

$$\exists x_1 = \mathrm{Str}\,(\emptyset) \forall y_1 \exists x_2 = \mathrm{Str}\,(x_1, y_1) \forall y_2 \exists x_3 = \mathrm{Str}\,(x_1, y_1, x_2, y_2) \forall y_3 \ldots$$

such that the First-Player wins (wins or has a draw).

Similarly, a winning (drawing) strategy Str for the Second-Player means

$$\forall x_1 \exists y_1 = \mathrm{Str}\,(x_1) \forall x_2 \exists y_2 = \mathrm{Str}\,(x_1, y_1, x_2) \forall x_3 \exists y_3 = \mathrm{Str}\,(x_1, y_1, x_2, y_2, x_3) \ldots$$

such that the Second-Player wins (wins or has a draw). Here $x_1, x_2, x_3, \ldots$ and $y_1, y_2, y_3, \ldots$ are the moves of the First-Player and the Second-Player, respectively.

### Complete analysis

Traditional game theory provided a simple algorithm for extending the **I**, **II**, **D**-labeling $F$ of the end-positions in a transparent (combinatorial) game $\mathbf{G}$. In fact, the label of the starting position answers the first ultimate question: "who wins".

Let $\mathbf{G} = (G, F)$ be an arbitrary transparent game, where $G = (V, E)$ is a finite rooted linear digraph, $V$ is the vertex-set, and $E$ is the edge-set. Let $Z \subset V$ be the set of end-positions (i.e. vertices with zero out-degree). The outcome-function $F$ cuts $Z$ into three parts ("I wins", "II wins", "draw"):

$$Z_I = \{u \in Z : F(u) = I\},$$

$$Z_{II} = \{u \in Z : F(u) = II\},$$

$$Z_D = \{u \in Z : F(u) = D\}.$$

We are going to extend this partition $Z = Z_I \cup Z_{II} \cup Z_D$ into a partition of $V = V_I \cup V_{II} \cup V_D$ satisfying $V_I \supseteq Z_I$, $V_{II} \supseteq Z_{II}$ and $V_D \supseteq Z_D$.

Let $u \in V$ be an arbitrary position (i.e. vertex), and its options are

$$\text{Options}(u) = \{v \in V : u \to v \in E\}.$$

By definition, all end-positions are labeled by **I**, **II**, and **D**. In a *rooted linear digraph* every vertex has a well-defined *distance* from the root. If the distance is *even* (*odd*) then the first (second) player moves next.

We use the natural rules (1)-(2)-(3) and (1')-(2')-(3') below to extend the **I**, **II**, **D**-labeling $F$ *backward* from $Z$ to the whole vertex-set $V$ in $O(|E|)$ steps:

(1) $\quad u \in V_I \quad \text{iff} \quad \text{Options}(u) \cap V_I \neq \emptyset;$

(2) $\quad u \in V_{II} \quad \text{iff} \quad \text{Options}(u) \subseteq V_{II};$

(3) $\quad u \in V_D \quad \text{iff} \quad \text{Options}(u) \cap V_I = \emptyset \quad \text{and} \quad \text{Options}(u) \cap V_D \neq \emptyset;$

if vertex $u$ has an *even* distance from the root; and

$$(1') \quad u \in V_{II} \quad \text{iff} \quad \text{Options}\,(u) \cap V_{II} \neq \emptyset;$$

$$(2') \quad u \in V_I \quad \text{iff} \quad \text{Options}\,(u) \subseteq V_I;$$

$$(3') \quad u \in V_D \quad \text{iff} \quad \text{Options}\,(u) \cap V_{II} = \emptyset \quad \text{and} \quad \text{Options}\,(u) \cap V_D \neq \emptyset;$$

if vertex $u$ has an *odd* distance from the root.

More precisely, let $s \in V$ denote the starting position. Let $V_0 = \{s\}$, $V_1 = \text{Options}\,(s)$, and in general,

$$V_{i+1} = \bigcup_{u \in V_i} \text{Options}\,(u).$$

$V_0, V_1, V_2, \ldots$ is a partition of $V$. Let $m$ be the *largest* index such that $V_m$ is *not* contained by set $Z$. Then depending on the *parity* of $m$ we use rules (1)-(2)-(3) if $m$ is *even* and rules (1')-(2')-(3') if $m$ is *odd* to extend the **I**, **II**, **D**-labeling to $V_m$, then to $V_{m-1}$, then to $V_{m-2}$, and so on back to $V_0$.

Let $\overline{F}$ denote this extension of labeling $F$ from the terminal positions $Z$ to the whole $V$. We call $\overline{\mathbf{G}} = (G, \overline{F})$ the **complete analysis** of game $\mathbf{G} = (G, F)$. The reason why we call $\overline{\mathbf{G}}$ the **complete analysis** is that it answers the two ultimate questions of "who wins" and "how to win". Rules (1), (2), (3), (1'), (2'), (3') guarantee that if a position is labeled **I** (**II**) then the first (second) player can force a win in the *subgame* starting from this particular position. Label **D** means that both players can maintain a draw in this *subgame.* If the position is the *starting position*, then the subgame is the whole game **G**. In other words, the label of the starting position answers the question of "who wins".

The next question is "how to win". The *complete analysis* $\overline{\mathbf{G}}$ provides an explicit way to *find* either players optimal strategy in game **G**.

(i) If the starting position is labeled by **I** (i.e. the first player has a winning strategy), then the first player's explicit winning strategy is to keep moving to any option with label **I** (he can always do that).

(ii) If the starting position is labeled by **II**, then the second player's explicit winning strategy is to keep moving to any option with label **II**.

(iii) If the starting position is labeled by **D**, then the first (second) player's explicit drawing strategy is to avoid label **II** (**I**).

Note that the complete analysis carries more information than merely answering the two ultimate questions. Indeed, by using the complete analysis one can easily obtain the *complete list* of all possible optimal strategies in a transparent game (although this list is usually *absurdly* long).

Furthermore, the complete analysis $\overline{\mathbf{G}}$ of **G** answers the two ultimate questions in every *subgame* where the players start from an *arbitrary* position, not necessarily the starting position.

We summarize the properties of the "backward labeling" in the following statement.

**Theorem 7** ("Backward Labeling Theorem"). *Let* $\mathbf{G} = (G, F)$ *be a transparent game. The "backward labeling algorithm" determines the complete analysis of* **G** *— in particular, it finds the outcome of the game, and provides either player's optimal strategy — in computing time* $\mathrm{O}\big(|T|\big)$, *that is, linear in size of the "game-tree".*

*In the special case of transparent games the computing time is linear in size of the "game-graph".* ∎

An important byproduct of the "backward labeling algorithm" is Zermelo's theorem.

## Outcomes of games

In 1912 Zermelo proved the fundamental result that every finite combinatorial game is "determined". Note that in our definition a combinatorial game is always finite.

**Theorem 8** ("Zermelo's Theorem"). *Let* $\mathbf{G} = (T, F)$ *be a combinatorial game with game tree* $T = (V, E)$. *Then* **G** *is determined which means that either one of the players has a winning strategy, or both of them have a drawing strategy.*

**Proof of Theorem 8.** The *existence* proof of Zermelo's theorem is a simple application of De Morgan's law. Indeed, there are three alternatives only.

(1) Either **I** has a winning strategy: $\exists x_1 \forall y_1 \exists x_2 \forall y_2 \ldots$ such that **I** wins;

(2) or **II** has a winning strategy: $\forall x_1 \exists y_1 \forall x_2 \exists y_2 \ldots$ such that **II** wins;

(3) or the negation of (1) $\vee$ (2):

$$\neg\big((\exists x_1 \forall y_1 \exists x_2 \forall y_2 \ldots \textbf{ I} \text{ wins}) \vee (\forall x_1 \exists y_1 \forall x_2 \exists y_2 \ldots \textbf{ II} \text{ wins})\big),$$

which, by De Morgan's law, is equal to

$$(\forall x_1 \exists y_1 \forall x_2 \exists y_2 \ldots \textbf{ I} \text{ loses or draw}) \wedge (\exists x_1 \forall y_1 \exists x_2 \forall y_2 \ldots \textbf{ II} \text{ loses or draw}).$$

So the third alternative is that both players have a drawing strategy, and this completes the existence proof.

The advantage of the "backward labeling" is that it gives an *algorithmic proof* of Zermelo's theorem. ■

**Remarks.** These three alternatives are the possible *outcomes of a combinatorial game.* Formally we distinguish this from the concept of the outcome of a *perfect play,* which means the outcome of a particular *play* where the players know and follow any of their optimal strategies. There is, however, an obvious connection between these two concepts. If the *outcome of a game* is that First-Player (Second-Player, neither player) has a winning strategy, then the *outcome of a perfect play* in the same game is First-Player's win (Second-Player's win, draw), and vice versa. So the loosely stated first question: "who wins a game", more precisely means "who has a winning strategy", or what is the same thing, "who wins a perfect play".

It is a completely different question whether real-world players know the best strategies or not. In advanced games like Chess and Go nobody knows any optimal strategy. Consequently it is only the *ignorance* of players that keeps a combinatorial game alive for competitive play. This is called the "paradox" of games of full skill.

A famous *mathematical* paradox is that by extending the concept of combinatorial games to *infinite* games, quite unexpectedly *undetermined* games show up. Historically the first example was the Banach-Mazur game. Around 1928 the Polish mathematician S. Mazur invented and studied the following game. Let $A$ be an arbitrary subset of the closed unit interval $[0, 1]$, and let $B$ be the complementary set $B = [0, 1] \setminus A$. The $(A, B)$-game is played by two players: player $A$ and player $B$, who move alternately. Player $A$ begins the play by choosing an arbitrary closed subinterval $I_1 \subset [0,1]$, then player $B$ chooses a closed subinterval $I_2 \subset I_1$, then again player $A$ chooses a closed subinterval $I_3 \subset I_2$, then player $B$ chooses a closed subinterval $I_4 \subset I_3$, and so on. At the end of a play the two players determine a nested infinite sequence of closed intervals $I_n$, $n = 1, 2, 3, \ldots$; player $A$ chooses those with odd indices, and player $B$ chooses those with even indices. If the intersection set $\bigcap_{n=1}^{\infty} I_n$ has at least one common point with the set $A$, then player $A$ wins; otherwise player $B$ wins.

Banach and Mazur proved the following perfect characterization of the $(A, B)$-game. Player $B$ has a winning strategy in the $(A, B)$-game if and only if $A$ is a *first category set.* On the other hand, player $A$ has a winning strategy in the $(A, B)$-game if and only if $B \cap I_1$ is a *first category set* for some proper subinterval $I_1 \subset [0,1]$.

Is it possible that neither of these two criteria hold? Yes! By using transfinite induction one can prove the *existence* of a subset $S$ of the real numbers such that both $S$ and its complement intersect all uncountable closed subsets of the reals. Such a set $S$ is called a *Bernstein set.* Now if $A$ is the intersection of the unit interval $[0,1]$ with a Bernstein set $S$, then the Banach-Mazur $(A, B)$-game is clearly *undetermined.*

We have the following relation between the classes of games discussed so far. We started with the class of Hales–Jewett $n^d$-games. The $n^d$-games all belong to the class of strong positional games. Strong positional games belong to the bigger class of transparent games. Finally, transparent games are all combinatorial games.

Let us return to the class of positional games. If the size of the board $|X| = N$, then the total number of positions, i.e. the number of vertices in the game-graph, is obviously $O(3^N)$. Indeed, in the course of a play every point on the board can have three "colors": marked by First-Player, marked by Second-Player, or unmarked. Therefore, the size of the game-graph, i.e. the number of edges, is clearly $O(N \cdot 3^N)$.

The size of the game-tree is clearly $O(N \cdot 3^N)$. Indeed, a *play* is a permutation of the board, so the total number of plays is clearly $N!$. Note that $O(N \cdot 3^N)$ is substantially less than $O(N!)$. This savings, $O(N \cdot 3^N)$ instead of $O(N!)$, is the very reason why we introduced the concept of transparent games.

## The main problem of combinatorial game theory

According to Theorem 7, the "backward labeling" algorithm describes the winner and provides an explicit winning or drawing strategy in linear time — in particular, for positional games, in $O(N \cdot 3^N)$ steps, where $N$ is the size of the board. This seems to be a pleasing answer. Well, not exactly. An obvious problem is that the "backward labeling" is hardly more than mindless computation: it lacks any kind of "understanding". But the main difficulty, the heart of the matter, is that a $3^N$-step algorithm is impractical. Of course, these two reasons are not independent from each other. The "backward labeling" is "intractable" *because* it lacks any kind of deeper understanding of the game.

To perform $3^N$ operations is far beyond the capacity of the fastest computer "possible" even for relatively small board-size like (say) N=100. As far as we understand it now, the fundamental laws of physics prevent us from building arbitrarily fast computers. There seems to be a universal bound on the speed and the speed much smaller than $3^{100}$ operations per second (or hour, or year, or century, it does not really matter). Right now $10^{10}$ operations per second is about the limit of what computers are capable to do. This looks a rather shaky foundation for a theory to build on. But this "complexity" viewpoint is very much the basis for Discrete Mathematics and Theoretical Computer Science. For a good survey paper about games and complexity, see Fraenkel [17].

This means that, unless one finds some substantial shortcut, the "brute force" complete analysis of a game with board-size at least 100 is beyond reach. The human brain can sometime diagnose shortcuts, but we cannot expect substantial shortcuts too often, certainly not in the *whole* class of positional games. In other words, even small positional games are so *complex* that it is well possible that the existing winning or drawing strategies may never be found.

Note that even the successes of the "brute force" search are discouraging. Every child "knows" that Tic-Tac-Toe is a draw game, but the precise mathematical proof is unpleasantly long – see the exhausting analysis on pp. 670–71 in the *Winning Ways* (see [11]). Another example is Tic-Toc-Tac-Toe. In 1977 Oren Patashnik [21] was able to find a First-Player's winning strategy in the 3-dimensional $4 \times 4 \times 4$ game (Tic-Toc-Tac-Toe). Patashnik's program required 1,500 hours of computing time, and the winning strategy contains 2,929 "strategic moves". Note that the board size in

Tic-Toc-Tac-Toe is $N = 4^3 = 64$, quite big, but the number of alternatives can substantially be cut down by excluding "obviously" bad moves. This is where Patashnik could use his comprehensive game-playing experience, and could cut down the size $3^{64} \approx 10^{30}$ of case study to a "realistic" size below $10^{10}$. We strongly recommend the reader to study Patashnik's fascinating paper [21].

We have to conclude, therefore, that the "backward labeling", although theoretically important, turns out to be a "dead-end" for the theory, at least from our complexity viewpoint.

Now we clearly see where the "catch" is in Theorem 7. The "backward labeling" answers the two ultimate questions in the fastest way in terms of the game-tree, or the game-graph for transparent games, *but* the game-tree or the game-graph are *not* the "natural" input data. The really interesting games always have much better "natural representations", and the game-tree or game-graph representations are terribly redundant. For example, a positional game $(X, \mathcal{F})$ in its "natural form" can be encoded by its winning sets, so the size of the input data is $O(|X| + |\mathcal{F}|)$, which is usually much smaller than $O(N \cdot 3^N)$, the size of the game-graph ($|X| = N$). If $|\mathcal{F}|$ is exponential in board-size $N = |X|$, then the "backward labeling" is *efficient*, but this is hardly ever the case. If $|\mathcal{F}|$ is polynomial, or ever just sub-exponential in $N$, which is almost always the case, then an *efficient* strategy should mean computing time $|\mathcal{F}|^{\text{const}}$, and the "backward labeling" becomes terribly inefficient.

If the winning sets form some kind of "complete structure" then the input data can be further reduced. For example, the Hales–Jewett $n^d$-game can be encoded simply by the binary forms of the numbers $n$ and $d$, that is, by a data of size $O(\log n + \log d)$.

After these preparations we are ready to formulate what we consider the "main problem" of combinatorial game theory. Theorem 7 is not the end, but it is rather the starting point of the theory. We consider the "backward labeling" the "worst case scenario", the "most complex" strategy. *The main problem in the "positive" direction is to describe wide subclasses of combinatorial games for which one can find efficient optimal strategies, or at least to substantially "beat" the "backward labeling".* This is exactly what we are going to do in the rest of this paper. We are going to describe subclasses of positional games for which there is an instant answer to the question of "who wins", and there is an efficient answer to "how to win". We recall that efficient strategy for positional games means that the *worst-case* running time is $|\mathcal{F}|^{\text{const}}$.

A complementary "negative" approach is to try to prove that some classes of positional games are *hard* in the sense that any *algorithm* which determines the "outcome" for *all* games in this class must necessarily take as much computations as the "backward labeling", or at least a very large amount of computations. Unfortunately we do not have the slightest idea of how to prove this kind of unconditional complexity lower bounds.

The bad news is the winning strategy in the strong positional game. Unfortunately, the Ramsey Criterion (Theorem 2) is the *only* criterion known for a strong win. This is a big roadblock for the theory. Equally bad is that the only way to find a strong winning strategy (whose existence is guaranteed by the Ramsey Criterion) is the "backward labeling" with its exponential $O(N \cdot 3^N)$ running time.

## §3. The quasiprobabilistic method

The quasiprobabilistic method is a desperate attempt to study complex games where all other methods fail to work. If we do not have time for the exhaustive search through all branches of the game-tree, then we can do no better than to assume that all options are equally likely. In other words, it is reasonable to study the *random walk* on the game-tree, that is, to study the *randomized game* where both players play randomly.

Of course, if by experience one has gained some knowledge about the game-tree, then it is enough to restrict the random walk to the unknown parts of the game-tree.

### Randomized combinatorial games

Let $\mathbf{G} = (T, F)$ be a combinatorial game, and consider the *symmetric random walk* on the tree $T = (V, E)$. This means that we begin the walk at the starting position and visit a sequence of positions as follows. If a position has (say) $d$ options (i.e. out-degree = d) then the probability of going to any of them is the same, namely $1/d$. This symmetric random walk defines a probability distribution on the set of all plays of the game $\mathbf{G}$. The probability of a particular play (a full branch) is clearly $\left(\prod_i d_i\right)^{-1}$ where $d_i$ is the out-degree of the $i$-th point on the dipath. The discrete random variable with this special distribution on the space of all plays of the game $\mathbf{G}$ is called the *random play* on $\mathbf{G}$.

Each particular play has a well-defined outcome: either **I** wins, or **II** wins, or the play ends in a draw. The outcome of the *random play* is a random variable which has these three possible values. We call this random variable the *random outcome* of the game $\mathbf{G}$. A random play together with the corresponding random outcome forms a *randomized game.*

It is rather surprising that the probabilistic analysis of a randomized positional game might help to find a *shortcut* answer to the ultimate questions of "who wins" and "how to win". *Shortcut* means a procedure which is much faster than "backward labeling".

The *quasiprobabilistic method* has two steps. The first step is the probabilistic analysis of the randomized game. The second step is *derandomization* by potential techniques.

For positional games the symmetric random walk distribution is particularly simple. If the board size is $N$, then the total number of plays is $N!$, assuming the simplifying condition that the players do not stop after occupying a winning set. Since the plays are equally likely, every play has the same probability $1/N!$.

The *majority principle*, a heuristic argument of the quasiprobabilistic method, goes as follow. Assume that in a weak game the overwhelming *majority* of the $N!$ plays have the same outcome, namely Maker's win. Then, according to the majority principle, there is a good chance that Maker has a winning strategy. Indeed, this is the case for large classes of weak games.

The same heuristic holds for Breaker's win in the *weak* version. If the overwhelming majority of the $N!$ plays end with Breaker's win, then according to the majority principle, the reason might be that Breaker has a winning strategy. And again, the principle hold for surprisingly many weak positional games.

By the way, this is exactly how we show that some strong games have a draw: by using the quasiprobabilistic method we prove the stronger statement that Breaker = **II** can win the weak version.

These results are very surprising, because in general there is no logical connection whatsoever between the "majority outcome" of all possible plays and the outcome of a single play with "perfect players".

The bad news concerns the strong games with First-Player's win. Unfortunately, in this case the quasiprobabilistic method is useless. Why is that? What is the crucial difference between weak and strong games? Well, strong positional games are symmetric. Therefore, when *draw is impossible*, the probability of First-Player's win is usually very close to $1/2$, like $1/2 + \varepsilon$. So we cannot speak about overwhelming majority at all. Strong games with First-Player's win are evenly balanced "head-to-head" games where one single mistake can be fatal. So the quasiprobabilistic method, a crude approximation of the complete analysis, does not have a chance of working.

On the other hand, for asymmetric games like the weak version, usually a sort of finite "zero-one law" holds. This "one-sidedness" gives the quasiprobabilistic method a good chance to work.

The pioneering result of the quasiprobabilistic method is the Erdős–Selfridge theorem. In 1973 Erdős and Selfridge [16] found a very useful sufficient condition for Breaker's win in the weak game (Breaker is the second player). Of course the same criterion automatically guarantees Second-Player's draw in the strong version. Quite contrary to the pairing strategy and other local approaches, the Erdős–Selfridge Theorem is a *global* criterion.

**Theorem 9** ("Erdős–Selfridge Theorem"). *If $\mathcal{F}$ is an $n$-uniform hypergraph and $|\mathcal{F}| < 2^{n-1}$, then* Breaker $=$ **II** *has a winning strategy in the weak game on $(X, \mathcal{F})$.*

**Remark.** The first step of the quasiprobabilistic method is *randomization.* The probabilistic analysis of the randomized game is very simple. Let $|X| = N$, and consider all possible $N!$ plays. It is an easy double-counting argument that under the Erdős–Selfridge condition there is at least one Breaker's win end-position. Indeed, the *expected number* of winning sets completely occupied by either player is precisely

$$2|\mathcal{F}| \frac{\binom{N-n}{[N/2]}}{\binom{N}{[N/2]}}.$$

Here, as usual, $[x]$ stands for the lower integral part of $x$. Since

$$\frac{\binom{N-n}{[N/2]}}{\binom{N}{[N/2]}} = \frac{[N/2]}{N}\frac{[N/2]-1}{N-1}\frac{[N/2]-2}{N-2}\cdots\frac{[N/2]-n+1}{N-n+1} \leq 2^{-n},$$

the *expected number* of the winning sets completely occupied by either player is less or equal to $2^{-n+1}|\mathcal{F}|$. By hypothesis, this number is less than one. This implies that there exist at least one draw end-position in the strong game, which is automatically a Breaker's win end-position in the weak game.

In general, if $2^{-n+1}|\mathcal{F}| = c$ is a large constant, then at least $(1-\frac{1}{c})N!$ of the $N!$ plays end with Breaker's win. According to the majority principle, the game-theoretic threshold should be around const $\cdot\, 2^n$.

Note that the calculation simplifies if we study the *random two-coloring* of the board with probability $p = 1/2$. This model is a little bit different from considering the *exactly* 50–50% two-colorings only — the case which corresponds to the randomized game. But the calculation in the random two-coloring model is more transparent. Indeed, in the random two-coloring model the *expected number* of monochromatic winning sets is clearly $2^{-n+1}|\mathcal{F}|$.

The second step is "derandomization". More precisely, "derandomization of the first moment method". We are going to turn the simple "counting argument" above into Breaker's winning strategy by using a *potential technique.* This part is the actual proof, but the probabilistic motivation was necessary for better understanding.

**Proof of Theorem 9.** Assume we are at the stage of the play where Maker = "first player" already occupied $x_1, x_2, \ldots, x_i$, Breaker = "second player" has occupied $y_1, y_2, \ldots, y_{i-1}$, and the question is how to choose Breaker's next point $y_i$. Those winning sets that contain at least one $y_j$ ($j \leq i-1$) are "harmless" — we call them "dead sets". The winning sets that are not "dead" are called "survivors". The "survivors" have a chance to be completely occupied by Maker at the end of the play, so they each represent some "danger". What is the total "danger" of the *whole* position? Consider the *randomized game* starting from this given position, i.e. the unoccupied elements of the board $X$ are colored randomly with $p = 1/2$ red for Maker and blue for Breaker. Let $A_s \in \mathcal{F}$ be a "survivor", and let $E_s$ denote the event that $A_s$ becomes monochromaticly red in the random game. The probability of $E_s$ is clearly $2^{-u_s}$ where $u_s$ is the number of unoccupied elements of $A_s$. What we are really interested in is the "loss-probability" $\text{Prob}\left\{\bigcup_{s\in S} E_s\right\}$. By the inclusion-exclusion formula

$$\text{Prob}\left\{\bigcup_{s\in S} E_s\right\} = \sum_{s\in S} \text{Prob}\{E_s\} - \sum_{s<t} \text{Prob}\{E_s \cap E_t\} +$$

$$+ \sum_{s<t<v} \text{Prob}\{E_s \cap E_t \cap E_v\} \mp \cdots .$$

This expression is obviously out of control, so we have to approximate it. A natural choice is "linear approximation", that is, to keep the first term $\sum_{s\in S} \text{Prob}\{E_s\}$ on the right-hand side only. This is no other than the *expected number* which was the key concept in the "global counting argument" above. This convinces us to evaluate the given position by the following expression, called the "danger-function": $D_i = \sum_{s\in S} 2^{-u_s}$, where $u_s$ is the number of unoccupied elements of the "survivor" $A_s$ ($s \in S =$ "*survivors*") and index $i$ indicates that we are at the stage of choosing the i-th point $y_i$ of Breaker. A natural choice for $y_i$ is to minimize the "danger" $D_{i+1}$ at the next stage. How to do that? Because of the simple linear structure of the danger-function $D_i$, it is quite easy to answer this question. Let $y_i$ and $x_{i+1}$ denote the next two moves. What is their effect on $D_{i+1}$? Well, $y_i$ "kills" all the "survivors" $A_s \ni y_i$, which means we have to subtract the sum

$$\sum_{A_s\,:\,y_i\in A_s} 2^{-u_s}$$

from $D_i$. On the other hand, $x_{i+1}$ doubles the "danger" of each "survivor" $A_s \ni x_{i+1}$, that is, we have to add the sum $\sum_{A_s\,:\,x_{i+1}\in A_s} 2^{-u_s}$ back. Warning: if some "survivor" $A_s$ contains both $y_i$ and $x_{i+1}$ then we do not have to give the corresponding term $2^{-u_s}$ back because that $A_s$ was previously "killed" by $y_i$.

Now the natural choice for $y_i$ is that unoccupied $z$ for which $\sum_{A_s : z \in A_s} 2^{-u_s}$ is maximum. Then clearly

$$D_{i+1} \le D_i - \sum_{A_s : y_i \in A_s} 2^{-u_s} + \sum_{A_s : x_{i+1} \in A_s} 2^{-u_s}$$

$$\le D_i - \sum_{A_s : y_i \in A_s} 2^{-u_s} + \sum_{A_s : y_i \in A_s} 2^{-u_s} = D_i.$$

In other words, Breaker can force the decreasing property $D_0 \ge D_1 \ge D_2 \ge \cdots \ge D_{\text{end}}$. This is exactly the *potential technique* that we were talking about before.

What Breaker really wants is to prevent his opponent from completely occupying some $A_0 \in \mathcal{F}$, that is, to avoid $u_0 = 0$. If $u_0 = 0$ then $D_{\text{end}} \ge 2^{-u_0} = 1$. On the other hand,

$$D_{\text{start}} = D_0 = \sum_{A : x_1 \in A \in \mathcal{F}} 2^{-n+1} + \sum_{A : x_1 \notin A \in \mathcal{F}} 2^{-n} \le |\mathcal{F}| 2^{-n+1} < 1,$$

so by the decreasing property of the danger-function, $D_{\text{end}} < 1$. This completes the proof of the Erdős–Selfridge theorem. ■

Theorem 9 is sharp. The full branches of a binary tree with $n$ levels form an $n$-uniform family of $2^{n-1}$ winning sets such that the first player can occupy a full branch in $n$ moves.

Another extremal system is the following. The board $X$ is a $(2n-1)$-elements set $\{w, x_1, x_2, \ldots, x_{n-1}, y_1, y_2, \ldots, y_{n-1}\}$. The family of winning sets consists of all possible $n$-elements subsets $A$ of $X$ with the following two properties: (1) $w \in A$, (2) $A$ contains exactly one point from each pair $\{x_i, y_i\}$, $i = 1, 2, \ldots, n-1$. The number of winning sets is $2^{n-1}$, and again the first player can occupy a winning set in $n$ moves.

It would be very interesting to describe all possible extremal systems of Theorem 9. For example, does there exist an extremal family (i.e. $n$-uniform of size $2^{n-1}$) in which it takes more than $n$ moves for the first player to occupy a winning set?

The Erdős–Selfridge potential technique resembles to Conway's beautiful resource counting argument in the *Solitaire Army* puzzle (see pp. 715–17 in the *Winning Ways* [11]).

*Solitaire Army* is a particular case of the class of *Solitaire* puzzles (*Peg Solitaire* in America). We do not consider them as games because there are no two players: they are one-player "games". The common features of these puzzles are as follows. Each puzzle is played with a board and men (pegs), and the board contains a number of holes each of which can hold one man. Each move consists of a jump by one man over one or more other men, the men jumped over being removed from the board. Each move therefore reduces the number of men on the board.

The Solitaire Army is played on the infinite plane (infinite chessboard), and the holes are at the lattice points. The permitted move is to jump a man horizontally or vertically but not diagonally. Let us draw a line across the infinite board parallel to a lattice point line, and start with all men behind this line. We consider this line the horizontal axis, so all men are in the lower half-plane. How many men do we need to send one man forward 1, 2, 3, 4, or 5 holes into the upper half-plane?

Two men are needed to send a man forward one hole, four men are needed to send a man forward two holes, eight men are needed to send a man forward three holes. The doubling pattern breaks for four holes: no less than twenty men are needed to send a man forward four holes. But the really surprising result is the case of five holes: it is *impossible* to send a man forward five holes into the upper half-plane. This striking result was discovered by Conway in 1961.

The idea behind Conway's resource counting is the following. We assign a value to each hole subject to the condition that if H1, H2, H3 are *any* three consecutive holes in a row or in a column, and w(H1), w(H2), w(H3) are the corresponding values, then w(H1) + w(H2) $\geq$ w(H3). We can evaluate a position by the sum of the values of those holes which are occupied by men.

The meaning of the inequality w(H1)+w(H2)$\geq$w(H3) is very simple. The effect of a move where a man in H1 jumps over another man in H2 and arrives at H3 is that we replace men with values w(H1) and w(H2) by a man with value w(H3). Since w(H1)+w(H2)$\geq$w(H3), this change cannot be an increase in the value of the new position.

The inequality w(H1)+w(H2)$\geq$w(H3) guarantees that no play is possible from an initial position to a target position if the target position has a higher value.

Let $w$ be the positive number which satisfies $w + w^2 = 1$. The actual value of $w$ is $\frac{\sqrt{5}-1}{2}$. Now Conway's resource counting goes as follows. Assume that one has succeeded in sending a man 5 holes forward into the upper half-plane by starting from a configuration of a finite number of men in the lower half-plane. Place a 1 where the man stands 5 holes forward into the upper half-plane, and extend it in the following way:

$$\begin{array}{ccccccccccc}
 & & & & & 1 & & & & & \\
 & & & & & w & & & & & \\
 & & & & & w^2 & & & & & \\
 & & & & & w^3 & & & & & \\
 & & & & & w^4 & & & & & \\
\cdots & w^9 & w^8 & w^7 & w^6 & w^5 & w^6 & w^7 & w^8 & w^9 & \cdots \\
\cdots & w^{10} & w^9 & w^8 & w^7 & w^6 & w^7 & w^8 & w^9 & w^{10} & \cdots \\
\cdots & w^{11} & w^{10} & w^9 & w^8 & w^7 & w^8 & w^9 & w^{10} & w^{11} & \cdots \\
\cdots & \cdots & \cdots & \cdots & \cdots & \cdots & \cdots & \cdots & \cdots & \cdots & \cdots
\end{array}$$

The value of the top line of the lower half-plane is

$$w^5 + 2w^6 + 2w^7 + 2w^8 + \cdots = w^5 + 2\frac{w^6}{1-w} =$$

$$= w^5 + 2\frac{w^6}{w^2} = w^5 + 2w^4 = w^3 + w^4 = w^2.$$

So the value of the whole lower half-plane is

$$w^2(1 + w + w^2 + w^3 + \cdots) = w^2\frac{1}{1-w} = w^2\frac{1}{w^2} = 1,$$

which is exactly the value of the target position. So no finite number of men in the lower half-plane will suffice to send a man forward five holes into the upper half-plane.

For *almost disjoint* families (i.e. any two sets have at most one point in common) Theorem 9 can be considerably improved. Roughly speaking, the upper bound $2^n$ can be replaced by $4^n$ (see Beck [3]).

**Theorem 10** ("Breaker's Win for Almost-Disjoint Families"). *There is an absolute constant $c > 0$ such that for every $n$-uniform almost disjoint family $\mathcal{F}$ of winning sets, if*

$$|\mathcal{F}| < 4^{n-c\sqrt{n}}$$

*then Breaker, as the second player, can force a win in the weak game on* $(X, \mathcal{F})$.

Note that Theorem 10 is asymptotically nearly best possible (see Erdős and Lovász [15]).

**Theorem 11** ("Erdős' Construction"). *There is an* $n$*-uniform almost-disjoint family* $\mathcal{F}^*$ *with no more than* $n^4 \cdot 4^n$ $n$*-sets which is 3-chromatic.*

Since the chromatic number is greater than 2, by Theorem 2 the strong game played on this $\mathcal{F}^*$ has a First-Player's win. By using the same strategy, Maker, as the first player, can force a win in the weak game.

What is the reason behind Theorem 10, i.e., why can we replace the size $2^n$ in Theorem 9 by its square $4^n$ in Theorem 10? Well, for almost disjoint families there is a surprising "degree reduction" result, due to Erdős and Lovász (see Erdős–Lovász [15]). But before formulating the Erdős–Lovász degree reduction theorem, let us make a simple observation. Let $\mathcal{F}$ be an arbitrary $n$-uniform almost-disjoint family, and let $X$ denote the union of the sets in $\mathcal{F}$. Since $\mathcal{F}$ is almost disjoint, we have $|\mathcal{F}|\binom{n}{2} \leq \binom{|X|}{2}$, which implies $(n-1)\sqrt{|\mathcal{F}|} < |X|$. This inequality implies that the *average degree* $\overline{d}$ of $\mathcal{F}$ is less than $2\sqrt{|\mathcal{F}|}$. Indeed,

$$\overline{d} = \frac{1}{|X|}\sum_{x \in X} d_x = \frac{n|\mathcal{F}|}{|X|} < \frac{n|\mathcal{F}|}{(n-1)\sqrt{|\mathcal{F}|}} \leq 2\sqrt{|\mathcal{F}|}.$$

Can one prove anything like this for the *maximum degree* instead of the average degree? The answer is *yes* if one throws out one point from each set $A \in \mathcal{F}$, i.e. by reducing the $n$-uniform family $\mathcal{F}$ into an $(n-1)$-uniform family of the same size.

**Theorem 12** ("Degree Reduction Theorem"). *Let* $\mathcal{F}$ *be an arbitrary* $n$*-uniform almost-disjoint family* $(n \geq 2)$. *Then for each* $n$*-elements set* $A \in \mathcal{F}$ *there is an* $(n-1)$*-elements subset* $\widetilde{A} \subset A$ *such that the* $(n-1)$*-uniform family* $\widetilde{\mathcal{F}} = \{\widetilde{A} : A \in \mathcal{F}\}$ *has maximum degree*

$$\text{max.degree}\,(\widetilde{\mathcal{F}}) \leq \sqrt{n|\mathcal{F}|}.$$

**Proof.** For each $A \in \mathcal{F}$ let $g(A) \in A$ be that point of $A$ which has the largest $\mathcal{F}$-degree. From each $A \in \mathcal{F}$ throw out the corresponding point $g(A)$, that is, let $\widetilde{A} = A \setminus \{g(A)\}$, and let $\widetilde{\mathcal{F}} = \{\widetilde{A} : A \in \mathcal{F}\}$. Let $d$ denote the maximum degree of $\widetilde{\mathcal{F}}$, and let $\widetilde{A_1}$, $\widetilde{A_2}$, $\widetilde{A_3}$, …, $\widetilde{A_d}$ be those sets of $\widetilde{\mathcal{F}}$ which cointain a point (say) $y$ of maximum degree in $\widetilde{\mathcal{F}}$. (The $\widetilde{\mathcal{F}}$-degree of $y$ is $d$.) The $d$ points $g(A_1)$, $g(A_2)$, $g(A_3)$, …, $g(A_d)$ are all different (because $y$ is a common point and $\mathcal{F}$ is almost disjoint), and all have $\mathcal{F}$-degree $\geq d$ (otherwise one would pick $y$ instead of $g(A_i)$). It follows that

$$n|\mathcal{F}| \geq \sum_{i=1}^{d} d_i \geq d^2,$$

where $d_i$ denotes the $\mathcal{F}$-degree of $g(A_i)$, $1 \leq i \leq d$. This completes the proof of Theorem 12. ■

In order to prove Theorem 10, we also need a "Neighborhood Principle for Almost Disjoint Families" (see Beck [3]).

**Theorem 13.** *Let $\mathcal{F}$ be an $n$-uniform family of almost disjoint sets. Assume that the maximum degree of $\mathcal{F}$ is at most $2^{n-c\sqrt{n}}$, that is, every point $x$ is contained by at most $2^{n-c\sqrt{n}}$ elements $A$ of $\mathcal{F}$ (local condition). Moreover assume that $|\mathcal{F}| < 2^{n^{3/2}}$ (global condition). If $n$ is sufficiently large, then Breaker, as the second player, has a winning strategy in the weak game on $\mathcal{F}$.*

Theorem 10 immediately follows from Theorems 12 and 13.

By using Theorem 13 one can easily settle the Hales–Jewett conjecture, at least for sufficiently large numbers.

But first let us try Theorem 9. The Erdős–Selfridge criterion applies to the weak $n^d$-game if

$$\frac{(n+2)^d - n^d}{2} < 2^{n-1}.$$

This means that Breaker can force a win if $n >$ const $\cdot\, d \cdot \log d$. Therefore, Second-Player can force a draw in the strong $n^d$-game when $n >$ const $\cdot\, d \cdot \log d$. This result just falls short of the Hales–Jewett conjecture which states that the strong game is a draw if $2n^d \geq (n+2)^d$, which is asymptotically equivalent to that $n \geq 2d/\log 2$ (see §1).

Next let us try Theorem 13 instead of Theorem 9. In the $n^d$-game every cell of the $d$-dimensional board $n \times \cdots \times n = n^d$ is contained by at most $(3^d - 1)/2$ ($n$ odd) or $2^d - 1$ ($n$ even) winning lines. Assume first that $n$ is odd. Theorem 13 applies here if

$$(1) \qquad \frac{3^d - 1}{2} \leq 2^{n - c\sqrt{n}} \quad \text{and} \quad \frac{(n+2)^d - n^d}{2} < 2^{n^{3/2}}.$$

Inequalities (1) trivially hold if

$$(2) \qquad n > \left(\frac{\log 3}{\log 2} + \varepsilon\right) d \quad \text{and} \quad n > n_0(\varepsilon).$$

If $n$ is even, the analogous bound is

$$(3) \qquad n > (1+\varepsilon)d \quad \text{and} \quad n > n_0(\varepsilon).$$

Observe that (2-3) are asymptotically better than the Hales–Jewett conjecture $n \geq 2d/\log 2$. This means that Second-Player can force a draw in the strong $n^d$-game even when a pairing strategy cannot exist.

In Theorem 13 the degree condition $2^{n-c\sqrt{n}}$ is just exponential. This local bound $2^{n-c\sqrt{n}}$ is a little bit smaller than the global bound $2^{n-1}$ in the Erdős–Selfridge Theorem (Theorem 9). But there is a big compensation: the "superexponential" global bound $2^{n^{3/2}}$. Note that $2^{n^{3/2}}$ is asymptotically much bigger than $2^n$.

In other words, in Theorem 13 the local condition is much more important than the global one. Results of this type are called the "Neighborhood Principle". So Theorem 13 is a justification of the "Neighborhood Principle," at least for almost disjoint families.

Note that the *neighborhood* of a set $A$ in $\mathcal{F}$ is the subfamily $\{B \in \mathcal{F} : A \cap B \neq \emptyset\}$. If $\mathcal{F}$ is $n$-uniform then the size of the neighborhood is at most $n$ times the maximum degree of $\mathcal{F}$. So the neighborhood-size and the maximum degree have the same order of magnitude.

The perfect justification of the "Neighborhood Principle" would be to completely drop the global condition $2^{n^{3/2}}$ in Theorem 13. Unfortunately this is an open problem. Note that the proof of Theorem 13 heavily uses the *almost disjointness* property. What happens in general, when we do not have any restriction on the pairwise intersections? Again we do not know the answer.

**Open Problem.** *Does there exist a constant $c > 1$ such that if $\mathcal{F}$ is an $n$-uniform family of winning sets with maximum degree less than $c^n$ and $n$ is sufficiently large depending on $c$, then Breaker as the second player can always force a win in the weak game on $\mathcal{F}$?*

In other words, can one replace the exponential *global* condition $2^{n-1}$ in the Erdős–Selfridge Theorem by any weaker but still exponential *local* (i.e. degree) condition. What we can prove here is a much weaker linear bound. Indeed, by the Matching Degree Criterion (Theorem 5), if max.degree $\leq n/2$ then Breaker can force a win. So the Open Problem is to jump from linear to exponential.

Even if we cannot solve the Open Problem, we can prove results in the "right direction".

**Theorem 14** ("Neighborhood Principle"). *Let $\mathcal{F}$ be an $n$-uniform family. Assume that the maximum degree of $\mathcal{F}$ is at most $2^{n/100}$ (local condition). Moreover assume that*

$$|\mathcal{F}| < 2^{2^n}$$

*(global condition). If $n$ is sufficiently large, then Breaker, as the second player, has a winning strategy in the weak game on $\mathcal{F}$.*

Note that the extremely weak double exponential global bound

$$|\mathcal{F}| < 2^{2^n}$$

can be replaced by the even weaker triple exponential bound

$$|\mathcal{F}| < 2^{2^{2^n}},$$

or by the the four-fold iterated exponential bound

$$|\mathcal{F}| < 2^{2^{2^{2^n}}},$$

and so on, if the local condition is replaced by stronger but still exponential bounds. We just formulate the triple exponential case.

**Theorem 15** ("Neighborhood Principle"). *Let $\mathcal{F}$ be an $n$-uniform family. Assume that the maximum degree of $\mathcal{F}$ is at most $2^{n/500}$ (local condition). Moreover assume that*

$$|\mathcal{F}| < 2^{2^{2^n}}$$

*(global condition). If $n$ is sufficiently large, then Breaker, as second player, has a winning strategy in the weak game on $\mathcal{F}$.*

By using Theorem 14 instead of Theorem 13 in the weak $n^d$-game one can improve the linear thresholds for Breaker's win $n > \left(\frac{\log 3}{\log 2} + \varepsilon\right) d$ ($n$ odd), and $n > (1+\varepsilon)d$ ($n$ even), to the nearly square-root threshold $n > \text{const} \cdot \sqrt{d \log d}$. But Theorem 14 alone is not enough, we also need a "degree reduction" result like Theorem 12, but a much stronger version. This "degree reduction" result is not as general as Theorem 12: it works for the family of the winning sets of the $n^d$-game only. Before formulating the result, we make some simple observations. The average degree of the family of winning sets in $n^d$ is

$$\text{average degree}\,(n^d) = \frac{n \cdot \text{familysize}}{\text{boardsize}} = \frac{n\left((n+2)^d - n^d\right)/2}{n^d} \approx \frac{n}{2}\left(e^{\frac{2d}{n}} - 1\right).$$

This is much smaller than the maximum degree $(3^d-1)/2$ ($n$ odd) and $2^d - 1$ ($n$ even).

Can we reduce the maximum degree of an $n$-uniform hypergraph to the order of magnitude of the average degree? The answer is an easy yes if we are allowed to throw out sets. But throwing out a set means that Breaker looses control over that set, and Maker might occupies it. So we cannot throw out whole sets, but we can throw out some points from each winning set just like in Theorem 12 we were throwing out one point from each set.

In other words, we can *truncate* the sets, but we cannot throw them out completely. Well, in this case the answer is *no* in general (we leave it to the reader), but it is *yes* for the special family of the winning sets in $n^d$. A substantial difference compared to Theorem 12 is that now we have to throw out as much as 75% from each winning set instead one single point.

**Theorem 16** ("Truncation Theorem for $n^d$"). *Let $\mathcal{F}_{n,d}$ denote the $n$-uniform family of winning sets of the Hales–Jewett $n^d$-game. Then for each $n$-set $A \in \mathcal{F}_{n,d}$ there is an $\frac{n}{4}$-element subset $\widetilde{A} \subset A$ such that the $\frac{n}{4}$-uniform truncated family $\widetilde{\mathcal{F}_{n,d}} = \{\widetilde{A} : A \in \mathcal{F}_{n,d}\}$ has maximum degree*

$$\text{max.degree}\,(\widetilde{\mathcal{F}_{n,d}}) < d^{\frac{4d}{n}}.$$

**Remark.** The upper bound $d^{\frac{4d}{n}}$ on the maximum degree of $\widetilde{\mathcal{F}_{n,d}}$ is not that far from the order of the magnitude of the average degree $\frac{n}{2}\left(e^{\frac{2d}{n}} - 1\right)$ of $\mathcal{F}_{n,d}$. Indeed, what really matters here are the exponents $\frac{4d}{n}$ and $\frac{2d}{n}$, and they are relatively close to each other.

Now the nearly square-root threshold $n > \text{const} \cdot \sqrt{d \log d}$ for a Breaker's win in the weak $n^d$-game is easy. It immediately follows from Theorems 14 and 16. Indeed, we apply Theorem 14 to the truncated $\frac{n}{4}$-uniform family $\widetilde{\mathcal{F}_{n,d}}$, where the truncated family $\widetilde{\mathcal{F}_{n,d}}$ comes from Theorem 16. This means two requirements:

$$d^{\frac{4d}{n}} < 2^{\frac{(n/4)}{100}} \quad \text{and} \quad n^d < 2^{2^n}.$$

These requirements are satisfied if $n > 40\sqrt{d \log d}$ and $n > c_0$.

How good is this nearly quadratic threshold $n > \text{const} \cdot \sqrt{d \log d}$. Well, it turns out to be best possible apart from the logarithmic factor $\sqrt{\log d}$. Indeed, we are going to show that if $n < \text{const} \cdot \sqrt{d}$ then Maker has a winning strategy in the weak $n^d$-game. It will follow from a general criterion for a weak win. For a *strong* win we do not know anything the Ramsey Criterion (Theorem 2). Our general criterion for a weak win usually performs far better than Theorem 3, see Beck [4].

**Theorem 17** ("Weak Win Criterion"). *Assume that $\mathcal{F}$ is an $n$-uniform hypergraph, and let $X$ be the union set (board). Moreover assume that, fixing any two points of the board $X$, there are no more than $d_2 = d_2(\mathcal{F})$ winning sets $A \in \mathcal{F}$ containing both points. If*

$$|\mathcal{F}| > 2^{n-3} \cdot d_2 \cdot |X|,$$

*then Maker, as the first player, has a winning strategy in the weak game on $(X, \mathcal{F})$.*

If $\mathcal{F}$ is almost disjoint, then fixing any two points of the board $X$, there is at most one winning set containing both of them. So we have the following.

**Corollary.** *If $\mathcal{F}$ is an $n$-uniform almost disjoint family of winning sets, and $|\mathcal{F}| > 2^{n-3}|X|$, then Maker, as the first player, has a weak win playing on $(X, \mathcal{F})$.* ■

We are going to discuss the quasiprobabilistic motivation of Theorem 17. For simplicity consider the "random two-coloring" model instead of the "exactly 50–50% two-coloring" model. We know that the *expected number* of the monochromatic winning sets is precisely $|\mathcal{F}|2^{-n+1}$. We can assume that roughly half of the monochromatic winning sets belong to Maker. We need a condition which guarantees that, with probability almost one, a random variable is close to its *expected value.* By Chebyshev's inequality, this is the case if the *standard deviation* is "small" compared to the expected value.

Let the random variable be the number of the monochromatic winning sets in the "random two-coloring" model. If the standard deviation is "small" compared to the expected value, and the expected value is "large", then Maker wins with probability almost one.

Since the standard deviation is square-root of the *variance*, we have to compute the variance of the number of monochromatic winning sets. Let $Y$ denote the number of monochromatic winning sets in the "random two-coloring" model. We have $Y = \sum_{A \in \mathcal{F}} Y_A$ where $Y_A$ is 1 if $A$ is monochromatic and 0 otherwise. The variance of $Y$ is $\mathrm{Var}(Y) = E(Y^2) - E^2(Y)$. The expected number is $E(Y) = |\mathcal{F}|2^{-n+1}$. To compute $E(Y^2)$ we have

$$E(Y^2) = E\Big(\sum_{A \in \mathcal{F}} \sum_{B \in \mathcal{F}} Y_A Y_B\Big) = \sum_{A \in \mathcal{F}} E(Y_A) + \sum_{A \in \mathcal{F}} \sum_{B \in \mathcal{F}:\, B \neq A} E(Y_A Y_B).$$

Clearly $\sum_{A\in\mathcal{F}} E(Y_A) = E(Y) = |\mathcal{F}|2^{-n+1}$. On the other hand, for every $A \in \mathcal{F}$,

$$\sum_{B\in\mathcal{F}\,:\,B\neq A} E(Y_AY_B) \leq \binom{n}{2} d_2 2^{-n+1} + \left(|\mathcal{F}| - \binom{n}{2} d_2\right) 2^{-2n+2}.$$

Indeed, by hypothesis, for every $A \in \mathcal{F}$ there are at most $\binom{n}{2}d_2$ other sets $B \in \mathcal{F}$ which have more than one common point with $A$. If $A$ and $B$ have at most one point in common, then $E(Y_AY_B) = 2^{-2n+2}$ if $|A \cap B| = 1$, and $E(Y_AY_B) = 2^{-2n+1}$ if $|A \cap B| = 0$.

We conclude that

$$\mathrm{Var}\,(Y) \leq |\mathcal{F}|\frac{n^2}{2} d_2 2^{-2n+2}.$$

On the other hand, we recall that

$$E^2(Y) = |\mathcal{F}|^2 2^{-2n+2}.$$

Now Var $(Y)$ is "small" compared to $E^2(Y)$ if $|\mathcal{F}|$ is "large" compared to $n^2d_22^n$. In this case, by Chebyshev's inequality, Maker wins with probability almost one. Therefore, the majority principle suggests that the game-theoretic "breaking point" for Maker's win comes around $n^2d_22^n$. But this is not the whole truth. The majority principle has to be combined with the *neighborhood principle*: the threshold $n^2d_22^n$, which comes from the majority principle, has to be the size of a typical neighborhood. The neighborhood of a winning set $A$ is the family of those other sets $B \in \mathcal{F}$ which intersect $A$. If almost every winning set $A$ intersects around $n^2d_22^n$ or more other sets $B \in \mathcal{F}$, then the degree of almost every point is at least $nd_22^n$. (Indeed, by hypothesis there are limited number of pairwise intersections of size $\geq 2$, so if two winning sets intersect, the common part is typically one-element.) Since there are $|X|$ points in the board, it means there are at least const $\cdot\, |X|d_22^n$ winning sets. This is exactly the condition in Theorem 17.

**Proof of Theorem 17.** This is a straightforward adaptation of the Erdős–Selfridge proof-technique. Assume we are at that stage of the play where Maker = "first player" has already occupied $x_1, x_2, \ldots, x_i$, Breaker = "second player" has occupied $y_1, y_2, \ldots, y_i$, and the question is how to choose Maker's next point $x_{i+1}$. Those winning sets which contain at least one $y_j$ $(j \le i-1)$ are "useless" for Maker. We call them "dead sets". The winning sets which are not "dead" are called "survivors". The "survivors" have a chance to be completely occupied by Maker at the end of the play. What is the total "chance" of the position? Consider the *randomized game* starting from this given position, i.e. we color the unoccupied elements of the board $X$ red and blue randomly with $p = 1/2$, red for Maker and blue for Breaker. Let $A_s \in \mathcal{F}$ be a "survivor", and let $E_s$ denote the event that $A_s$ becomes monochromatically red in the random game. The probability of $E_s$ is clearly $2^{-u_s}$ where $u_s$ is the number of unoccupied elements of $A_s$. We are interested in the "win-probability" Prob $\left\{\bigcup_{s\in S} E_s\right\}$. By the inclusion-exclusion formula

$$\text{Prob}\left\{\bigcup_{s\in S} E_s\right\} = \sum_{s\in S} \text{Prob}\{E_s\} - \sum_{s<t} \text{Prob}\{E_s \cap E_t\} +$$

$$+ \sum_{s<t<v} \text{Prob}\{E_s \cap E_t \cap E_v\} \mp \cdots .$$

We consider the "linear approximation", that is, we keep the first term $\sum_{s\in S} \text{Prob}\{E_s\}$ on the right-hand side. This is nothing else than the *expected number* of monochromaticly red sets. We evaluate the given position by the following "chance-function": $C_i = \sum_{s\in S} 2^{-u_s}$ where $u_s$ is the number of unoccupied points of the "survivor" $A_s$ ($s \in S$ = "*survivors*") and index $i$ indicates that we are at the stage of choosing the $i+1$-st point $x_{i+1}$ of Maker. A natural choice for $x_{i+1}$ is to maximize the "chance" $C_{i+1}$ at the next stage. Let $x_{i+1}$ and $y_{i+1}$ denote the next moves of the two players. What is their effect on $C_{i+1}$? Well, first $x_{i+1}$ doubles the "chance" of each "survivor" $A_s \ni x_{i+1}$, that is, we have to add the sum $\sum_{A_s : x_{i+1}\in A_s} 2^{-u_s}$ to $C_i$.

On the other hand, $y_{i+1}$ "kills" all the "survivors" $A_s \ni y_{i+1}$, which means we have to subtract the sum

$$\sum_{A_s : y_{i+1}\in A_s} 2^{-u_s}$$

from $C_i$.

Warning: we have to make a correction to those "survivors" $A_s$ which contain both $x_{i+1}$ and $y_{i+1}$. These 'survivors" $A_s$ were "doubled" first and "killed" second. So what we have subtract from $C_i$ is not

$$\sum_{A_s:\{x_{i+1},y_{i+1}\}\subset A_s} 2^{-u_s}$$

but the twice as large

$$\sum_{A_s:\{x_{i+1},y_{i+1}\}\subset A_s} 2^{-u_s+1}.$$

It follows that

$$C_{i+1} = C_i + \sum_{A_s:x_{i+1}\in A_s} 2^{-u_s} - \sum_{A_s:y_{i+1}\in A_s} 2^{-u_s} - \sum_{A_s:\{x_{i+1},y_{i+1}\}\subset A_s} 2^{-u_s}.$$

Now the natural choice for $x_{i+1}$ is that unoccupied $z$ for which $\sum_{A_s:z\in A_s} 2^{-u_s}$ is maximum. Then clearly

$$C_{i+1} \geq C_i - \sum_{A_s:\{x_{i+1},y_{i+1}\}\subset A_s} 2^{-u_s}.$$

We trivally have

$$\sum_{A_s:\{x_{i+1},y_{i+1}\}\subset A_s} 2^{-u_s} \leq \frac{d_2}{4}.$$

Indeed, there are at most $d_2$ winning sets $A_s$ containing the given two points $\{x_{i+1}, y_{i+1}\}$, and $2^{-u_s} \leq 2^{-2}$ since $x_{i+1}$ and $y_{i+1}$ were definitely unoccupied points at the previous stage. Therefore,

$$C_{i+1} \geq C_i - \frac{d_2}{4}.$$

What happens at the end? Let $l$ denote the number of stages, i.e. the $l$th stage is the last one. Clearly $l = |X|/2$. The inequality $C_l = C_{\text{last}} > 0$ means that Breaker was not able to "kill" (block) all the winning sets. Because at the last stage all the points are occupied, it follows that Maker was able to completely occupy a winning set, i.e. Maker wins.

So all what we have to check is that $C_l = C_{\text{last}} > 0$. But this is trivial. Indeed, $C_{\text{first}} = C_0 = |\mathcal{F}|2^{-n}$, so we have,

$$C_{\text{last}} \geq |\mathcal{F}|2^{-n} - \frac{|X|}{2}\frac{d_2}{4}.$$

It follows that $C_{\text{last}} > 0$ if $|\mathcal{F}| > 2^{n-3}|X|d_2$. This completes the proof of Theorem 17. ∎

Now let us apply the Corollary of Theorem 17 to the weak $n^d$-game. Maker has a weak win, i.e. a winning strategy in the weak $n^d$-game, if

$$|\mathcal{F}_{n,d}| = \frac{(n+2)^d - n^d}{2} > 2^{n-3}n^d,$$

which is equivalent to

(4) $$\left(1 + \frac{2}{n}\right)^d > 2^{n-2} + 1.$$

Inequality (4) holds for the weak $3^3, 4^4, 5^7, 6^{10}, 7^{14}, 8^{19}, 9^{24}, 10^{30}, \ldots$ games. (By exhaustive search we know that the strong $3^2$ (Tic-Tac-Toe) is a draw game, and the strong $4^3$-game (Tic-Toc-Tac-Toe) is a First-Player's win according to Patashnik.)

Asymptotically inequality (4) holds if $n < \sqrt{2d/\log 2}$, or equivalently, if $d > (\log 2) \cdot n^2/2$ ($n$ is sufficiently large).

Let us summarize what we know about the $n^d$-games. The best known bounds for strong and weak wins are at an "astronomical distance" from each other.

**Theorem 18.** *In the weak wersion of the Hales–Jewett $n^d$-game,*

(i) *if $d < \frac{3n^2}{10^4 \log n}$ then Breaker has a winning strategy (if $n$ is sufficiently large);*

(ii) *if* $d > \frac{(\log 2)\cdot n^2}{2}$ *then Maker has a winning strategy (if* $n$ *is sufficiently large). On the other hand, in the strong version of the* $n^d$*-game,*

(iii) *if* $d < \frac{3n^2}{10^4 \log n}$ *then it is a drawn game;*

(iv) *if* $d >$ *"Shelah's supertower function of* $n$*"*

*then First-Player has a winning strategy (if* $n$ *is sufficiently large).*

The probabilistic motivation of the "Neighborhood Principle" is the well-known Lovász Local Lemma (a probabilistic sieve argument), see Erdős and Lovász [15]. To explain it in detail, including the proofs of Theorems 14, 15 and 16, requires another long paper, which will appear elsewhere very soon. Note that the proofs of Theorems 14 and 15 are game-theoretic adaptations of the proof-technique in Beck [6].

## APPENDIX: ON THE $n \times n$ GAMES

The simplest $n^d$-games are the 2-dimensional $n^2$-games, that is, the $n$-in-a-row game on an $n \times n$ board. The most difficult cases turn out the be the smallest values $n = 3$ and 4. For $n = 3$, i.e. in Tic-Tac-Toe, every child "knows" that Second-Player can force a draw. A precise mathematical proof of this fact can be found in Chapter 22 of the *Winning Ways* (see [11]). This is nothing else than a "brute force" complete analysis, making use of the symmetries of the board.

A good reason why Tic-Tac-Toe is not as trivial as it seems is that Second-Player cannot prevent First-Player from occupying a winning triplet. That is, Breaker = **II** loses the weak game.

If $n = 4$ then again the Second-Player can force a draw in the strong game. What is more, in contrast to $n = 3$, Breaker, as second player, can even win the weak game. We are going to prove this fact later by an elegant argument of *David Galvin* who is a graduate student at Rutgers University.

If $n \geq 5$ then again the Second-Player can force a draw, but the situation is much simpler since we can employ pairing strategies. This automatically implies the stronger statement that Breaker, as second player, can even win the weak game.

First we give a quick proof of the cases $n = 5$ and 6 by applying the following pairing stategy: whenever the first player occupies a numbered cell, the second player takes the other cell of the same number.

$$n = 5 : \qquad \begin{bmatrix} 2 & 10 & 5 & 5 & 1 \\ 6 & 9 & 12 & 8 & 9 \\ 6 & 11 & * & 11 & 4 \\ 7 & 10 & 12 & 7 & 4 \\ 1 & 3 & 3 & 8 & 2 \end{bmatrix}$$

If the first player takes the *-marked center, the second player may take any cell, and if the cell he is required to take by the pairing strategy is occupied, he may play anywhere.

The following figure gives exactly the same pairing strategy: all you have to do is to make sure that for every move of your opponent in a marked square you take the similarly marked square in the direction indicated by the mark.

$$\begin{bmatrix} \diagdown & - & \mid & - & \diagup \\ \mid & - & - & \mid & \mid \\ - & \mid & * & \mid & - \\ \mid & \mid & - & - & \mid \\ \diagup & - & \mid & - & \diagdown \end{bmatrix}$$

Similarly, for $n = 6$ the numbering below is equivalent to the direction-marks next to it:

$$n = 6 : \qquad \begin{bmatrix} 1 & 13 & 2 & 13 & 3 & 12 \\ 6 & 14 & 5 & 14 & 4 & 12 \\ 7 & 8 & 15 & 9 & 10 & 15 \\ 16 & 3 & 11 & 1 & 16 & 2 \\ 17 & 4 & 11 & 6 & 17 & 5 \\ 7 & 8 & 18 & 9 & 10 & 18 \end{bmatrix} \begin{bmatrix} \diagdown & - & \diagdown & - & \diagup & \mid \\ \diagdown & - & \diagdown & - & \diagup & \mid \\ \mid & \mid & - & \mid & \mid & - \\ - & \diagup & \mid & \diagdown & - & \diagdown \\ - & \diagup & \mid & \diagdown & - & \diagdown \\ \mid & \mid & - & \mid & \mid & - \end{bmatrix}$$

The following pairing for $n = 6$ is even more symmetric. You may reply to a move on a diagonal with *any* other move on the same diagonal.

$$\begin{bmatrix} \diagdown & - & \mid & \mid & - & \diagup \\ - & \diagdown & - & - & \diagup & \mid \\ - & \mid & \diagdown & \diagup & \mid & - \\ - & \mid & \diagup & \diagdown & \mid & - \\ - & \diagup & - & - & \diagdown & \mid \\ \diagup & - & \mid & \mid & - & \diagdown \end{bmatrix}$$

These last three examples were elegant special cases of Theorem 4 ("Matching Criterion").

Observe that Theorem 5 ("Matching Degree Condition") immediately settles the $n \times n$ game for all $n \geq 8$: the game is a draw by a pairing strategy. The missing case $n = 7$ follows from Theorem 4 since the center is the only cell with 4 winning lines passing through it.

Now let us return to the first numbering (direction-marking) of the $6 \times 6$ board. This numbering leads to an elegant proof that the *unrestricted* 9-in-a-row is a draw game. *Unrestricted* 9-in-a-row means that the board is an infinite chess board, and the object of the game is to find 9-in-a-row in horizontal, in vertical, or in the two diagonal directions. Indeed, cover the infinite board with copies of that 6 by 6 matrix. The second player can force a draw by always taking the *nearest* cell with the same number as of the previous play. It is easy to see that the first player can obtain no line longer than 8.

Finally we prove that the 4-in-a-row on a $4 \times 4$ board is a draw game. What is more, Breaker, as the second player, can force a win in the weak game. Even if pairing strategy *cannot* work (because the number of cells is less than twice the number of winning lines: $16 = 4^2 < 2(4 + 4 + 2) = 20$), after his first move the Second-Player can always employ a pairing strategy which depends on the opening move of the opponent. The proof is based on the following three pictures:

$$\begin{bmatrix} O & - & \mid & - \\ \mid & X & \diagup & \mid \\ \mid & \diagup & - & - \\ - & - & \mid & \mid \end{bmatrix} \quad \begin{bmatrix} \mid & O & - & - \\ \mid & X & \diagup & \mid \\ - & \diagup & \mid & - \\ - & - & \mid & \mid \end{bmatrix} \quad \begin{bmatrix} \diagdown & \mid & - & - \\ \mid & O & X & \mid \\ - & - & \diagdown & \mid \\ \mid & \mid & - & - \end{bmatrix}$$

The explanation is as follows. Apart from symmetry there are three different opening moves, marked by $O$. Second-Player's reply is $X$, and for the rest of the play he employs the pairing strategy direction-marked in the corresponding picture (for every move of the opponent in a marked square he takes the similarly marked square in the direction indicated by the mark). This elegant argument is due to *David Galvin.*

## References


[1] N. Alon, and J. Spencer, *The Probabilistic Method,* Academic Press (New York, 1992).

[2] J. Baumgartner, F. Galvin, R. Laver and R. McKenzie, Game theoretic versions of partition relations, in: *Colloq. Math. Soc. János Bolyai "Infinite and Finite sets"* (Keszthely, Hungary, 1973), 131–135.

[3] J. Beck, On positional games, *Journal of Combinatorial Theory, ser. A,* **30** (1981), 117–133.

[4] J. Beck, Van der Waerden and Ramsey type games, *Combinatorica,* **2** (1981), 103–116.

[5] J. Beck, Random graphs and positional games on the complete graph, *Annals of Discrete Math.,* **28** (1985), 7–13.

[6] J. Beck, An algorithmic approach to the Lovász Local Lemma. I., *Random Structures and Algorithms,* **2** (1991), 343–365.

[7] J. Beck, Deterministic graph games and a probabilistic intuition, *Combinat. Probab. Comput.,* **3** (1993), 13–26.

[8] J. Beck, Foundations of positional games, *Random Structures and Algorithms,* **9** (1996), 15–47.

[9] J. Beck and L. Csirmaz, Variations on a game, *Journal of Combinatorial Theory, ser. A,* **33** (1982), 297–315.

[10] C. Berge, Sur les jeux positionelles, *Cahiers Centre Études Rech. Opér.,* **18** (1976).

[11] E. R. Berlekamp, J. H. Conway and R. K. Guy, *Winning Ways,* Academic Press (London, 1982).

[12] B. Bollobás, *Random Graphs,* Academic Press (London, 1985).

[13] J. H. Conway, *On Numbers and Games,* Academic Press (London, 1976).

[14] P. Erdős and V. Chvátal, Biased positional games, *Annals of Discrete Math.*, **2** (1978), 221–228.

[15] P. Erdős and L. Lovász, Problems and results on 3-chromatic hypergraphs and some related questions, in: *Infinite and Finite Sets* (eds.: A. Hajnal et al.), *Colloq. Math. Soc. J. Bolyai*, **11**, North-Holland, Amsterdam, 1975, 609–627.

[16] P. Erdős and J. Selfridge, On a combinatorial game, *Journal of Combinatorial Theory, ser. A*, **14** (1973), 298–301.

[17] A. S. Fraenkel, Complexity of games, *Combinatorial Games, Proc. Symposia in Applied Mathematics*, AMS, Providence, RI, Vol. 43, 111–153.

[18] R. L. Graham, B. L. Rothschild and J. H. Spencer, *Ramsey Theory*, Wiley-Interscience Ser. in Discrete Math. (New York, 1980).

[19] A. W. Hales and R. I. Jewett, On regularity and positional games, *Trans. Amer. Math. Soc.*, **106** (1963), 222–229.

[20] L. Lovász, *Combinatorial Problems and Exercises*, North-Holland and Akadémiai Kiadó (Budapest, 1979).

[21] O. Patashnik, Qubic: $4 \times 4 \times 4$ Tic-Tac-Toe, *Mathematics Magazine*, **53** Sept. (1980), 202–216.

[22] A. Pluhár, Generalized Harary Games, *Acta Cybernetica*, **13** (1997), 77–83.

[23] S. Shelah, Primitive recursive bounds for Van der Waerden numbers, *Journal of the American Math. Soc.*, **1** 3 (1988), 683–697.

József Beck

*Department of Mathematics*
*Rutgers University*
*New Brunswick, New Jersey 08903*
*U.S.A.*

e-mail: jbeck@math.rutgers.edu

BOLYAI SOCIETY
MATHEMATICAL STUDIES, 10

**Contemporary Combinatorics**
**Budapest, 2002,** pp. 139–184.

# A State Space Definition of the HOMFLY Invariant

B. BOLLOBÁS, L. PEBODY, and D. WEINREICH

Polynomial invariants are powerful tools in knot theory. One of the most useful invariants is the HOMFLY polynomial, discovered in 1986. Taking our lead from Kauffman's explicit expression for the Jones polynomial, we provide a more explicit definition of the HOMFLY invariant based on the state space of the link in question. In fact, we define a polynomial on basepointed, component-ordered link diagrams and provide a complete proof that it has the properties of, and is thus a new representation of, the HOMFLY polynomial.

## 1. Introduction

The study of knots and links as mathematical objects was introduced by Listing in 1847 and, independently, by Thomson in 1869. In their work, and in much subsequent research, knots and links were considered as geometric or topological objects. In their formulation, which we shall use here, a *link* of $n$ components is a subset of $\mathbf{R}^3$ consisting of $n$ disjoint piecewise linear simple closed curves. A *knot* is a link with a single component.

The question of which links are the 'same' is far from trivial. Regardless of your definition of 'sameness,' the fact that curves can be transformed (non-trivially) in space without changing their nature causes problems. This is clear to anyone who has held a seemingly knotted string and pulled it straight. More precisely, we say two links are equivalent when there is an orientation-preserving homeomorphism from $\mathbf{R}^3$ to itself taking one link to the other. Two such links are said to be *ambient isotopic.* A link which is ambient isotopic to a circle is called an *unknot.* Similarly, a link which is ambient isotopic to a collection of $n$ disjoint circles in a plane is called *$n$ unlinked unknots.* A representation of a link in $\mathbf{R}^2$ is called a *diagram* corresponding to the link.

Ideally, one would like to be able to completely classify knots and links via their ambient isotopy classes. Hence we would like some tools to investigate link equivalence. The work of Reidemeister in the 1920s showed that link-equivalence may be viewed as a purely combinatorial property of link diagrams (see, for example, [16]). He proved that two links are ambient isotopic if and only if their diagrams can be transformed into each other by planar isotopy and a finite combination of the three so-called Reidemeister moves. In Section 3, we describe these moves and describe a result involving the Reidemeister moves and unknots.

Even with this useful result, finding a complete characterization of ambient isotopy classes turns out to be quite difficult, so researchers turned their attention instead to creating link invariants which can be computed easily and yet can distinguish many classes of links. In particular, such invariants were introduced by (among others) Alexander [1], Conway [3], and Reidemeister and more recently by Kauffman [10], Jones [9] and the HOMFLY authors [4]. These are mathematical objects (groups/polynomials, etc.) which are constructible from the link diagrams and which are preserved by ambient isotopy.

In this paper, we shall consider knots and links as combinatorial objects by considering a multigraph based on a link. We start from a link diagram, the projection of the link into $\mathbf{R}^2$ with no triple points and with any double point corresponding to a 'clean' crossing of the two strands involved. We then create a 4-regular multigraph based on this projection. This multigraph is called the *universe* of the link. Each vertex corresponds to a crossing of the link, a place where the projection crosses itself, and each pair of non-consecutive edges at a vertex corresponds to a continuous curve in the link. By labeling each vertex according to which of the pairs of non-consecutive edges is higher, we retain all of the information given by the link. We call this extra structure (and the link diagram itself) a *state* of the multigraph. For a given 4-regular multigraph, we call the set of all states the *state space* of the graph, and the set of all such link diagrams the universe.

In 1987, Kauffman [11] gave a state space definition of his own polynomial and related it to the Jones polynomial. Previously, polynomial invariants had been defined by way of a recurrence relation. By describing the states of the crossings in a link diagram and resolving them in a certain way, he was able to concretely define the Jones polynomial of any link. A state model has the advantage of being closely related to models in statistical physics. Also, in the case of Kauffman's invariant, a state space model made verification of its existence and uniqueness properties easy. We describe how this was done in Section 4. Our main aim in this paper is to give a similar state space definition of the HOMFLY polynomial.

In particular, we define a new polynomial in terms of an explicit summation based on the states of the nodes of the link's multigraph. We then show that this is an invariant of links, and moreover, it is precisely a two-variable form of the HOMFLY invariant. This polynomial was discovered simultaneously in 1985 by Freyd and Yetter [5], Hoste [6], Lickorish and Millett [13], Ocneanu [14], and Przytycki and Traczyk [15] (see also [4]), the last pair of authors only narrowly missing having their names included in the acronym. Unfortunately, the state space definition only simplifies the proof of existence and uniqueness slightly. The method we use is similar to that of Hoste, Lickorish and Millett, and Przytycki and Traczyk, while the work of Freyd and Yetter and Ocneanu is quite different. However, the state-space definition does have the advantage that it streamlines some of the arguments and enables us to use some clearer language. More importantly, the focus of the present work is on the definition of the polynomial as an explicit sum, and we prove many useful lemmas which do not appear elsewhere.

Our result shows that the HOMFLY polynomial, and hence the polynomials of Alexander, Conway, and Jones, can be defined in terms of state spaces. Furthermore, the state spaces do not need to be described in terms of a diagram of unlinked unknots, but may be described in terms of any diagram in the universe of the link in question. These state space definitions alone are enough to prove the existence and nature of the HOMFLY polynomial, adding to the usefulness of the invariant.

## 2. Preliminary definitions

Unfortunately, there are many terms which are used in the theory of knots which need formal definitions for mathematical rigor. In this section, we make explicit the concepts discussed implicitly in the introduction. Throughout, we mix formal definitions with assertions about the object defined. While this sometimes seems cumbersome, it is in fact the most seamless way to introduce the concepts. Those familiar with the theory of knots may want to skim the definitions, included for precision.

**Definition 1.** A *link* $L$ *of* $n$ *components* is a subset of $\mathbf{R}^3$ consisting of $n$ disjoint piecewise linear simple closed curves. The number of components of a link will be denoted $cL$. A *knot* is a connected link (a link $L$ such that $cL = 1$). If we give each of the components of $L$ an orientation, we call $L$ an *oriented link.*

In fact, we view links not as defined above, but as purely combinatorial objects. To do this, we project a link into $\mathbf{R}^2$ to form a graph. As geometric deformation does not affect the nature of a link, we may deform the link slightly so that the projection is easy to interpret. For example, if in the projection there is a point at which three strands cross, we can pull aside one of the strands so that only two strands cross at each point. Furthermore, we can bend the strands slightly, so that at every crossing the edges 'cross cleanly.' That is, there is a positive angle between them at their point of intersection in the projection plane, and they do indeed cross; they can never coincide or merely touch. Doing this, we project a link into $\mathbf{R}^2$ to form a 4-regular plane multigraph with (perhaps) loops and disjoint circles (representing unknot components). We then give our diagrams some extra structure, in that we say which of the two crossing pairs of edges is above the other. Any such diagram can be transformed back into the appropriate link in $\mathbf{R}^3$ by traversing the plane graph by following its edges, at each vertex passing from an edge to the opposite (half)edge, up to geometric isomorphism as described above. More precisely, we have the following definition.

**Definition 2.** Let a *graph* be a multigraph with loops, where we allow a graph to have degenerate graph components each consisting of one loop and no vertices. We shall call such a graph component a 0-*cycle* of the graph.

A 0-cycle corresponds to an unknot in the link which is not intertwined with any other component of the link. A *link diagram* is a 4-regular plane graph, with the extra structure that at each vertex we specify which pair of opposite edges is higher. In this additional structure, the higher of the two pairs of edges is called the *overcrossing* pair and the other is called the *undercrossing* pair. A *strand* of a link diagram is a trail in the diagram which uses the overcrossing pair at each interior vertex, and one edge of an undercrossing pair at its endpoints. An *arc* of a link diagram is the image of a curve in the link itself.

Confusion could arise over the use of the word "component" between link components and graph components. If we use the term component, we mean a link component. When we mean graph components, we shall say so explicitly. Note that the discussion before the definition implies that any 4-regular plane graph is a diagram of some link, with the appropriate structure added. Each such graph of order $n$ gives rise to $2^n$ link diagrams, corresponding to the choice of overcrossing or undercrossing at each vertex.

**Definition 3.** If $U$ is a 4-regular plane graph, the graph is called the *universal diagram* of the link diagrams that arise from it. A particular link diagram is called a *state* of the universal diagram. The set of all link diagrams with the same universal diagram is called the *state space* or the *universe* of any of the diagrams. The set of all link diagrams is denoted by $\mathcal{L}$, and the collection of link diagrams with at most $n$ crossings is denoted by $\mathcal{L}_n$.

For technical reasons, we will find it useful to consider not link diagrams, but diagrams with yet another additional structure. This will not change the link that the diagram represents, but will enable us to describe in a concrete way particular links in a diagram's universe, as the next three definitions show. Given a link diagram, we will need to determine the number of components contained in the link it represents. In fact this is easy to do, by traversing the edges of the graph in order according to their pairing and counting the number of circuits traversed in this way.

**Definition 4.** Given a link diagram, we can provide each of the link components with an orientation, place a *basepoint* on each component (at a place other than a crossing), and give the components an ordering. We will call a link diagram that has been thus prepared a *bod.* If $L$ is a bod, we will denote the underlying oriented link diagram by $\overline{L}$. The set of all bods is denoted by $\mathcal{B}$, and the collection of bods with at most $n$ crossings is denoted by $\mathcal{B}_n$.

Analogous to Definition 3, we define a universe for bods. If we use the term universe without modification, however, we will mean the link universe.

**Definition 5.** Given a bod $L$, the *universal bod* of $L$ is the universal diagram of $\overline{L}$ where each component is given the same basepoints and component ordering as $L$. A particular bod is called a *state* of the universal bod. The set of all bods with the same universal bod is called the *bod universe* (which we denote $\mathcal{U}_L$) of any of the bods.

The next definition shows why the extra information on a bod is useful, in that it allows us to identify a particular bod in each bod universe. Note that this special bod is a diagram of a collection of unlinked unknots.

**Definition 6.** Given a bod $L$, consider the following trail: start at the basepoint of the first (link) component, proceed as dictated by the orientation, and traverse each component in turn according to the component ordering. The bod is *descending* if each vertex is first visited via the over-crossing pair of edges. For any bod $L \in \mathcal{B}_n$, we define $\alpha L$ to be the unique descending bod in the bod universe of $L$. If $x$ is a crossing of $L$ that agrees with $\alpha L$, we say the crossing at $x$ is *descending*. Otherwise we say that the crossing is *ascending*.

Note that we have broadened the collection of objects we will consider from links to link diagrams to bods, in each case exponentially increasing the number of objects we consider. We wish to describe two types of equivalence relations between bods that help control this proliferation.

**Definition 7.** Given two bods $L$ and $M$, we say that $L$ and $M$ are *siblings* if they differ only in basepoints and component ordering. This means that $L$ and $M$ have the same underlying oriented link diagram, or $\overline{L} = \overline{M}$.

**Definition 8.** If $M$ is in $\mathcal{U}_L$, we call $L$ and $M$ *similar* and define the set $T(L, M)$ to be the set of crossings on which they differ. If $\left|T(L, M)\right| = 1$, we say that $L$ and $M$ are *adjacent*.

Now we use these types of equivalence relations to define related links, providing a method to move from one universe to another.

**Definition 9.** If $L$ and $M$ are bods whose underlying links are in the same universe, then we denote by $M_{\mathcal{U}_L}$ the bod which is similar to $L$ but a sibling of $M$. That is, $M_{\mathcal{U}_L}$ is $\overline{M}$ with basepoints and ordering as in $L$.

The following lemma is an easy consequence of the definitions.

**Lemma 1.** *The set $T(L, M)$ is not affected by the component ordering or the choice of basepoints. That is, if $N$ is a sibling of $L$ ($N = L_{\mathcal{U}_N}$), then $T(L_{\mathcal{U}_N}, M_{\mathcal{U}_N}) = T(L, M)$.* ∎

Oriented crossings can be assigned values of $\pm 1$, called *signs*, according to the rules shown in Figure 1.

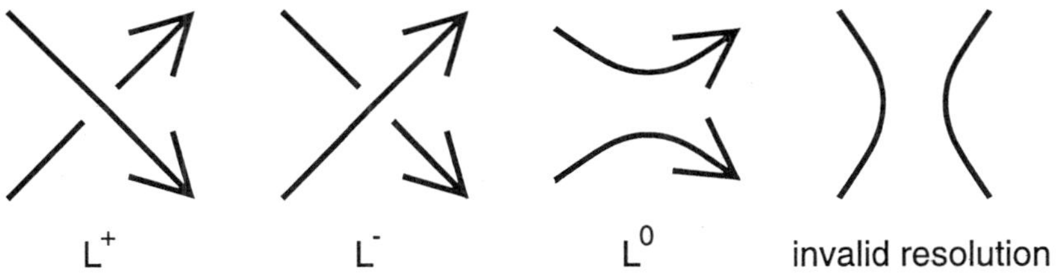


*Fig. 1.* A positive and negative crossing, their oriented resolution, and the other resolution of the non-oriented link.

**Definition 10.** Given an oriented link diagram (or bod) $L$, and a crossing $v$ in $L$, rotate the arc composed of the overcrossing edge-pair counterclockwise to overlap the arc made up of the undercrossing edge-pair. We say that $v$ is *positive* if the two arcs are then oriented in the same direction and *negative* otherwise. The *sign* of $v$, $s(v, L)$, is $+1$ if $v$ is positive, and $-1$ if $v$ is negative. Given a set of crossings $S$ of $L$, let $p_L(S)$ be the number of crossings in a positive state and $n_L(S)$ be the number of crossings in a negative state. We will suppress the subscript if it is clear from context. For the indicator function of a crossing, we will write $p(v)$ or $n(v)$ rather than $p(\{v\})$ or $n(\{v\})$.

**Definition 11.** Given a (non-oriented) link diagram $L$ and a crossing $v$ in $L$, we say link diagram $L'$ is a *resolution* of $L$ at $v$ if it is obtained by removing $v$ and joining the loose ends pairwise so as not to intersect.

See Figure 1 for a visual explanation of these definitions. We see that there are two resolutions of each crossing in a non-oriented link diagram. However, it is clear that a crossing in an oriented link diagram can only be resolved in one direction that maintains the orientation. This resolution is independent of the sign, and depends only on the orientation.

**Definition 12.** Given an oriented link diagram $L$, and a crossing $v$ in $L$, we denote by $L_v^0$, $L_v^+$ and $L_v^-$ the diagram $L$ with the crossing at $v$ resolved, positive, and negative, respectively. Where there is no chance of confusion we suppress $v$ and denote these link diagrams by $L^0$, $L^+$ and $L^-$.

**Definition 13.** Given a bod $M$, and a set $S$ of crossings in $M$, we define $[M]_S$ to be the bod we get from $M$ by resolving the crossings in $S$ in the order they appear in the bod (starting at the basepoint of the first component, following the orientation, and visiting the components in order) with the following rules:

- If, in resolving a crossing, we combine two components then the new component replaces the earlier of them. It also keeps that component's basepoint.
- If, in resolving a crossing, we split one component into two, then whichever component contained the old basepoint keeps it, and stays at the same position in the ordering. The other component gets a basepoint where the crossing was, and immediately precedes the other component in the ordering.

Following the rules above, $M^+$, $M^-$, and $M^0$ are defined for bods as in Definition 12.

Now that the preliminary details of definitions are taken care of, we may start discussing some properties of links and bods.

## 3. Reidemeister moves

We shall explore polynomial invariants of links in this paper, so we need some tools to verify invariance. These tools are methods for manipulating a bod or link diagram, while maintaining the equivalence class of the link represented. In this section, we develop those tools and explore some of their properties.

It is clear that distorting a line of a diagram within its plane region does not change the link represented by a link diagram, and hence we will not distinguish between planar isotopic link diagrams: two diagrams are considered the same if they come from isotopic plane graphs in which the isotopy preserves the extra structure of overcrossings and undercrossings. Thus planar isotopy will be one of the tools we use.

Another tool will be the *Reidemeister moves* illustrated in Figure 2. While these transformations on a link diagram change the diagram, they clearly do not change the represented link. As mentioned in the introduction, Reidemeister [16] proved that in fact these three moves are sufficient for ambient isotopy. That is to say, two links are ambient isotopic if and only if there is a finite sequence of Reidemeister moves which transforms one diagram to the other.

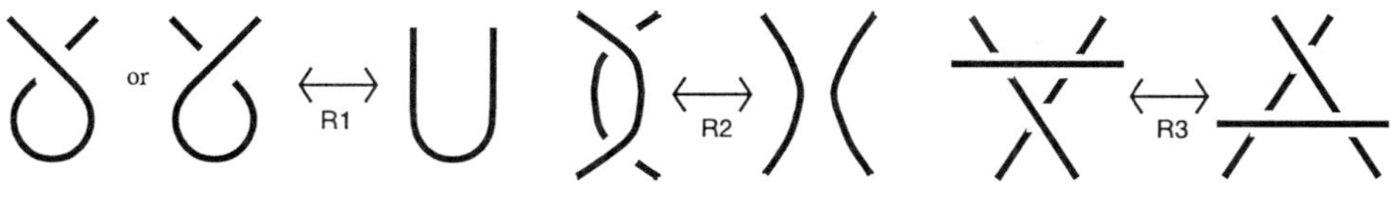


*Fig. 2.* The three Reidemeister moves, in which two links differ only in the regions shown. A move of the first type (R1) adds or removes a kink, a move of the second type (R2) adds or removes a pair of consecutive undercrossings or overcrossings, and a move of the third type (R3), a triangle move, changes the position of two crossings. Note that there are two moves of the third type depending on the sign of the middle crossing. The triangles shown before and after the move of the third type are called *Reidemeister triangles.*

Reidemeister's result provides us with a tool to examine the validity of a conjectured invariant. That is, a parameter is invariant under ambient isotopy exactly when it is preserved under the Reidemeister moves and plane isotopy.

As noted in Section 2, we will be especially interested in descending bods, which are a particular class of bods representing unlinked unknots. Although Reidemeister's result implies that a descending bod will remain equivalent to the appropriate collection of unlinked unknots after any collection of Reidemeister moves, it will not necessarily remain descending. This is easy to see if you apply a Reidemeister move of the first type, creating a twist in the non-descending direction. Even if we only allow Reidemeister moves which do not increase the number of crossings, it is not true, as is shown in Figure 3.

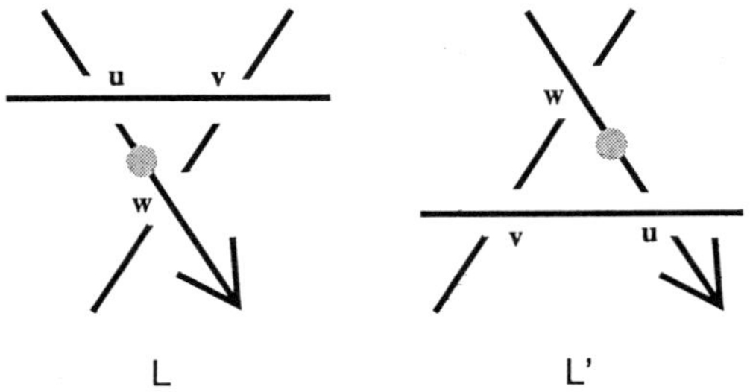


*Fig. 3.* A Reidemeister move of the third type which changes a bod from descending to non-descending: all strands are from the same component and we traverse the strands shown from the arrow to the top strand, and then to the bottom strand.

As long as we avoid this situation and do not increase the number of crossings, simple inspection verifies that the link will remain descending. In view of this, the following statement is true but cannot be made significantly stronger.

**Lemma 2.** *Let $L$ be a descending bod. If $L'$ is obtained from $L$ by a Reidemeister move which decreases the number of crossings or by a Reidemeister move of the third type in which no basepoint appears inside the Reidemeister triangle, then $L'$ is descending.* ■

We will use an additional result, due to Przytycki and Traczyk [15], that if a bod has no 0-cycle, there is a descending diagram with the same universal link diagram, which can be transformed to a diagram with fewer crossings by Reidemeister moves which do not ever increase the number of crossings. As they note, this is not difficult and was probably known to

Reidemeister as well. While it seems rather esoteric, the result is critical for proving the invariance of our polynomial. The proof we present is due to Przytycki and Traczyk, though their terminology is quite different.

We would like a result that is much stronger than this. It seems highly likely that, given any descending bod $L$, we can eradicate $L$ completely if we are allowed only to use Reidemeister moves which do not increase the number of crossings, and we are allowed to remove 0-cycles whenever they appear. However, we have yet to find a proof of this assertion. In [6], Hoste proves an equivalent statement for bods with two components, but the result does not extend in any obvious way. Hoste's result, however, could replace the result we present here and, with slight changes in Section 6.5, our proof would still be valid. Lickorish and Millett seem to imply the full assertion in their proof (for bods with any number of components), but they make some unstated assumptions which do not hold.

Given a bod, we will focus our attention on certain areas of the plane bounded by the strands of the bod. Figure 4 and Definition 14 describe the areas in which we will be interested.

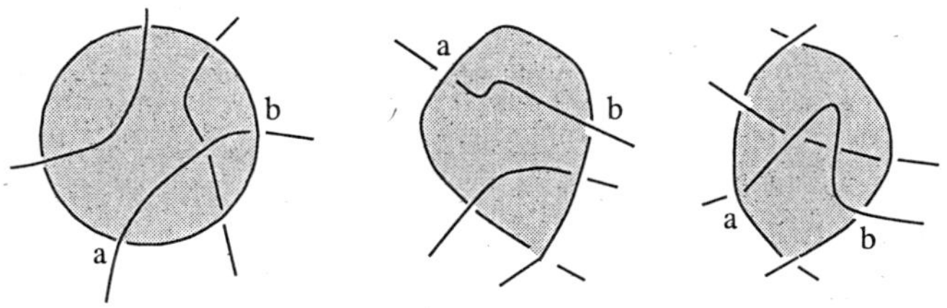


*Fig. 4.* Good regions: from left to right, $b(R) = 0, 1, 2$, respectively. For each region, the arc $ab$ is one example of a transversal.

**Definition 14.** A *region* $R$ of a bod is a closed region of the plane bounded by arcs of the bod. Each bounding arc is called a *side* of the region. The minimal number of arcs which can be used to bound a particular region is called its *bounding number* and is denoted $b(R)$, unless $R$ is bounded by an entire component with no self-crossings, in which case we say $b(R) = 0$. If $b(R) \leq 2$, we say the region is *good*. If a region is good and it contains no other good region, we say that the region is a *best* region. An arc is a *transversal* of $R$ if it joins two distinct points outside $R$, crosses the boundary of $R$ exactly twice, and is involved in no crossings outside of $R$.

Note that the existence of a region is independent of the state of the crossings of the bod, so a region is the same whether we consider the bod, the underlying link diagram, or even the universal bod or universal diagram. Hence we consider regions on *link-objects*, that is, on any of the aforementioned objects. We denote regions by the labels on their crossings; that is, a region $R$ with sides $ab$, $bc$, $cd$, and $da$ may be denoted by $R = (a, b, c, d)$.

It is clear that every link-object has a good region: starting at any point on the link-object, we either come to a self-crossing, forming a 1-sided region, or we complete the component, which forms a region $R$ with $b(R) = 0$. Note also that every good region contains a best region, so every link-object has a best region.

We first prove a technical lemma, which says that if a 3-sided region in the link-object is cut cleanly (from one side to another, as shown in Figure 5) then the link-object has a 3-sided region with empty interior. This is a correction of the lemma that appears in [15], which claims more than is proven there.

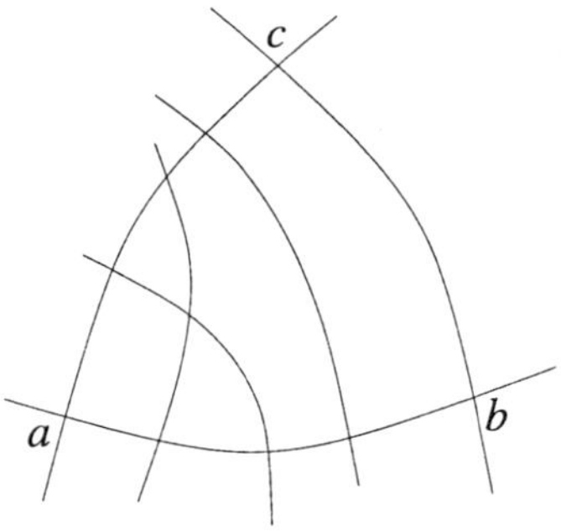


*Fig. 5.* A region satisfying the conditions of Lemma 3.

**Lemma 3.** *Let $R$ be a region of a link-object $L$ with $b(R) \in \{2, 3\}$ having sides $ab$, $ac$, and $bc$, where $bc$ may be a point, such that every transversal of $R$ crosses first $ab$ and then $ac$ without any self-crossings and no pair of transversals cross each other twice. If $R$ has a transversal, there is a region $R' \subseteq R$ such that $b(R') = 3$, one side of $R'$ lies along $ab$, and $R'$ has no transversal.*

**Proof.** We proceed by induction on the number $k$ of transversals of $R$. If $k = 1$, then the region bounded by the unique transversal and having third corner $a$ is a 3-sided region with a side along $ab$ having no transversal. So assume we can find such a region for all 2- or 3-sided regions with $k - 1$ transversals. Let $xy$ be a transversal of $R$. Remove $xy$ and, using the induction hypothesis, find a region $R'' = (a', b', c')$ with side $a'b' \subseteq ab$ in what remains. If $xy$ does not cross $R''$ or $xy$ crosses $a'b'$, we are done, in the latter case by applying the lemma to $R''$. Therefore $xy$ must cut $R''$ on sides $a'c'$ and $b'c'$. Traversing $xy$ from $x$ to $y$, let $x'$ and $y'$ be the crossings of $xy$ with the sides of $R''$ in order (see Figure 6). The cases $x'$ is on $a'c'$ and $x'$ is on $b'c'$ are symmetric, so we consider only the former, illustrated in Figure 6. Hence $y'y \cap \mathbf{Int}(R'') = \emptyset$. Consider the region $R'''$ bordered by $yy'$, $b'y'$, and $b'y$. We claim that $R'''$ contains a 3-sided region with no transversals.

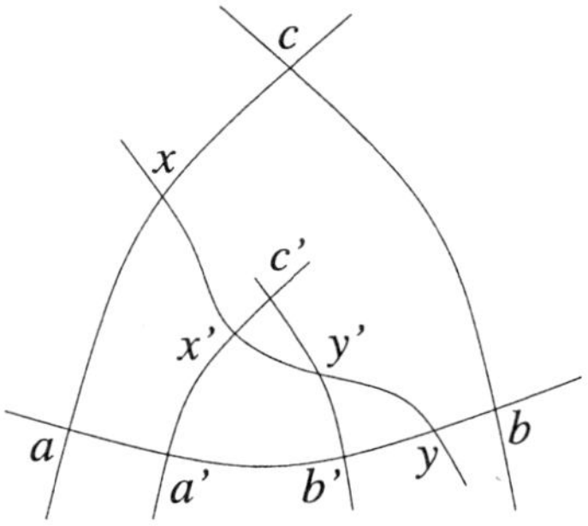


*Fig. 6.* The regions $R = (a, b, c)$, $R'' = (a', b', c')$, and $R''' = (b', y', y)$ in the proof of Lemma 3.

Since no transversal of $R$ (and hence no arc of $L$) other than $xy$ crosses $R''$ and $xy$ only crosses $b'c'$ once, no arc of $L$ crosses $b'y'$. Hence every transversal of $R'''$ must cross first $yy'$ and then $b'y'$ without any self-crossings. Furthermore, since $xy$ does not cross itself, no subarc of $xy$ is a transversal of $R'''$. Hence, either $R'''$ has no transversals, in which case it will do for $R'$, or $R'''$ has fewer than $k$ transversals, in which case the proof is complete by the induction hypothesis. ■

Our proof of the main result of this section relies on this lemma and the concept of a best region.

**Theorem 4.** *Let $U$ be a universal diagram without a 0-cycle. There exists a choice of basepoints for each of the components such that, given any descending bod $L$ with universal diagram $U$ and these basepoints, we can apply a finite sequence of Reidemeister moves to $L$ and obtain a descending bod with fewer crossings without increasing the number of crossings at any intermediate step.*

**Proof.** Let $R$ be a best region in $U$. If $b(R) \leq 1$, then $R$ has no transversals, as it would otherwise contain a region $R' \neq R$ with $b(R') \leq 2$, contradicting the fact that $R$ is best. Then, since $U$ has no 0-cycles, $b(R) \neq 0$. Hence $b(R) = 1$, and $R$ has no transversals, so a Reidemeister move of the first type will give us a universe with fewer crossings.

Thus we may assume that $b(R) = 2$. As noted earlier, every component of a link contains a region with $b(R) \leq 1$, so no component of $U$ lies fully within $R$. Furthermore, no transversal of $R$ has self-crossings, and no pair of transversals intersect twice, as in each case $R$ would contain a good region, and would thus not be best. Hence $R$ satisfies the conditions of Lemma 3.

Pick a bod in $U$ all of whose basepoints lie outside of the region $R$. From such a choice, create the descending bod $L$. Note that our choice of basepoints guarantees that any transversal of $R$ passes either above or below both of the sides of $R$. Furthermore, one side of $R$ passes above the other side of $R$ at both crossings.

We claim that we can apply a finite sequence of Reidemeister moves to the transversals of $R$ without increasing the number of crossings and obtain a 2-sided region with no transversals in a descending link diagram.

We prove this claim by induction on $c$, the number of times the transversals of $R$ cross each other. This will be referred to as the number of crossings in $R$. If $c = 0$, then we can move the transversals out of the region by Reidemeister moves of the third type, obtaining an empty region as desired. We will still have a descending bod by Lemma 2, as all basepoints lie outside $R$ and hence outside of the Reidemeister triangles.

So assume that the result is true if $R$ has fewer than $c$ crossings and $c \geq 1$. We can apply Lemma 3 to $R$ to find a triangle $R'$ inside $R$ with no transversals and a side along $R$. This will be a Reidemeister triangle since there are no basepoints inside $R$. As shown in Figure 7, a Reidemeister move of the third type will reduce the number of crossings in $R$, and the resulting bod will still be descending by Lemma 2. Thus we can apply the induction hypothesis to complete the proof of the claim.

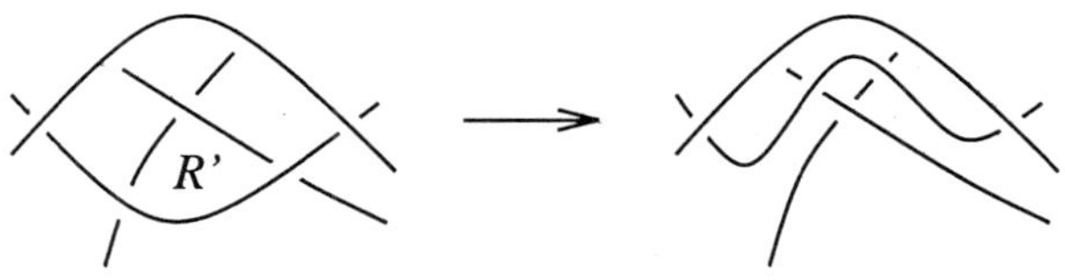


*Fig. 7.* Reducing the number of crossings in a region with 2 sides.

As the link we obtain is still descending, one side of the 2-sided region we obtain will still pass above the other side at both crossings. Hence we may apply a Reidemeister move of the second type to the sides of the region to reduce the number of crossings of $L$. In this process we have not increased the crossings at any step. ■

One might naively assume that this result, applied inductively on the number of crossings, would give us the strengthening of Hoste's result mentioned earlier. However, it is important to note that Theorem 4 is much weaker than it may appear. It says only that, given a bod, there is a descending bod in the *link* universe whose crossings can be reduced by Reidemeister moves. Hence even after obtaining the bod with fewer crossings, we might need to move the basepoints again to reduce crossings further. The more general result remains elusive.

## 4. Definition of the Kauffman bracket

With the preliminaries finally out of the way, we shall now shift focus and embark on a discussion of how a polynomial of links or bods can be defined from the state space of the links or bods. The details we have worked so hard to develop in the previous two sections will not be needed until Sections 5 and 6. The main result of this paper will be to give a state-space definition of a polynomial on bods that we shall show to have certain useful features. To facilitate understanding of our methods, we shall describe various polynomials of links and here present a state-space definition, discovered by Kauffman [10], of a collection of simpler polynomials of link diagrams (as opposed to bods). We shall follow Kauffman's proof that these polynomials have certain other useful properties. For this section we shall need little of the extra machinery we have created. In particular, link diagrams are not given orientation, basepoints or a component ordering. The methods we shall use are much more simple than those we shall employ later, but should be instructive towards the general idea.

Without orientation, as noted earlier, there are two ways of resolving a crossing in an (unoriented) link diagram. We want to be able to tell them apart. Locally, the two strands at a crossing split the nearby area into 4 regions. It is, of course, possible for two opposite local regions to be in fact the same global region; for example consider the link diagram with only one crossing. However, by the Jordan curve theorem, no other pair of regions may be part of the same global region.

**Definition 15.** Viewed locally, there are four regions created at every crossing of a link diagram. At any crossing of an unoriented link diagram, rotate the over-crossing arc counterclockwise until the under-crossing arc is reached. We call the two regions swept out by this rotation the two *A regions.* The other two regions are called the *B-regions* (see Figure 8). The two ways of resolving a crossing are called an *A-slice* (where we unite the $A$ regions, keeping the $B$ regions apart) and a *B-slice* (where we unite the $B$ regions, keeping the $A$ regions apart). Given a link diagram $L$ with a crossing at $v$, we write $L_v^A$ for the link diagram obtained from $L$ by an $A$-slice at $v$, and $L_v^B$ for the link diagram obtained from $L$ by a $B$-slice at $v$ (see Figure 8).

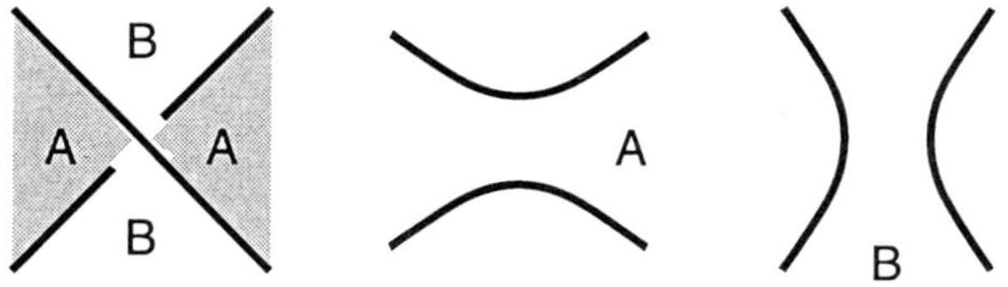


*Fig. 8.* A crossing $v$ with $A$ and $B$ regions, and their respective slices $L_v^A$ and $L_v^B$.

In Section 2 we defined a state of a universe to be a link diagram obtained by making a choice from $\{+, -\}$ for each crossing. Now we want to define a state of a link diagram, analogously, to be a choice from $\{A, B\}$ for each crossing. Based on this choice, we will then resolve every crossing.

Clearly we will be left with a link diagram of no crossings, which must therefore be a collection of unlinked trivial diagrams of unknots.

**Definition 16.** A *state* $S$ of a link diagram is a choice of direction for slicing at each crossing of the diagram. The $S$*-splitting* of link diagram $L$ is then the result of slicing all crossings according to the state $S$. We write $a_L(S)$ for the number of $A$-slices in $S$, $b_L(S)$ for the number of $B$-slices, and $c_L(S)$ for the number of components of the $S$-splitting of $L$.

For example, the following diagram shows a simple link and its two possible splittings.

*Fig. 9.* A twist, its {A}-split and its {B}-split.

Now we are ready to present Kauffman's definition of his square bracket of a link diagram.

**Definition 17.** The *Kauffman square bracket* $[L]$ of a link diagram $L$, with value in $\mathbf{Z}[A, B, d]$ is defined as follows:

$$[L] = \sum_S A^{a_L(S)} B^{b_L(S)} d^{c_L(S)-1},$$

with the summation being over all states $S$ of the diagram $L$.

The basic properties of the Kauffman square bracket are given in the following theorem. Let us write $\bigcirc$ for a connected link diagram without crossings (which we shall call *a trivial diagram of an unknot*), and $L_1 \cup L_2$ for a diagram which is a disjoint union of two link diagrams $L_1$ and $L_2$.

**Theorem 5.** *There is a unique map* $\phi : \mathcal{L} \to \mathbf{Z}[A, B, d]$ *such that*

1. *if* $L$ *and* $L'$ *are plane homotopic link diagrams then* $\phi(L) = \phi(L')$;

2. $\phi(\bigcirc) = 1$;

3. $\phi(L \cup \bigcirc) = d\phi(L)$ *for every link diagram* $L$;

4. $\phi(L) = A\phi(L_v^A) + B\phi(L_v^B)$ *for every crossing* $v$ *of every link diagram* $L$.

*Furthermore,* $\phi(L) = [L]$.

**Proof.** Suppose $\phi$ and $\phi'$ are two distinct maps with all four properties. As they are distinct, there are link diagrams on which they differ. Take such a diagram $L$ with a minimal number $n$ of crossings. If $n = 0$, then $\phi(L) = \phi'(L) = d^{k-1}$ by Properties 3 and 2, and by induction on the number $k$ of components of $L$. If $n > 0$, let $v$ be a crossing of $L$. Then as $L_v^A$ and $L_v^B$ have only $n-1$ crossings, $\phi$ and $\phi'$ agree at both diagrams. Thus, by Property 4, $\phi(L) = \phi'(L)$. This contradiction proves uniqueness.

Existence will follow from the fact that $[L]$ has all the required properties. Properties 1–3 follow immediately from the definition. To prove Property 4, let $v$ be a crossing of $L$. Let $L' = L_v^A$ and $L'' = L_v^B$. Then

$$
\begin{aligned}
[L] &= \sum_S A^{a_L(S)} B^{b_L(S)} d^{c_L(S)-1} \\
&= \sum_{S\,:\,S(v)=A} A^{a_L(S)} B^{b_L(S)} d^{c_L(S)-1} + \sum_{S\,:\,S(v)=B} A^{a_L(S)} B^{b_L(S)} d^{c_L(S)-1} \\
&= A \sum_{S'} A^{a_{L'}(S')} B^{b_{L'}(S')} d^{c_{L'}(S')-1} + B \sum_{S''} A^{a_{L''}(S'')} B^{b_{L''}(S'')} d^{c_{L''}(S'')-1} \\
&= A[L'] + B[L'']
\end{aligned}
$$

as was required, where $\sum_{S'}$ denotes summation over all states $S'$ of $L'$ and $\sum_{S''}$ denotes summation over all states $S''$ of $L''$. ∎

In its present form, the Kauffman square bracket is certainly not the same on all equivalent links. Consider, for example, the unknot and the equivalent twist, shown in Figure 9. The Kauffman bracket of the twist is $A + Bd$ while the Kauffman bracket of the unknot is 1. However, we can change the bracket so that it is invariant on equivalent links by weakening Reidemeister's definition of link equivalence. We therefore define another equivalence relation on link diagrams, called *regular isotopy*.

**Definition 18.** If two link diagrams can be transformed into one another by a finite sequence of Reidemeister moves of the second and third type (see Diagram 3), they are said to be *regular isotopic* to each other.

A careful choice of values for $A$, $B$, and $d$ creates a version of the Kauffman square bracket that is invariant on links which are regular isotopic to each other. This one-variable version of the Kauffman square bracket is called the Kauffman angle bracket.

**Definition 19.** The *Kauffman angle bracket* $\langle L\rangle$ of a link $L$ is a Laurent polynomial in one variable, defined by setting $\langle L\rangle(A) = [L](A, A^{-1}, -A^2 - A^{-2})$.

Note that Property 4 of Theorem 5 says $\langle L\rangle = A\langle L_v^A\rangle + A^{-1}\langle L_v^B\rangle$ for every crossing $v$ and Property 3 says $\langle L \cup \bigcirc\rangle = (-A^2 - A^{-2})\langle L\rangle$. These relations will be used in the next proof.

**Lemma 6.** *The Kauffman angle bracket is invariant under regular isotopy.*

**Proof.** Let $B$ and $d$ be such that $AB = 1$ and $d = -A^2 - A^{-2}$, so that $\langle L\rangle(A) = [L](A, B, d)$.

Consider first the effect of a Reidemeister move of the second type. Let $L$ and $L'$ be links that differ only in the region involved in the Reidemeister move. Let $L$ and $L'$ be as in Figure 10, so that applying to $L$ a Reidemeister move of the second type will leave $L$ plane isotopic to $L'$. Furthermore, $L_v^A = L_1$, $L_v^B = L_2$, $(L_1)_w^A = (L_2)_w^B = L_3$, $(L_1)_w^B = L_4$, and $(L_2)_w^A = L'$.

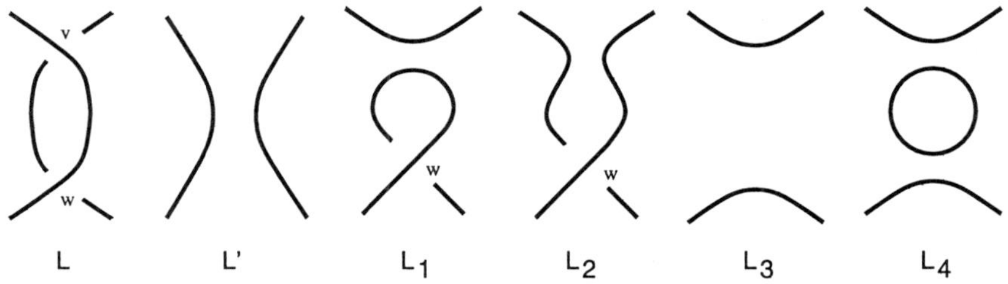


*Fig. 10.* Figures for invariance under a Reidemeister move of the second type in the proof of Lemma 6.

Then $[L] = A[L_1]+B[L_2] = A\big(A[L_3]+B[L_4]\big) + B\big(A[L']+B[L_3]\big)$. But $L_4$ is simply $L_3 \cup \bigcirc$. Thus $[L] = (A^2+ABd+B^2)[L_3]+AB[L']$. Then by the above values of $B$ and $d$, this reduces to $\langle L\rangle = \langle L'\rangle$. Thus a Reidemeister move of the second type preserves the angle bracket.

To complete the proof, we shall show that invariance under a Reidemeister move of the second type implies invariance under a Reidemeister move of the third type.

*Fig. 11.* Figures for invariance under a Reidemeister move of the third type in the proof of Lemma 6. Figures for a Reidemeister move of the third type with the sign of its middle crossing switched may be obtained by reflecting each diagram.

Let $L$ and $L'$ differ only in the region affected by the Reidemeister move and let $L$ and $L'$ be as in Figure 11, so that applying to $L$ a Reidemeister move of the third type will leave it planar isotopic to $L'$. Furthermore, $L_v^A = L_1$, $(L')_v^A = L_3$, and $L_v^B = (L')_v^B = L_2$. Furthermore, $L_1$ is transformable to $L_3$ by a pair of Reidemeister moves of the second type. Hence, by the first part of this proof, $\langle L_1\rangle = \langle L_3\rangle$. Property 4 of Theorem 5 gives us $\langle L\rangle = A\langle L_1\rangle + A^{-1}\langle L_2\rangle = A\langle L_3\rangle + A^{-1}\langle L_2\rangle = \langle L'\rangle$. ∎

As it happens, it is easy to alter slightly the Kauffman (angle) bracket $\langle L\rangle$ to turn it into an ambient isotopy invariant of links. To do this we need to make use of another regular isotopy invariant of oriented links, this time related to the signs of the crossings.

**Definition 20.** The sum of the signs of all the crossings of a link diagram is called the *writhe* $w(L)$ of the link diagram. A crossing $v$ is called a *self-crossing* if the two strands crossing at $v$ are part of the same component. The *self-writhe* $s(L)$ of a link diagram is the sum of the signs of all the self-crossings.

It is easy to see that the self-writhe of a link diagram is independent of orientation. Thus we can speak of the self-writhe of an unoriented link diagram. Furthermore, writhe and self-writhe are clearly regular-isotopy invariants, as a Reidemeister move of the third type does not change the signs of crossings, and a Reidemeister move of the second type removes or adds two crossings of opposite signs. The first Reidemeister move changes writhe and self-writhe by the same amount, so that $w(L) - s(L)$ is an ambient-isotopy invariant of oriented link diagrams. We will now describe an ambient-isotopy invariant of unoriented links and an ambient-isotopy invariant of oriented links, both constructed from the Kauffman bracket.

**Definition 21.** For oriented link diagram $L$, let $\langle L \rangle$ be the Kauffman angle bracket defined on the unoriented diagram based on $L$.

**Theorem 7.** *The following Laurent polynomials are ambient isotopy invariants on their respective link-objects.*

1. $f(L) = (-A)^{-3s(L)}\langle L \rangle$ *on unoriented links.*
2. $f[L] = (-A)^{-3w(L)}\langle L \rangle$ *on oriented links.*

**Proof.** Since $w(L)$ (for oriented link diagrams), $s(L)$ and $\langle L \rangle$ are all regular-isotopy invariants, so are $f(L)$ and $f[L]$. Hence we only need to show that the two polynomials are invariant under a Reidemeister move of the first type.

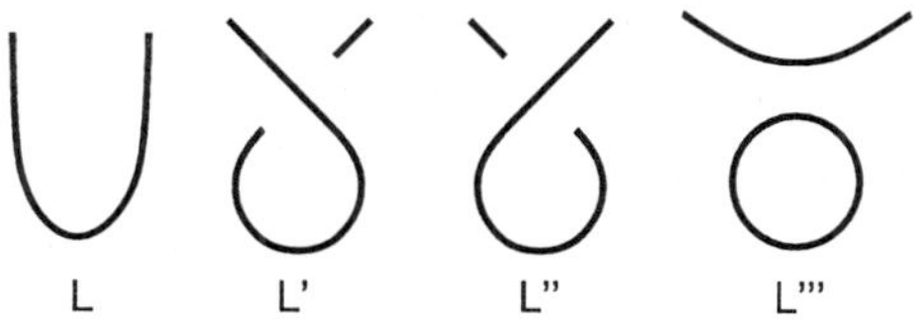


*Fig. 12.* $L$, $L'$, $L''$ and $L'''$ for the proof of Theorem 7.

We first show that $f(L) = (-A)^{-3s(L)}\langle L \rangle$ is invariant on unoriented links. Let $L$, $L'$, and $L''$ be so that they agree everywhere except in a specific region, each appearing there as either $L$, $L'$, or $L''$ of Figure 12.

It is clear that $(L')_v^A = L'''$, $(L')_v^B = L$, $(L'')_v^A = L$ and $(L'')_v^B = L'''$. Furthermore, $L''' = L \cup \bigcirc$. So

$$\begin{aligned}\langle L' \rangle &= A\langle L''' \rangle + A^{-1}\langle L \rangle \\ &= A(-A^2 - A^{-2})\langle L \rangle + A^{-1}\langle L \rangle \\ &= (-A^3 - A^{-1} + A^{-1})\langle L \rangle = (-A)^3\langle L \rangle.\end{aligned}$$

Also,

$$\begin{aligned}\langle L'' \rangle &= A\langle L \rangle + A^{-1}\langle L''' \rangle \\ &= \left(A + A^{-1}(-A^2 - A^{-2})\right)\langle L \rangle = (-A)^{-3}\langle L \rangle.\end{aligned}$$

Noting that $s(L') = s(L) + 1$ and $s(L'') = s(L) - 1$ completes the proof of part 1.

Recall that $w(L) - s(L)$ is an ambient-isotopy invariant of oriented link diagrams. Hence if $f(L) = (-A)^{-3s(L)}\langle L \rangle$ is maintained under a Reidemeister move of the first type, $f[L] = (-A)^{-3w(L)}\langle L \rangle$ is also maintained under a Reidemeister move of the first type and thus is also an invariant of ambient isotopy. ∎

Finally, we find that for oriented links, $f[L]$ satisfies a new recurrence relation, and is in fact a variant of the *Jones polynomial* [9], which was an important polynomial invariant defined on links, and the one which started the flurry of recent research into the subject. In fact, this was a major strength of Kauffman's polynomials. Modified in the appropriate way, his definition became a state-space definition of the Jones polynomial. It should of course be noted that Jones' work preceded that of Kauffman.

**Theorem 8.** *There is a unique polynomial in a single variable $t$, $V_L(t)$, defined for each oriented link $L$ and satisfying the properties*

1. *if $L$ is a trivial diagram of an unknot $V_L = 1$,*
2. *for every crossing $v$ in $L$,*

$$t^{-1}V_{L^+} - tV_{L^-} = (t^{1/2} - t^{-1/2})V_{L^0}, \tag{1}$$

3. *the polynomial $V$ is preserved by ambient isotopy.*

*Furthermore,* $V_L(t) = f[L](t^{-1/4})$ *(where* $f[L] = (-A)^{-3w(L)}\langle L\rangle(A)$*).*

**Proof.** Given existence, uniqueness follows a standard argument. We first show that if a link diagram $L$ represents $k$ unlinked unknots, then the conditions specify $V_L(t)$. Then given two different polynomials, say $V_L(t)$ and $V'_L(t)$ there is a link $L$ with minimal numbers of crossings such that $V_L(t) \neq V'_L(t)$. Then since $L$ is linked, there must be at least one crossing $v$. Also, since $L^0_v$ has fewer crossings, if we change the sign of $v$ to obtain $L'$, condition 2 gives $V_{L'}(t) \neq V'_{L'}(t)$. Hence we can change as many crossings as we like and obtain the result that for any sibling $M$ of $L$, $V_M(t) \neq V'_M(t)$. But every link diagram $L$ has an unlinked sibling (add basepoints, order the components and form the descending bod), and the polynomials must agree on unlinked diagrams. This contradiction proves uniqueness.

To prove existence, we merely show that the given $V_L(t) = f[L](t^{-1/4})$ has the required properties. Property 1 is trivial: If $L$ is a trivial diagram of an unknot, $\langle L\rangle = 1$ and $w(L) = 0$. Property 3 follows from Theorem 7.

Given crossing $v$ in $L$, let $L_\infty$ be the unoriented link diagram formed by resolving $v$ opposite to orientation. Also write $n = w(L^0)$, so that $w(L^+) = n + 1$ and $w(L^-) = n - 1$. Then

$$\langle L^+\rangle = A\langle L^0\rangle + A^{-1}\langle L_\infty\rangle \quad \text{and} \quad \langle L^-\rangle = A\langle L_\infty\rangle + A^{-1}\langle L^0\rangle.$$

Hence

$$\begin{aligned} A\langle L^+\rangle - A^{-1}\langle L^-\rangle &= A\big(A\langle L^0\rangle + A^{-1}\langle L_\infty\rangle\big) - A^{-1}\big(A\langle L_\infty\rangle + A^{-1}\langle L^0\rangle\big) \\ &= (A^2 - A^{-2})\langle L^0\rangle. \end{aligned}$$

We wish to show equation (1) holds. Let $V_L(t) = f[L](t^{-1/4})$ as suggested. Then $t = A^{-4}$ and the left hand side of (1),

$$\begin{aligned} t^{-1}V_{L^+} - tV_{L^-} &= A^4 f[L^+] - A^{-4} f[L^-] \\ &= A^4(-A)^{-3(n+1)}\langle L^+\rangle - A^{-4}(-A)^{-3(n-1)}\langle L^-\rangle \\ &= (-A)^{-3n}\big( - A\langle L^+\rangle + A^{-1}\langle L^-\rangle\big) \end{aligned}$$

$$= (-A)^{-3n}(A^{-2} - A^2)\langle L^0 \rangle = (A^{-2} - A^2)f[L^0]$$

$$= (t^{1/2} - t^{-1/2})V_{L^0}. \qquad \blacksquare$$

In the next section we extend such observations to a new polynomial, defined on oriented links. In fact, we shall first define our polynomial on bods rather than oriented link diagrams. The state spaces we consider are the actual state spaces of the link diagrams, described in terms of positive and negative crossings, rather than in terms of $A$ and $B$ splittings. We shall show that the polynomial is the same on all bods with the same underlying oriented link diagrams, and in fact is the same when they represent the same oriented link, as is desired.

## 5. Definition of the polynomial

We have seen that one can define a polynomial of links which is invariant under ambient isotopy via an explicit definition involving $A$ and $B$ splittings. By defining the Jones polynomial in this way, the proof of its existence, uniqueness, and fundamental properties becomes almost trivial. We found that a new polynomial that generalizes the Jones polynomial also can be represented by a state-space summation. This polynomial is the so-called HOMFLY polynomial, discovered simultaneously by five teams of researchers and expressed by a recurrence relationship much as the Alexander and Jones polynomials were originally. Although our state-space representation is not nearly as simple as the one which Kauffman developed for the Jones polynomial, expressing the polynomial in this way yields some new observations and provides notational benefits to various proofs.

As defined, our polynomial is a two-variable version of the HOMFLY polynomial. We shall show that it has the required properties of the HOMFLY, and uniqueness will give the fact that the polynomials are the same. Keep in mind, however, that our goal is to explicitly describe the polynomial in terms of a state-space representation, and not merely to prove the existence of HOMFLY. Although our definition is written as a recursion, note that with the boundary conditions it is actually an explicit sum.

**Definition 22.** We inductively define a map $P$ on $\mathcal{B}_n$ as follows: Let $L \in \mathcal{B}_n$, $T = T(L, \alpha L)$, $p = p_{\alpha L}$, and $n = n_{\alpha L}$. Recursively define $P$ so that it satisfies the two relations

$$P(\alpha L) = (x+y)^{cL-1}, \tag{2}$$

$$P(L) = (-1)^{|T|} \left(\frac{x}{y}\right)^{p(T)-n(T)} \sum_{S \subseteq T} (-1)^{|S|} x^{-p(S)} y^{-n(S)} P\big([\alpha L]_S\big). \tag{3}$$

Note that equations (2) and (3) are consistent, as for a descending bod $L$, (3) reduces to $P(L) = P(\alpha L)$. The induction in the definition works as follows. Equation (2) defines $P$ on any descending diagram, as if $L$ is descending $L = \alpha L$ and $P(L) = (x+y)^{cL-1}$. To build the polynomial on non-descending bods, we look at equation (3), and perform induction on the number of crossings of $L$. This induction is simple, since each summand uses a polynomial defined either on a descending bod (when $S = \emptyset$) or on a bod with fewer crossings (otherwise). Hence it is clear that the polynomial is well defined.

What is not so clear is that the polynomial has any use. To be useful in a standard setting, we would first like the polynomial to be defined on link diagrams as opposed to bods. In other words, we would like all bods with the same underlying diagram to have the same polynomial. Also, we would like the polynomial to be ambient isotopy invariant. Furthermore, as has been done classically, we would like polynomials of links which agree everywhere but on one crossing to be related. In fact, this polynomial does satisfy all of these requirements, as is stated in the following theorem, which we shall prove in Section 6.

**Theorem 9.** *The defined polynomial has the following properties:*

1. $P(L^0) = xP(L^+) + yP(L^-)$ *for each crossing of every bod* $L$*;*
2. $P(L)$ *is independent of basepoints;*
3. $P(L)$ *is preserved by Reidemeister moves;*
4. $P(L)$ *is independent of component ordering.*

Upon proving this theorem, we will have the following corollary immediately, which is basically the main theorem of the HOMFLY paper [4].

More precisely, if we set $x = -X/Z$ and $y = -Y/Z$ in Corollary 10, we get the unique polynomial of links satisfying $XP(L^+)+YP(L^-)+ZP(L^0) = 0$ and the same boundary conditions. This unique polynomial is precisely the HOMFLY polynomial.

**Corollary 10.** *There exists a unique ambient isotopy invariant (Laurent) polynomial of oriented links satisfying $P(L^0) = xP(L^+) + yP(L^-)$ and $P(\bigcirc) = 1$. Furthermore, this polynomial is the polynomial of Definition 22.*

**Proof.** Consider the polynomial $P$ of Definition 22 and Theorem 9. Properties 2 and 4 tell us that $P$ can be viewed as a polynomial of oriented link diagrams rather than bods, as they imply that all bods with the same underlying oriented link diagram have the same polynomial. Property 3 tells us that $P$ is an ambient isotopy invariant, while Property 1 and equation (2) (with $L = \bigcirc$) give us the relations we wish to have. Hence such a polynomial does exist.

The uniqueness of the polynomial is guaranteed by the recurrence relation $P(L^0) = xP(L^+) + yP(L^-)$. Suppose $Q$ is another polynomial satisfying this relation and the boundary condition. Consider Figure 13, and note that if $\bigcirc^k$ is $k$ unlinked unknots then $L^0 = \bigcirc^k$. Also, since $Q$ is ambient isotopy invariant, $Q(L^+) = Q(L^-) = Q(\bigcirc^{k-1})$. Therefore the boundary condition $Q(\bigcirc) = 1$ and the recurrence relation implies that $Q(\bigcirc^k) = (x+y)^{k-1}$. But, any descending bod $L$ is ambient isotopic to $\bigcirc^{cL}$, so this is exactly the relationship in equation (2) and $Q$ agrees with $P$ on descending bods.

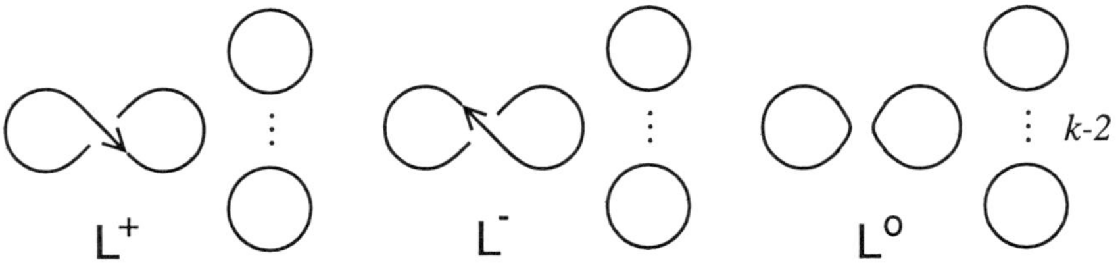


*Fig. 13.* Resolving a twist to make an unknot.

We apply induction, first on the number of crossings and then on $|T(L, \alpha L)|$. Suppose that $Q(L') = P(L')$ for all links $L'$ with fewer than $n$ crossings. As noted above, this is true for links with no crossings, so our induction is rooted. Let $L$ be a link with $n$ crossings such that $\left|T(L, \alpha L)\right| = k \geq 1$. Assume that $Q(L') = P(L')$ for links $L'$ with $n$ crossings and $\left|T(L', \alpha L')\right| < k$. This induction is rooted since $Q = P$ on all descending diagrams (where $\left|T(L, \alpha L)\right| = 0$). Since $\left|T(L, \alpha L)\right| \geq 1$, there is a crossing $v$ in $L$ such that either $L = L^+$ and $\left|T(L^-, \alpha L)\right| < \left|T(L, \alpha L)\right|$ or $L = L^-$ and $\left|T(L^+, \alpha L)\right| < \left|T(L, \alpha L)\right|$. In either case, the recurrence relation and the induction hypothesis imply that $Q(L) = P(L)$. ■

We now look at the polynomial from a different viewpoint, and provide another formulation for the polynomial which will actually be the formulation we primarily use. We start with an identity that will prove useful in the work to come. It may be easily verified by substituting $\pm 1$ for $s(v, L)$.

$$(4) \qquad \left(-\frac{x}{y}\right)^{s(v,L)} + (x+y)(xy)^{-1/2}\left(\frac{x}{y}\right)^{s(v,L)/2} = 1$$

Now, let $L$ and $M$ be similar adjacent bods with $T(L, M) = \{v\}$. Then the reader should convince himself that Property 1 of Theorem 9 can be written as

$$(5) \quad P(M) = -\left(\frac{x}{y}\right)^{s(v,L)}\left(P(L) - (xy)^{-1/2}\left(\frac{x}{y}\right)^{-s(v,L)/2} P(L^0)\right).$$

For $T = T(L, M) = \{v\}$, this is of a form similar to (3), as $L^0 = [L]_{\{v\}}$ and $L = [L]_{\emptyset}$. That is, we have precisely

$$P(M) = (-1)^{|T|}\left(\frac{x}{y}\right)^{p(T)-n(T)} \sum_{S \subseteq T} (-1)^{|S|} x^{-p(S)} y^{-n(S)} P\big([L]_S\big).$$

Motivated by this we make an auxiliary definition.

**Definition 23.** Given two similar bods $L$ and $M$, let $T = T(L, M)$, $p = p_L$ and $n = n_L$. We define a polynomial

$$(6) \quad P_L(M) = (-1)^{|T|}\left(\frac{x}{y}\right)^{p(T)-n(T)} \sum_{S \subseteq T} (-1)^{|S|} x^{-p(S)} y^{-n(S)} P\big([L]_S\big).$$

Note that for $L = \alpha M$, this definition is precisely equation (3) of Definition 22. In other words, $P_{\alpha L}(L) \equiv P(L)$. For $L = M$, equation (6) reduces to $P_L(L) = P(L)$.

With this new family of polynomials defined, we can restate Property 1 of Theorem 9. The right-hand side of (5) is exactly $P_L(M)$. Hence Property 1 becomes simply: If $L$ is adjacent to $M$, $P(M) = P_L(M)$. We note that the statement $P(M) = P_L(M)$ is symmetric in $M$ and $L$ for adjacent $L$ and $M$.

Our next lemma extends the argument in Figure 13 and the condition $P(L) = (x+y)^{cL-1}$ for descending $L$ to another situation.

**Lemma 11.** *If $\beta L = L \cup \bigcirc$, the disjoint union of a bod $L$ and a trivial diagram of an unknot, then $P(\beta L) = (x+y)P(L)$.*

**Proof.** First we note that if $L$ is a descending diagram with $k$ components, then $\beta L$ is a descending diagram with $k+1$ components, so $P(\beta L) = (x+y)^k = (x+y)P(L)$.

To prove the assertion we apply induction on the number of crossings. The result is clear if $L$ has no crossings. As we have shown, we are done if $L$ is descending as well. So assume $L$ has $n \geq 1$ crossings and is not descending, that is $T = T(L, \alpha L) \neq \emptyset$. Note that $T = T(\beta L, \alpha\beta L)$, as there are no crossings between $L$ and the unknot we added to obtain $\beta L$.

By the definition of the polynomial,

$$P(\beta L) = (-1)^{|T|} \left(\frac{x}{y}\right)^{p(T)-n(T)} \sum_{S \subseteq T} (-1)^{|S|} x^{-p(S)} y^{-n(S)} P\big([\alpha\beta L]_S\big),$$

where $p = p_{\alpha\beta L} = p_{\alpha L}$ and $n = n_{\alpha\beta L} = n_{\alpha L}$.

Now, for all $S$, $[\alpha\beta L]_S$ is of the form $\beta[\alpha L]_S$. Thus, for any set $S \neq \emptyset$, $P\big([\alpha\beta L]_S\big) = (x+y)P\big([\alpha L]_S\big)$ by the induction hypothesis. If $S = \emptyset$, $P\big([\alpha\beta L]_S\big) = (x+y)P\big([\alpha L]_S\big)$ from the first comment above, so factoring out $(x+y)$ from each term of the summand leaves the definition of $P(L)$. ∎

In fact, the statement of Lemma 11 can be generalized from a single trivial diagram of the unknot to any descending bod. With nearly the same proof, one can show that if a descending bod of $k$ components is placed above and before any other bod $L$ (so that each of its crossings with $L$ is an overcrossing, and each component precedes each component of $L$ in the component ordering) to create a new bod $L'$, then $P(L') = (x+y)^k P(L)$. As we only need to use the more limited statement above, the proof of this more general statement is left to the interested reader.

## 6. Proof of Theorem 9

We shall prove Theorem 9 through a series of lemmas. Our basic approach shall be induction on $n$, the number of crossings in the bod $L$. For Property 3, $n$ is taken as the number of crossings before or after the move, whichever is larger.

### 6.1. Property 1

We will now prove that the recurrence relation holds for the polynomial as defined. We begin with a lemma, which, as noted immediately prior to Definition 23, will easily imply the the first property in the theorem.

**Lemma 12.** *Let $L$ and $M$ be similar bods. Then $P_L(M) = P(M)$.*

We will need the following claim in order to prove the lemma.

**Claim 13.** *Let $K$ and $L$ be similar adjacent bods. Then one of the following is true:*

1. *$P_K(M) = P_L(M)$ for all $M$ in $\mathcal{U}_L$.*
2. *$P_K(M) \neq P_L(M)$ for all $M$ in $\mathcal{U}_L$.*

**Proof.** For similar bods $K$, $L$, and $M$ with $K$ and $L$ adjacent, we will prove the equivalence of $P_K(M) = P_L(M)$ and $P_K(L) = P_L(L)$. Since $M$ is arbitrary, this will give us our result.

Let $T(K, L) = \{v\}$. We consider two cases, depending on whether $M$ agrees with $K$ or $L$ at $v$.

**Case 1:** $v \in T(K, M)$.

Then let $T = T(L, M)$ and note $T(K, M) = T \cup \{v\}$. Note also that for any set $S$ not containing $v$, $p_L(S) = p_K(S)$ and $n_L(S) = n_K(S)$. Also, $p_K\big(T \cup \{v\}\big) - n_K\big(T \cup \{v\}\big) = p_K(T) - n_K(T) + s(v, K)$ (similarly for $L$). So we will use $p(S)$ and $n(S)$ for both when $v \notin S$, considering the sign of $v$ separately when $v \in S$. With that in mind, we write

$$P_K(M) = (-1)^{|T|+1} \left(\frac{x}{y}\right)^{p(T)-n(T)+s(v,K)} \times \sum_{S \subseteq T \cup \{v\}} (-1)^{|S|} x^{-p_K(S)} y^{-n_K(S)} P\big([K]_S\big).$$

Therefore, shifting the normalizing factor to the left-hand side and splitting the sum according to whether $v \in S$, we obtain

$$P_K(M) \left((-1)^{|T|} \left(\frac{y}{x}\right)^{p(T)-n(T)}\right) = (-1) \left(\frac{x}{y}\right)^{s(v,K)} \sum_{S \subseteq T} (-1)^{|S|} x^{-p(S)} y^{-n(S)} P\big([K]_S\big) + (-1) \left(\frac{x}{y}\right)^{s(v,K)} \sum_{S \subseteq T} (-1)^{|S|+1} x^{-p(S)-p_K(v)} y^{-n(S)-n_K(v)} P\big([K]_{S \cup \{v\}}\big),$$

Rearranging the terms, we get

$$P_K(M) \left((-1)^{|T|} \left(\frac{y}{x}\right)^{p(T)-n(T)}\right) = (-1) \left(\frac{x}{y}\right)^{s(v,K)} \sum_{S \subseteq T} (-1)^{|S|} x^{-p(S)} y^{-n(S)} \Big(P\big([K]_S\big) - x^{-p_K(v)} y^{-n_K(v)} P\big([K]_{S \cup \{v\}}\big)\Big).$$

However, $v \notin S$ so $T([L]_S, [K]_S) = \{v\}$ and $s(v, [K]_S) = s(v, K)$. Thus we can say

$$P_{[K]_S}([L]_S) = (-1)\left(\frac{x}{y}\right)^{s(v,K)}\left(P([K]_S) - x^{-p_K(v)}y^{-n_K(v)}P\big(([K]_S)_{\{v\}}\big)\right).$$

But as $[K]_{S\cup\{v\}}$ and $([K]_S)_{\{v\}}$ differ only in component ordering and base-points and have fewer crossings than $K$, by our inductive hypothesis we see that $P([K]_{S\cup\{v\}}) = P\big(([K]_S)_{\{v\}}\big)$. This allows us to combine the last two equations to obtain

$$P_K(M) = (-1)^{|T|}\left(\frac{x}{y}\right)^{p(T)-n(T)} \sum_{S\subseteq T} (-1)^{|S|} x^{-p(S)} y^{-n(S)} P_{[K]_S}([L]_S).$$

Now, by the definition,

$$P_L(M) = (-1)^{|T|}\left(\frac{x}{y}\right)^{p(T)-n(T)} \sum_{S\subseteq T} (-1)^{|S|} x^{-p(S)} y^{-n(S)} P([L]_S).$$

But by induction, if $S$ is non-empty, the fact that Lemma 12 holds on bods with fewer crossings gives us $P_{[K]_S}([L]_S) = P([L]_S)$. Therefore

$$P_K(M) - P_L(M) = (-1)^{|T|}\left(\frac{x}{y}\right)^{p(T)-n(T)} \big(P_K(L) - P_L(L)\big).$$

Hence we know that $P_K(M) = P_L(M)$ if and only if $P_K(L) = P_L(L)$.

**Case 2:** $v \notin T(K, M)$.

Then $v \in T(L, M)$ and the same argument as in Case 1 with $K$ and $L$ interchanged proves $P_K(M) = P_L(M)$ if and only if $P_K(K) = P_L(K)$. We previously noted that the latter is symmetric in $K$ and $L$. Therefore we get $P_K(M) = P_L(M)$ if and only if $P_K(L) = P_L(L)$. ■

We can now prove our initial lemma.

**Proof of Lemma 12.** For the purposes of our proof, we see this lemma primarily as a statement about bod $L$. Let us call a bod $L$ *agreeable* if for all similar $M$, $P_L(M) = P(M)$.

We claim that if $L$ is adjacent to an agreeable bod $N$, then $L$ is also agreeable. This is because $P_L(L) = P(L) = P_N(L)$ by the agreeability of $N$, and so by Claim 13 applied to $L$ and $N$, for all similar $M$, $P_L(M) = P_N(M) = P(M)$.

Furthermore for any link $L$, the link $\alpha L$ is agreeable by the definition of the polynomials. We complete the proof by induction on $|T(L, \alpha L)|$. If $|T(L, \alpha L)| = 0$, then $L = \alpha L$ and we are trivially done. Otherwise, let $L'$ be obtained from $L$ by changing one of the crossings in $T(L, \alpha L)$. Then $\alpha L = \alpha L'$ and $|T(L', \alpha L')| = |T(L, \alpha L)| - 1$. So the induction hypothesis implies $L'$ is agreeable. But then $L$ is adjacent to an agreeable bod, so $L$ is agreeable as well. ∎

As promised, Lemma 12 implies the recurrence relation. The result follows from a simple calculation involving the definition of the polynomial.

**Corollary 14.** *$P(L^0) = xP(L^+) + yP(L^-)$ for each crossing of every bod $L$.*

**Proof.** Note that if $v$ is the crossing we are interested in, $L^0 = [L^+]_{\{v\}} = [L^-]_{\{v\}}$ and $T = T(L^+, L^-) = \{v\}$. By Lemma 12, $P(L^-) = P_{L^+}(L^-)$. Then by (6) in the second polynomial definition,

$$P(L^-) = (-1)\left(\frac{x}{y}\right)\left((-1)^1 x^{-1} P(L^0) + P(L^+)\right) = -\frac{x}{y}P(L^+) + \frac{1}{y}P(L^0).$$

Multiplying through by $y$ gives the desired result. ∎

## 6.2. A critical result

The remaining properties require us to compare the polynomial on two related bods. A basic technique used to prove each property shall be to find a pair of bods we know to have the same polynomial (for example, descending bods) and which have the same relation to the bods in question. The following result allows us to then say that the polynomials of the original pair of bods are also the same. This is a direct corollary of Lemma 12 and again follows with a simple calculation, the details of which are included for completeness.

**Lemma 15.** *Let $L$, $M$ and $L'$, $M'$ be two pairs of similar bods with a bijection $\phi : T(L,M) \to T(L',M')$ satisfying*

1. *for all $v \in T(L,M)$, the sign of $v$ in $M$ is the same as the sign of $\phi v$ in $M'$,*

2. *for all non-empty $S \subseteq T(L,M)$, $P\big([M]_S\big) = P\big([M']_{\phi S}\big)$.*

*Then $P(L) = P(L')$ if and only if $P(M) = P(M')$.*

**Proof.** Note that, by Lemma 12, $P(L) = P_M(L)$ and $P(L') = P_{M'}(L')$. Let $T = T(L,M)$, $p = p_M$ and $n = n_M$. Then

$$P(L) = (-1)^{|T|}\left(\frac{x}{y}\right)^{p(T)-n(T)} \sum_{S\subseteq T} (-1)^{|S|} x^{-p(S)} y^{-n(S)} P\big([M]_S\big),$$

and so

$$P(L)(-1)^{|T|}\left(\frac{y}{x}\right)^{p(T)-n(T)} = P(M) + \sum_{\emptyset\neq S\subseteq T} (-1)^{|S|} x^{-p(S)} y^{-n(S)} P\big([M]_S\big).$$

Then $\phi T = T(L',M')$ and, if we let $p' = p_{M'}$ and $n' = n_{M'}$ similarly,

$$P(L')(-1)^{|\phi T|}\left(\frac{y}{x}\right)^{p'(\phi T)-n'(\phi T)}$$

$$= P(M') + \sum_{\emptyset\neq \phi S\subseteq \phi T} (-1)^{|\phi S|} x^{-p'(\phi S)} y^{-n'(\phi S)} P\big([M']_{\phi S}\big).$$

The first condition of the corollary guarantees that $p'(S) = p(\phi S)$ and $n'(S) = n(\phi S)$ for all $S \subseteq T$. The bijection guarantees that the sum in each equation has the same number of terms and that the terms correspond as desired. Finally, the second condition of the corollary yields the equation

$$P(L) - P(L') = (-1)^{|T|} \left(\frac{x}{y}\right)^{p(T)-n(T)} \left(P(M) - P(M')\right).$$

Thus the result is shown. ■

There are many cases where we apply this result to siblings, so we state the following corollary. Clearly the conditions of Lemma 15 apply to the situation described. The second condition applies due to the induction hypothesis.

**Corollary 16.** *If $L$ and $L'$ are siblings, then $P(L) = P(L')$ if and only if $P(\alpha L') = P(\alpha' L)$, where $\alpha' L = (\alpha L')_{\mathcal{U}_L}$.* ■

Now we will proceed as suggested before the statement of Lemma 15. The property we are interested in will suggest the pair of bods to examine. Finding a pair related in the same way but which can be shown directly to have the same polynomial will complete our proofs.

## 6.3. Property 2

Our next property is invariance under choice of basepoints. We begin with bods which differ only at one basepoint, and that only with respect to one crossing.

**Lemma 17.** *Let $L$ be a bod. If we move one basepoint of $L$ forward past one crossing $v$ of $L$ to form a new bod $L'$, then $P(L) = P(L')$.*

**Proof.** Let $\alpha' L = (\alpha L')_{\mathcal{U}_L}$, that is, $\alpha' L$ is the bod obtained from $\alpha L'$ by moving the identified basepoint back to its original position in $L$.

Then by Corollary 16, it is sufficient to prove that $P(\alpha' L) = P(\alpha L')$. But $P(\alpha L') = (x+y)^{cL-1} = P(\alpha L)$, so in fact it is sufficient to show $P(\alpha' L) = P(\alpha L)$.

If $v$ is a crossing of two distinct components, then $\alpha' L = \alpha L$, and we are done, as both $\alpha' L$ and $\alpha L'$ are descending with $cL$ components.

If $v$ is a self-crossing of a component, then $T(\alpha' L, \alpha L) = \{v\}$ and $[\alpha L]_{\{v\}}$ is descending with one more component than $\alpha L$. Thus, setting $p = p_{\alpha L}$ and $n = n_{\alpha L}$,

$$(7) \qquad P(\alpha' L) = -\left(\frac{x}{y}\right)^{p(v)-n(v)} \left(P(\alpha L) - x^{-p(v)} y^{-n(v)} P\big([\alpha L]_{\{v\}}\big)\right)$$

$$= -\left(\frac{x}{y}\right)^{p(v)-n(v)} P(\alpha L)\big(1 - (x+y) x^{-p(v)} y^{-n(v)}\big).$$

But either $p(v) = 1$ and $n(v) = 0$, in which case

$$1 - (x+y) x^{-p(v)} y^{-n(v)} = 1 - \frac{x+y}{x} = -\left(\frac{y}{x}\right)^{1},$$

or $p(v) = 0$ and $n(v) = 1$, and then

$$1 - (x+y) x^{-p(v)} y^{-n(v)} = 1 - \frac{x+y}{y} = -\left(\frac{y}{x}\right)^{-1}.$$

Either way, we have

$$1 - (x+y) x^{-p(v)} y^{-n(v)} = -\left(\frac{y}{x}\right)^{(p(v)-n(v))}.$$

Therefore, equation (7) becomes $P(\alpha' L) = P(\alpha L)$. ∎

If we have two bods $L$ and $M$ that differ only in basepoints, we can convert $L$ to $M$ by moving basepoints over each crossing, one by one, until we achieve the configuration in $M$. At each stage, Lemma 17 guarantees that the polynomial remains unchanged. Thus we have the following corollary.

**Corollary 18.** *$P(L)$ is independent of basepoints.* ∎

We will use this corollary quite heavily in the future. In particular, it implies that we can move the basepoints of a bod as we see fit without changing the polynomial. So given a bod with a potentially bad placement of basepoints, we will move the basepoints as best fits our needs. This is a great help in proving the invariance of the polynomial under Reidemeister moves.

## 6.4. Property 3

To prove $P$ is invariant under the three Reidemeister moves, we shall consider each move individually and in order. Recall that we proceed by induction: we assume that we can perform any Reidemeister move on a bod with fewer crossings without changing its polynomial. Additionally, as we prove invariance under a Reidemeister move of each type in succession, we will use the Reidemeister moves of previous types to prove each new move. Although the proof techniques in each case are quite similar, there are subtle differences worth noting.

**6.4.1. The first Reidemeister move.** The first Reidemeister move reduces the number of crossings of a bod, allowing us to deal easily with the change.

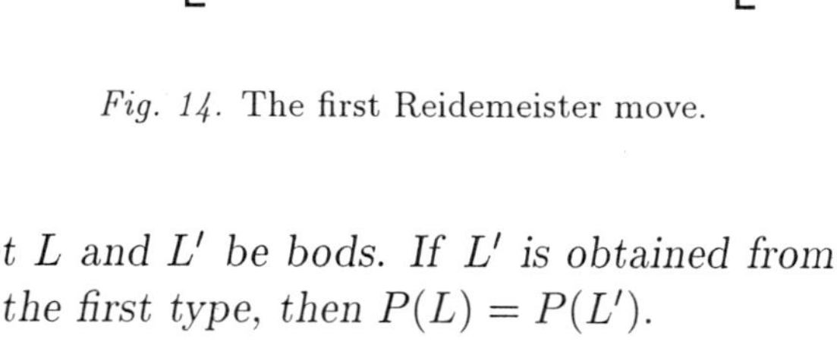


*Fig. 14.* The first Reidemeister move.

**Lemma 19.** *Let $L$ and $L'$ be bods. If $L'$ is obtained from $L$ by a Reidemeister move of the first type, then $P(L) = P(L')$.*

**Proof.** Let $L$ and $L'$ be as in Figure 14, where strand $s$ crosses over $t$ at $x$.

As mentioned at the end of Section 6.3, our first step is to make the bod easier to deal with by moving basepoints. In both $L$ and $L'$, move the basepoint on the component containing strand $s$ onto the strand $s$ itself.

On $L$, place the basepoint just before (with respect to the orientation) the crossing at $x$, so that $x$ is a descending crossing. In this way, we can ensure that $T(L, \alpha L) = T(L', \alpha L')$, as $x$ is descending in $L$ and does not exist in $L'$.

Then the conditions of Lemma 15 are satisfied for the pairs $L, \alpha L$ and $L', \alpha L'$, due to Lemma 19 on bods of fewer crossings. As $L$ and $L'$ have the same number of components, $P(\alpha L) = P(\alpha L') = (x+y)^{cL-1}$. Hence, by Lemma 15, $P(L) = P(L')$. ∎

**6.4.2. The second Reidemeister move.** In the previous section, the removal of a crossing by a Reidemeister move of the first type allowed us to consider one less crossing, providing easy access to induction. Although the second Reidemeister move removes two crossings, it is not as easy to get to the descending diagram here as with a Reidemeister move of the first type, as the component order may prevent us from changing the crossings to descending ones simply by moving basepoints.

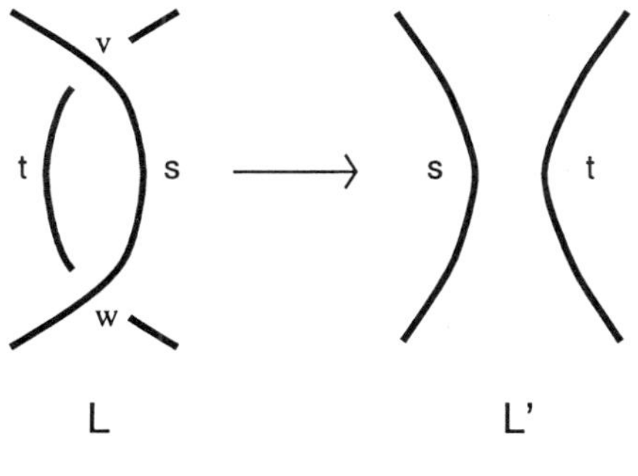


*Fig. 15.* The second Reidemeister move.

**Lemma 20.** *Let $L$ and $L'$ be bods. If $L'$ is obtained from $L$ by a Reidemeister move of the second type, then $P(L) = P(L')$.*

**Proof.** Let the situation be as in Figure 15, where $v$ and $w$ are the crossings removed by the Reidemeister move.

Again we move the basepoint on the component containing strand $s$ onto strand $s$. On bod $L$, we further position the basepoint so that it is before the first of $v$ and $w$ in the orientation of the component of $s$.

Let $\alpha' L$ be the diagram that agrees with $\alpha L$ outside of $\{v, w\}$ and agrees with $L$ on $\{v, w\}$. Note that in both $\alpha L$ and $\alpha' L$, one of $v$ and $w$ is a negative crossing and the other is a positive crossing. Also note that by our choice of the basepoint, it is easy to see that $T(L', \alpha L') = T(L, \alpha' L)$. Furthermore $[\alpha L']_S$ and $[\alpha' L]_S$ differ by a Reidemeister move of the second type. Therefore, by the induction hypothesis, Lemma 20 applies on bods of fewer crossings, and thus the conditions of Lemma 15 apply for pairs $L, \alpha' L$ and $L', \alpha L'$. Then it is sufficient to show that $P(\alpha' L) = P(\alpha L')$. Once again we note that $\alpha L$ and $\alpha L'$ are descending bods with the same number of components, so $P(\alpha L) = P(\alpha L') = (x+y)^{cL-1}$.

If $v$ and $w$ are descending in $L$, then $\alpha' L = \alpha L$ and we are done.

Otherwise, the component containing strand $t$ precedes that of $s$, and both $v$ and $w$ are ascending crossings. Hence the sign of each crossing in $\alpha L$ is the opposite of its sign in $\alpha' L$.

Let $M = [\alpha L]_{\{v\}}$, $M' = [\alpha L]_{\{w\}}$, $M'' = [\alpha L]_{\{v,w\}}$. By the definition of the polynomial (as $\alpha L = \alpha(\alpha' L)$),

$$P(\alpha' L) = P_{\alpha L}(\alpha' L) = (-1)^2 \left(\frac{x}{y}\right)^0 \sum_{S \subseteq \{v,w\}} (-1)^{|S|} x^{-p(S)} y^{-n(S)} P\big([\alpha L]_S\big).$$

Here $p = p_{\alpha L}$ and $n = n_{\alpha L}$. Simplifying terms and substituting the appropriate $M$, $M'$, or $M''$ for our resolved bods, we get

$$P(\alpha' L) = P(\alpha L) - x^{-p(v)} y^{-n(v)} P(M) - x^{-p(w)} y^{-n(w)} P(M') + x^{-1} y^{-1} P(M'').$$

We need to show

$$-x^{-p(v)} y^{-n(v)} P(M) - x^{-p(w)} y^{-n(w)} P(M') + x^{-1} y^{-1} P(M'') = 0.$$

Rearranging the terms, and noting that $1 - p(w) = p(v), 1 - n(w) = n(v)$, we see that this is equivalent to

$$(8) \qquad P(M) = -\left(\frac{x}{y}\right)^{p(v)-n(v)} \big(P(M') - x^{-p(v)} y^{-n(v)} P(M'')\big).$$

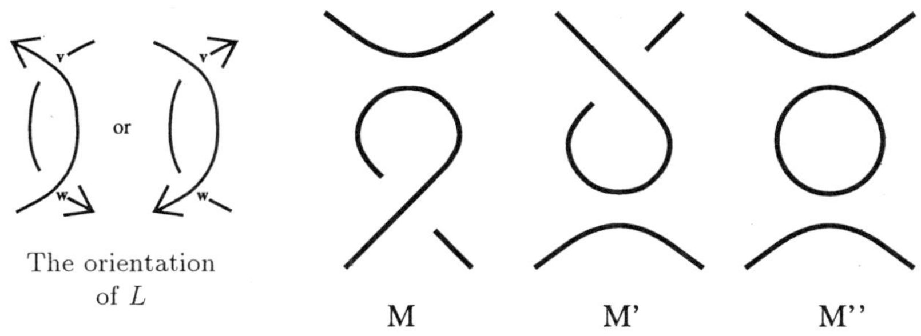


*Fig. 16.* Strands oriented in the opposite directions.

Hence if we show that (8) holds, we are done. Depending on the directions in the strands, $M$, $M'$, and $M''$ either look like Figure 16 or 17. We consider these two cases separately.

**Case 1:** The orientation is as in Figure 16.

We can use the first Reidemeister move to show that $P(M) = P(M')$. Furthermore, $P(M'') = (x+y)P(M)$ by Lemma 11. Recalling identity (4), we see that equation (8) holds.

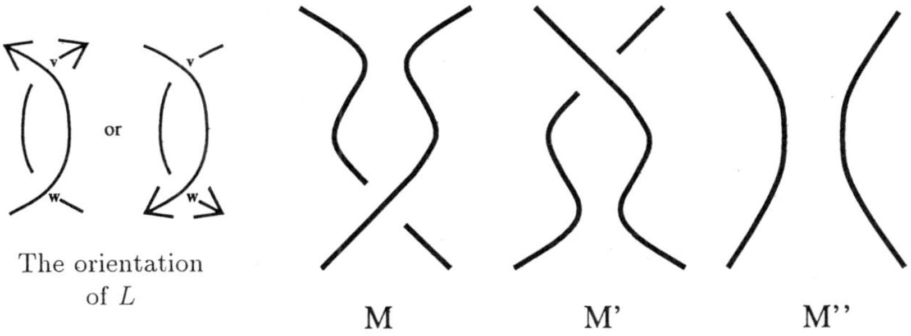


*Fig. 17.* Strands oriented in the same directions.

**Case 2:** The orientation is as in Figure 17.

We observe that $M$, $M'$, and $M''$ differ only at one crossing, where one of $M$ or $M'$ has the crossing positive, the other has the crossing negative, and $M''$ is the same as $M$ and $M'$ with the crossing resolved. Then, by

Property 1, we have $P(M) = P_{M'}(M)$, which is exactly equation (8) as

$$P_{M'}(M) = -\left(\frac{x}{y}\right)^{p(v)-n(v)} \left(P(M') - x^{-p(v)}y^{-n(v)}P(M'')\right). \qquad \blacksquare$$

**6.4.3. The third Reidemeister move.** The third Reidemeister move wreaks havoc with the crossings, moving but not removing three different crossings. By focusing our attention on only one of the crossings, however, we can make our work considerably easier. In the Reidemeister triangle (that is, $L$ in Figure 18 or its reflection (equivalently, or with the sign of the middle crossing ($w$) changed)), we identify the three strands as the top strand, the bottom strand, and the interweaved strand.

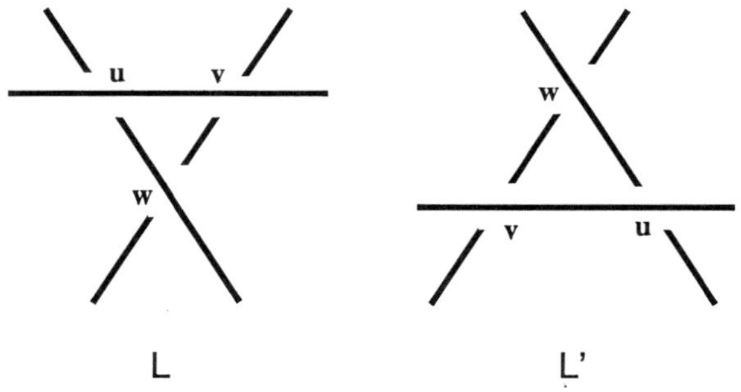


*Fig. 18.* The third Reidemeister move.

**Lemma 21.** *Let $L$ and $L'$ only differ by Reidemeister move of the third type. Let $x$ be one of the Reidemeister triangle crossings involving the interweaved strand. Then* $P([L]_x) = P([L']_x)$.

**Proof.** Note that in Figure 18, $x$ can be crossing $u$ or $w$. Suppose first that $x = u$. Figures 19(a) and 19(b) show the two possible outcomes when we resolve $u$ in $L$, depending on the directions of the strands which cross.

In the case of diagram 19(a), $[L]_u$ and $[L']_u$ are siblings with one fewer crossing than $L$ (they may be different bods because of base-points or ordering, in particular if a basepoint appears between crossings). Using the main induction hypothesis and applying Properties 2 and 4,

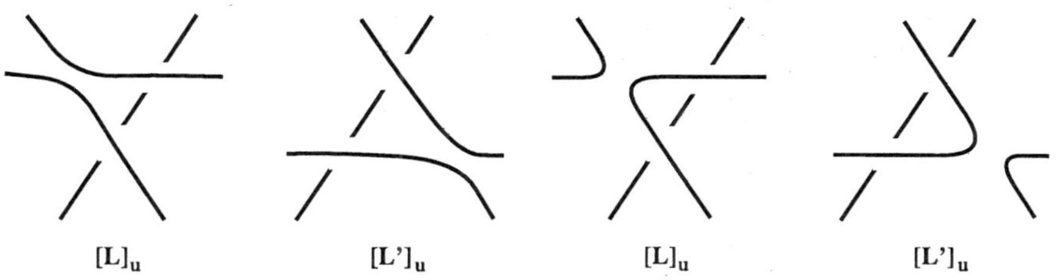


(a) Strands in same direction (b) Opposite directions

*Fig. 19.* Possible resolutions of a crossing of the triangle.

$P\big([L]_u\big) = P\big([L']_u\big)$. In the case of diagram 19(b), applying a Reidemeister move of the second type to $[L]_u$ and $[L']_u$ yields a pair of diagrams that are siblings and have fewer crossings than $L$, so again we are done by the induction hypothesis. Analogous arguments (reflected vertically) give the result when $x = w$. ■

We now develop a special case of Lemma 15 that applies to this situation.

**Corollary 22.** *Let $M$ and $N$ be similar and adjacent bods, with $M'$ and $N'$ obtained from the same Reidemeister move of the third type on $M$ and $N$, respectively. Then $M'$ and $N'$ are similar and adjacent, and $P(M) = P(M')$ if and only if $P(N) = P(N')$.*

**Proof.** We wish to apply Lemma 15. It is clear that $M'$ and $N'$ are similar, and we fix the bijection $\phi$ made by the Reidemeister move. Since a Reidemeister move of the third type does not change the sign of any crossing, $s(x, N) = s(\phi(x), N')$ for all crossings $x$.

Now, since $M$ and $N$ are adjacent, they only differ at one crossing, say $v$. We claim that if $v$ is one of the crossings of the triangle of the Reidemeister move, then it is a crossing of the interweaved strand, since otherwise, the Reidemeister move would not be available in both $M$ and $N$. Thus Lemma 21 guarantees the second condition of Lemma 15. If $v$ is not a crossing of the triangle, the second condition of Lemma 15 is given by the main induction hypothesis involving Reidemeister moves.

Therefore the hypotheses of Lemma 15 are satisfied. We apply Lemma 15 for the result. ■

After all of this preparation, we are ready to prove that $P$ is invariant under a Reidemeister move of the third type.

**Lemma 23.** *Let $L$ and $L'$ be bods. If $L'$ is obtained from $L$ by a Reidemeister move of the third type, $P(L) = P(L')$.*

**Proof.** We assume for the time being that the Reidemeister triangle can be rotated so as to resemble Figure 18 up to planar isotopy.

Move the basepoints so that they are outside the Reidemeister triangle. This does not affect the polynomial by Lemma 18. Consider the application of the Reidemeister move of the third type that transforms $L$ to $L'$. Since no basepoints are in the region, we can obtain $\alpha L'$ by applying the Reidemeister move of the third type to $\alpha L$ in the same way.

It is easier for our purposes if we know the nature of the crossings in $\alpha L$ and $\alpha L'$, as opposed to those in $L$ and $L'$. We rotate and apply planar isotopy so that the relevant crossings of $\alpha L$ and $\alpha L'$ to be labelled as suggested by Figure 18. Then we wish to examine the possible states of crossings of the triangle of the Reidemeister triangle in $L$ and $L'$. In fact, it is easily shown that there are six possibilities for the configurations of the triangles in $L$ and $L'$, as displayed in Figure 20. We will call graphs with these configurations $L_1 - L_6$ and $L'_1 - L'_6$, respectively. Note that the pair $L_1$, $L'_1$ has the same configuration as in Figure 18.

Let $\alpha' L$ be the diagram that agrees with $L$ on $\{u, v, w\}$ and with $\alpha L$ elsewhere. Analogously we define $\alpha' L'$ to be the diagram that agrees with $L'$ on $\{u, v, w\}$ and with $\alpha L'$ elsewhere. The triangles in these graphs also fall into the six cases of Figure 20.

By the inductive hypothesis, Lemma 23 holds on graphs with fewer crossings, so the conditions of Lemma 15 apply for the pairs $L, \alpha' L$ and $L', \alpha' L'$. Thus it will be sufficient to show that $P(\alpha' L) = P(\alpha' L')$. Then since for some $1 \leq i \leq 6$, $L = L_i$ and $L' = L'_i$, we are done if we show for each $1 \leq i \leq 6$ that $P(\alpha' L_i) = P(\alpha' L'_i)$.

Note that $\alpha' L_1 = \alpha L$ and $\alpha' L'_1 = \alpha L'$. Since $\alpha L$ and $\alpha L'$ are descending bods differing only in a Reidemeister move of the third type, they have the same number of components and hence the same polynomial. Thus $P(\alpha' L_1) = P(\alpha' L'_1)$. Furthermore, $\alpha' L_{i+1}$ is adjacent to $\alpha' L_i$, so by Corollary 22, $P(\alpha' L_{i+1}) = P(\alpha' L'_{i+1})$ if and only if $P(\alpha' L_i) = P(\alpha' L'_i)$. Applying induction on $i$, the proof is complete.

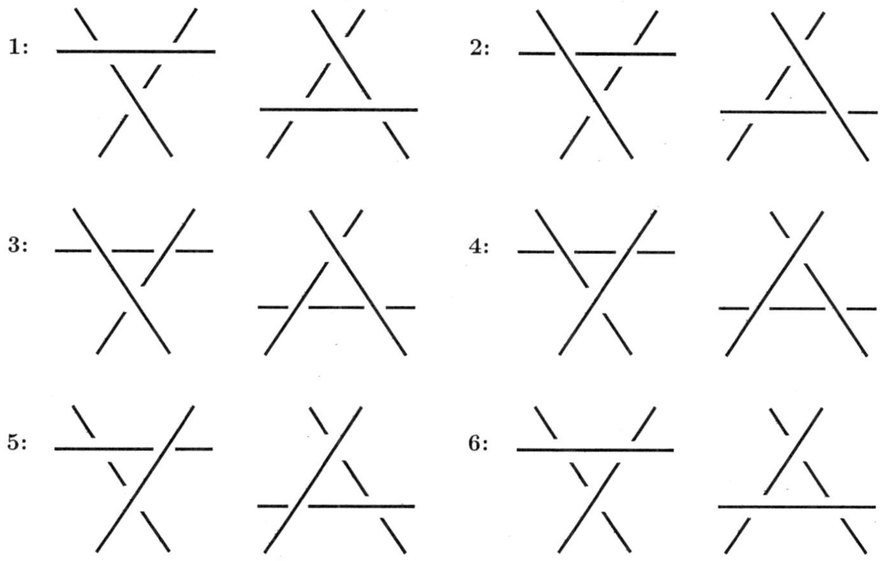


*Fig. 20.* The possibilities for $L$ and $L'$.

Identical arguments apply if the Reidemeister triangle resembles the reflection of Figure 18 (when the middle crossing ($w$) is of opposite sign). ■

We have now shown, in Lemmas 19, 20, and 23, that two bods which differ in a single Reidemeister move of any type have the same polynomial. We can apply Reidemeister moves in any number and any combination, at each step computing the polynomial, to see that any bods related by a sequence of Reidemeister moves have the same polynomial. Thus we have the following lemma.

**Lemma 24.** *If $L'$ can be obtained from $L$ by a finite sequence of Reidemeister moves, staying within $\mathcal{B}_n$, then $P(L) = P(L')$.* ■

To complete the proof of Theorem 9, we need to show that the polynomial is invariant under component ordering. We will use Reidemeister moves to reorder the bod and thereby have the result.

### 6.5. Property 4

In this section we will need the machinery developed in Section 3. Property 4 can be rephrased as follows.

**Lemma 25.** *If $L, L' \in \mathcal{B}_n$ are siblings, then $P(L) = P(L')$.*

**Proof.** Let $n$ be the number of crossings of $L$. If $n = 0$, $L$ and $L'$ are both descending, with the same number of components. Thus by definition, we are done.

If $n \geq 1$, and $L$ contains a 0-cycle, then $L'$ also contains a 0-cycle in the same place. Let $M$ be $L$ with all 0-cycles removed, and $M'$ be $L'$ with all 0-cycles removed. Then if $L$ originally had $k$ 0-cycles, $P(L) = (x+y)^k P(M)$ and $P(L') = (x+y)^k P(M')$ by Corollary 11. Then it would be sufficient to prove that $P(M) = P(M')$. Therefore we may assume that there are no 0-cycles.

Let $U$ be the universal diagram of $\overline{L}$ (and hence also of $\overline{L'}$) and $U'$ be the universal bod that agrees in structure with $U$ but with basepoints placed as dictated by Theorem 4 and any component ordering.

By property 2, we can move the basepoints of $L$ and $L'$ to match this choice without affecting the polynomial of $L$ or $L'$, so without loss of generality assume that $L$ and $L'$ have basepoints in the same place and as given by $U'$. Thus $L$ and $L'$ differ only in component ordering.

Let $\alpha' L = (\alpha L')_{\mathcal{U}_L}$.

By Corollary 16, it is sufficient to show that $P(\alpha L') = P(\alpha' L)$. That is, we may assume that $L$ is descending, as $\alpha L'$ and $\alpha' L$ are siblings with all the properties we have stated for $L$ and $L'$.

So let us assume that $L$ and $L'$ are siblings with no 0-cycles and at least one crossing, with $L$ descending. By Theorem 4, we can apply a finite sequence of Reidemeister moves to reduce the number of crossings of $L$ without increasing the number of crossings at any intermediate step. That is to say, we can apply a finite sequence of Reidemeister moves to $L$, at each step getting a bod with no more than $n$ crossings and ending with a bod $M$ lying in $\mathcal{B}_{n-1}$.

Basepoints, orientation, and component ordering play no role in any of the Reidemeister moves, so applying a Reidemeister move (of any type) to a bod and its sibling yields another pair of siblings. In particular, we can perform the sequence of Reidemeister moves that transform $L$ to $M$ to the bod $L'$. In doing so, we again remain within $\mathcal{B}_n$ and end with a bod $M'$ lying in $\mathcal{B}_{n-1}$. As noted, $M$ and $M'$ will be siblings, and the induction hypothesis says $P(M) = P(M')$.

Furthermore, since all links in the sequence from $L$ to $M$ and $L'$ to $M'$ lie in $\mathcal{B}_n$, Property 3 implies $P(L) = P(M)$ and $P(L') = P(M')$. Therefore, $P(L) = P(L')$. ■

## 6.6. Conclusion of proof of Theorem 9

We have now proved that, given that Theorem 9 is true for each bod in $\mathcal{B}_{n-1}$, each of the stated properties in Theorem 9 holds for each bod in $\mathcal{B}_n$. Therefore to complete the proof we only need to show that each property holds in $\mathcal{B}_0$.

If $L \in \mathcal{B}_0$, then $L$ consists of disjoint 0-cycles, ordered in some way and each given a basepoint. It is clear that $L$ is a descending bod, so by definition $P(L) = (x+y)^{cL-1}$, regardless of basepoint position and component order. Thus Properties 2 and 4 hold. Furthermore, Properties 1 and 3 hold vacuously, as there are no crossings in $L$ and no Reidemeister moves stay within $\mathcal{B}_0$.

Therefore our induction is rooted and the proof is complete. ■

## References

[1] J. W. Alexander, Topological invariants of knots and links, *Trans. Amer. Math. Soc.*, **20** (1923), 275–306.

[2] B. Bollobas, *Modern Graph Theory,* Springer-Verlag (New York, 1998).

[3] J. H. Conway, An enumeration of knots and links, and some of their algebraic properties. *1970 Computational Problems in Abstract Algebra* (Proc. Conf., Oxford, 1967), Pergamon, Oxford, 329–358.

[4] P. Freyd, J. Hoste, W. B. R. Lickorish, K. Millett, A. Ocneanu, and D. Yetter, A new polynomial invariant of knots and links, *Bull. Am. Math. Soc.*, **12** 2 (1985), 239–246.

[5] P. Freyd and D. Yetter, A new invariant for knots and links, *preprint.*

[6] J. Hoste, A polynomial invariant of knots and links, *Pac. J. Math.*, **124** 2 (1986), 295–320.

[7] F. Jaeger, A combinatorial model for the Homfly polynomial, *Europ. J. Comb.*, **11** 6 (1990), 549–558.

[8] F. Jaeger, Erratum: A combinatorial model for the Homfly polynomial, *Europ. J. Comb.*, **12** 1 (1991).

[9] V. Jones, A polynomial invariant for knots via Von Neumann algebras, *Bull. Amer. Math. Soc*, **12** 1 (1985), 101–108.

[10] L. Kauffman An invariant of regular isotopy, *Trans. Amer. Math. Soc.*, **318** 2 (1990), 417–471.

[11] L. Kauffman, State models and the Jones polynomial, *Topology,* **26** (1987), 395–407.

[12] L. Kauffman, State models for link polynomials, *Enseign. Math.*, **36** (1990), 1–37.

[13] W. B. R. Lickorish and K. Millett, A polynomial invariant of oriented links, *Topology,* **26** 1 (1987), 107–141.

[14] A. Ocneanu, A polynomial invariant for knots: A combinatorial and an algebraic approach, *preprint.*

[15] J. Przytycki and P. Traczyk, Invariants of links of Conway type, *Kobe J. Math.*, **4** (1987), 113–139.

[16] K. Reidemeister, *Knotentheorie,* Chelsea Publishing Company, New York (1948). Copyright 1932, Julius Springer, Berlin.

Béla Bollobás

*Department of Mathematical Sciences*
*University of Memphis*
*Memphis, TN 38152*
*U.S.A.*

and

*Trinity College*
*Cambridge CB2 1TQ*
*England*

e-mail: bollobas@msci.memphis.edu

Luke Pebody

*Department of Mathematical Sciences*
*University of Memphis*
*Memphis, TN 38152*
*U.S.A.*

e-mail: pebodyl@msci.memphis.edu

David Weinreich

*Department of Mathematics*
*University of Wisconsin - La Crosse*
*La Crosse, WI 54601*
*U.S.A.*

e-mail: weinreid@msci.memphis.edu

BOLYAI SOCIETY
MATHEMATICAL STUDIES, 10

**Contemporary Combinatorics**
**Budapest, 2002,** pp. 185–246.

# Better Bounds for Max Cut

B. BOLLOBÁS and A. D. SCOTT

For a multigraph $G$, let $f(G)$ be the size of a largest cut of $G$. We define $f(m)$ to be the minimum of $f(G)$ over graphs of size $m$, and $f_W(m)$ to be the minimum over multigraphs of size $m$. For $n$ sufficiently large, and $0 \le \binom{k}{2} \le n$, we determine $f(m)$ for $m = \binom{n}{2} + \binom{k}{2}$ and give the extremal graphs. Furthermore, by considering the weighted problem, we determine $f(m)$ and $f_W(m)$ to within an additive constant for every $m$, and find the extremal graphs for many values of $m$. This extends independent work of Alon and Halperin.

In the second part of the paper, we turn to the problem of finding efficient algorithms for obtaining large bipartite subgraphs. We give a linear time algorithm that, for a multigraph $G$ with $m$ edges and $n$ vertices, finds a bipartite subgraph with $f_W(m)$ edges. We give an algorithm running in time $O(2^{ck^4} + m + n)$ that finds a bipartite subgraph with at least $m/2 + \sqrt{m/8} + k$ edges if one exists and otherwise provides an optimal partition. We also provide a linear time weak approximation algorithm for $f(m) - m/2 - \sqrt{m/8}$.

In the final part of the paper, we generalize our results to the related problems Max $k$-Cut and Max Directed Cut.

## 1. Introduction

The well-known Max Cut problem asks for a largest cut of a graph $G$. A cut of maximal size clearly corresponds to a bipartite subgraph of maximal size, and we shall use both formulations. Max Cut is NP-hard and has been the focus of extensive study, both from the algorithmic perspective in computer science and the extremal perspective in combinatorics. The extremal problem asks how small a largest bipartite subgraph of a graph with $m$ edges can be, and which graphs achieve this bound. The algorithmic problem asks for efficient algorithms that determine or approximate the maximal size of a bipartite subgraph and that provide large bipartite subgraphs. An important survey of the Max Cut problem is given by Poljak and Tuza [32]; an excellent bibliography from the perspective of combinatorial optimization is given by Laurent [26].

This article is a combination of survey and research paper. We shall indicate some recent progress on the Max Cut problem, from both combinatorial and algorithmic perspectives, and prove a number of new results. The article was originally written in the autumn of 1997 for the Erdős Workshop in Budapest in the summer of 1998. Early in 1998, we became aware of the work of Alon and Halperin [2], who also addressed the extremal Max Cut problem. They determined the recurrence (5) for the extremal function $f_w(m)$; however, they did not determine the extremal graphs or consider the algorithmic aspects of the problem. Although our approach is similar to Alon and Halperin's, we give the details for clarity of exposition, and also so that we can obtain the extremal graphs for (4) and for some cases of (5). In addition, the arguments are used in later sections on algorithms.

For a graph $G$, let $f(G)$ be the maximal number of edges in a bipartite subgraph of $G$. For $m > 0$, we define $f(m)$ to be the minimum value of $f(G)$ for graphs $G$ with $m$ edges. As observed by Erdős, $f(m) \geq m/2$. This can be seen by noting that a random bipartition of a graph $G$ gives a cut with expected weight $e(G)/2$. In 1973, answering a question of Erdős, Edwards ([9], see also [10]) proved that

$$f(m) \geq \left\lceil \frac{m}{2} + \sqrt{\frac{m}{8} + \frac{1}{64}} - \frac{1}{8} \right\rceil . \tag{1}$$

We remark that, in fact, we can demand significantly more from a bipartition: it was shown in [6] that every graph $G$ with $m$ edges has a bipartition $V(G) = V_1 \cup V_2$ such that (1) is satisfied and, in addition, for $i = 1, 2$,

$$e(V_i) \leq \frac{m}{4} + \sqrt{\frac{m}{32} + \frac{1}{256}} - \frac{1}{16}. \tag{2}$$

This is an example of a *judicious partitioning result,* in which we demand that *every* class in a vertex partition satisfies some inequality. We remark that the result is best possible when $m = \binom{2l+1}{2}$, in which case $K_{2l+1}$ is the unique extremal graph (for $l \geq 2$). We shall not consider judicious partitions further in this paper; however, a discussion of related results and problems can be found in [7].

The Edwards bound (1) is exact for complete graphs. Thus if $m = \binom{n}{2}$ then $f(m) = f(K_n) = \lfloor n^2/4 \rfloor$. In fact, if $n \neq 4$ then $K_n$ is the unique extremal graph (for $n = 3$, two additional extremal graphs are obtained by taking an edge-disjoint union of two copies of $K_3$). For other values of $m$, however, the situation is less simple. Indeed, Erdős [11] conjectured that the difference between $f(m)$ and (1) can be arbitrarily large. This was proved by Alon [1], who showed that there exist $c, c' > 0$ such that, if $n$ is sufficiently large and $m = 2n^2$,

$$f(m) \geq n^2 + \frac{n}{2} + c\sqrt{n} \geq \frac{m}{2} + \sqrt{\frac{m}{8}} + c'm^{1/4}. \tag{3}$$

Alon also showed that, for some $c'' > 0$ and every $m > 0$,

$$f(m) \leq \frac{m}{2} + \sqrt{\frac{m}{8}} + c''m^{1/4}.$$

Thus the Edwards formula is exact for $m = \binom{n}{2}$ and out by $O(m^{1/4})$ when $m$ is about halfway between $\binom{n}{2}$ and $\binom{n+1}{2}$.

Our first aim in this paper is to determine $f(m)$ exactly for a range of values between $\binom{n}{2}$ and $\binom{n+1}{2}$. Indeed, suppose $m = \binom{n}{2} + \binom{k}{2}$, where $n$ and $k$ are non-negative integers with $0 \leq \binom{k}{2} \leq n$. The (two) graphs obtained by taking an edge-disjoint union of $K_n$ and $K_k$ show that

$$f(m) \leq \left\lfloor \frac{n^2}{4} \right\rfloor + \left\lfloor \frac{k^2}{4} \right\rfloor$$

while any graph obtained by deleting $\binom{n+1}{2} - \binom{n}{2} - \binom{k}{2}$ edges from $K_{m+1}$ shows that

$$f(m) \leq \left\lfloor \frac{(n+1)^2}{4} \right\rfloor.$$

In § 2 we shall prove that, provided $m$ is sufficiently large,

$$f(m) = \min\left\{\left\lfloor\frac{n^2}{4}\right\rfloor + \left\lfloor\frac{k^2}{4}\right\rfloor, \left\lfloor\frac{(n+1)^2}{4}\right\rfloor\right\}. \tag{4}$$

Furthermore, the graphs we have described include all the extremal graphs unless $k = 4$, when the graphs that can be obtained from an edge-disjoint union of $K_n$ and two copies of $K_3$ are also extremal.

Note that $\lfloor (n+1)^2/4 \rfloor$ is smaller than $\lfloor n^2/4\rfloor + \lfloor k^2/4\rfloor$ if

$$n - \sqrt{\frac{n}{2}} + \varepsilon(n,k) \le \binom{k}{2} \le n$$

where $|\varepsilon(n,k)|$ is $O(1)$. If there is some integer $k$ with $\binom{k}{2}$ in this range, then since $f(m)$ is monotonic increasing and $f\left(\binom{n+1}{2}\right) = \lfloor (n+1)^2/4 \rfloor$, it follows that, for $\binom{n}{2} + \binom{k}{2} \le m \le \binom{n+1}{2}$,

$$f(m) = \left\lfloor\frac{(n+1)^2}{4}\right\rfloor.$$

We obtain the surprising consequence that $f(m)$ is constant on intervals of length up to about $\sqrt{n/2} \approx (m/2)^{1/4}$.

In §3 we turn to the problem of determining $f(m)$ for arbitrary values of $m$. Our arguments are similar to those of §2, but we are faced with some additional technical difficulties, which make it necessary to consider graphs in which the edges are weighted. For a graph $G$ with edge-weighting $w$, let $f(G)$ be the maximal weight of a cut of $G$. Let $f_w(m)$ be the minimum of $f(G)$ over graphs whose edges are weighted with nonnegative integers and have total weight $m$ (or, equivalently, over multigraphs with $m$ edges). It is easily seen that $f_w(m) \le f(m)$. We prove that, for sufficiently large $m$,

$$f_w(m) = \min\left\{\left\lfloor\frac{(n+1)^2}{4}\right\rfloor, \left\lfloor\frac{n^2}{4}\right\rfloor + f_w\left(m - \binom{n}{2}\right)\right\}, \tag{5}$$

where the integer $n$ is defined by $\binom{n}{2} \le m < \binom{n+1}{2}$. This provides a recursive formula for $f_w(m)$. (This recurrence was found independently by Alon and

Halperin [2] and implies (4).) To prove a recursion, note that if we were to know $f_w(m)$ for every $m \le m_0$, where $m_0$ is sufficiently large, then (5) determines $f_w(m)$ for all $m$. In any case, (5) determines $f_w(m)$ to within an additive constant. Furthermore, by considering $K_{n+1}$ and graphs of form $K_n \cup H$ with $e(H) = m - \binom{n}{2}$, it is easy to see that

$$f(m) \le \min\left\{ \left\lfloor \frac{(n+1)^2}{4} \right\rfloor, \left\lfloor \frac{n^2}{4} \right\rfloor + f\left(m - \binom{n}{2}\right) \right\}.$$

Setting $C = \max_{m<m_0} \left| f(m) - f_w(m) \right|$, it follows that for every $m > 0$,

$$\left| f(m) - f_w(m) \right| \le C.$$

It seems likely that $f(m) = f_w(m)$ for every $m > 0$.

We use the results of §3 in §4, where we return to the problem of finding extremal graphs for Max Cut. Writing $m = \binom{n_1}{2} + \cdots + \binom{n_k}{2}$, where the $n_i$ are nonnegative integers with $\binom{n_i}{2} < n_{i+1}$ for $i < k$, we determine $f(m)$ provided $n_{k-1}$ is sufficiently large: if $\lfloor n_i^2/4 \rfloor + \cdots + \lfloor n_k^2/4 \rfloor < \left\lfloor (n_i+1)^2/4 \right\rfloor$ for each $i$, then $f(m) = \sum_{i=1}^{k} \lfloor n_i^2/4 \rfloor$; we also give the extremal graphs.

In the second part of the paper we concentrate on algorithmic results. Max Cut is NP-hard (see [24], [16]), even for quite restricted classes of graphs ([21], [5]), and it is therefore of interest to find polynomial time algorithms that give large bipartite subgraphs. A number of authors have given algorithms that yield a bipartite subgraph at least as large as that guaranteed by the Edwards bound (1) (see remarks in Section 2, where several algorithms are also discussed; see also [35]). In Section 5 we give a linear time algorithm that is optimal in terms of edge-weight: for any graph with integer edge-weights and total weight $m$, the algorithm yields a cut of weight at least $f_w(m)$.

Much recent progress on the Max Cut problem has concerned the existence of good approximation algorithms. Building on results in the theory of probabilistically checkable proofs, Håstad [18] has shown that, for any $\varepsilon > 0$, it is NP-hard to approximate Max Cut within a factor $17/16 - \varepsilon$. On the positive side, Goemans and Williamson [17] have given a 1.1383-approximation algorithm. (For the Max $k$-Cut problem Kann, Khanna, Lagergren and Panconesi [23] have shown that it is NP-hard to approximate within a factor $1 + 1/34k$; while Frieze and Jerrum [15] have given an algorithm that approximates within a factor $(1-1/k+2\ln k/k^2)^{-1}$.) Furthermore, it is known that good approximation algorithms exist for dense graphs (see [4], [25], [13]). Note that the difficulty for these algorithms lies in recognising and partitioning graphs $G$ for which $f(G)$ is large. Graphs for which $f(G)$ is small can be partitioned by the trivial greedy algorithm that yields a cut of weight at least $w(G)/2$.

In Sections 6 and 7 we concentrate on graphs of weight $m$ for which $f(G)$ is close to $f_w(m)$. Mahajan and Raman [29] have shown that there are algorithms running in time $O(n^3+m2^{4k})$ and $O(2^{ck^2}+m+n)$ that find a cut of size at least $\lceil m/2\rceil + k$ in a graph with $m$ edges and $n$ vertices, if such a cut exists. In Section 6, we show that, for any fixed integer $k$, there is an algorithm running in time $O(2^{ck^4}+m+e)$ that finds a cut of weight at least $m/2+\sqrt{m/8}+k$ in a graph with integer edge-weights, $e$ edges and total weight $m$ if such a cut exists and otherwise finds an optimal cut. In Section 7 we concentrate on the quantity $f(G)-m/2-\sqrt{m/8}$: we note that it is NP-hard to approximate this quantity within a factor $(9/8-\varepsilon)$, but provide a linear time algorithm that approximates its logarithm.

In the final part of the paper we consider two related problems. In Section 8, we consider the Max $k$-Cut problem for $k>2$. We prove versions of our results on bipartitions for the $k$-partite case. Finally, in Section 9, we consider the problem of finding large bipartite subgraphs of a directed graph, and give some extremal results.

Throughout the paper, we use $w$ for an integer-valued edge-weighting. For disjoint sets of vertices we write $E(X,Y)$ for the set of edges between $X$ and $Y$, $e(X,Y)=\left|E(X,Y)\right|$ and $w(X,Y)=\sum_{e\in E(X,Y)} w(e)$. We will also sometimes write $e(x,Y)$ and $w(x,Y)$ for $e\big(\{x\},Y\big)$ and $w\big(\{x\},Y\big)$.

# Part I: The Extremal Problem

## 2. Max Cut for graphs

Our main aim in this section is to find the exact value of $f(m)$ for every sufficiently large $m$ of form $\binom{n}{2}\le m=\binom{n}{2}+\binom{k}{2}\le\binom{n+1}{2}$, and determine the extremal graphs. The value of $f(m)$ can also be obtained from the results of the next section and from Alon and Halperin [2]. However, our aim in this section is also to determine the extremal graphs, which turn out to be surprisingly varied.

Note that, for any $m$, we can obtain an upper bound for $f(m)$ by writing $m = \binom{n_1}{2} + \binom{n_2}{2} + \cdots \binom{n_r}{2}$, where each $n_i$ in turn is chosen to be as large as possible; then by considering $K_{n_1} \cup \cdots \cup K_{n_r}$, it is clear that $f(m) \le \lfloor n_1^2/4 \rfloor + \cdots + \lfloor n_r^2/4 \rfloor$. A straightforward calculation shows that, for every $m$,

$$f(m) \le \frac{m}{2} + \sqrt{\frac{m}{8}} + (8m)^{1/4} + O(m^{1/8}), \tag{6}$$

while taking $k \approx \sqrt{2n} - 1$ in the theorem below shows that

$$f(m) \ge \frac{m}{2} + \sqrt{\frac{m}{8}} + \big(1 + o(1)\big)\,(8m)^{1/4}$$

for infinitely many values of $m$.

**Theorem 1.** *For $n > 5 \times 10^8$, every graph $G$ with*

$$e(G) = \binom{n}{2} + \binom{k}{2}$$

*where*

$$0 \le \binom{k}{2} \le n - 1$$

*satisfies*

$$f(G) \ge \min\left\{ \left\lfloor \frac{n^2}{4} \right\rfloor + \left\lfloor \frac{k^2}{4} \right\rfloor, \left\lfloor \frac{(n+1)^2}{4} \right\rfloor \right\}. \tag{7}$$

*Furthermore, the extremal graphs are the two graphs obtained by taking an edge disjoint union of $K_n$ and $K_k$ if $\lfloor n^2/4 \rfloor + \lfloor k^2/4 \rfloor \le \lfloor (n+1)^2/4 \rfloor$ and $k \ne 4$; and all graphs obtained by deleting $\binom{n+1}{2} - \binom{n}{2} - \binom{k}{2}$ edges from $K_{n+1}$ if $\lfloor n^2/4 \rfloor + \lfloor k^2/4 \rfloor \ge \lfloor (n+1)^2/4 \rfloor$. If $k = 4$ then the extremal graphs* *are obtained by taking an edge-disjoint union of $K_n$ and $K_4$ (two graphs) or $K_n$ and two copies of $K_3$ (seven graphs).*

We will make use of several lemmas in our proof of Theorem 1. Lemma 2 is due to Edwards [9]; recently a short proof was given by Erdős, Gyárfás and Kohayakawa [12]. Poljak and Turzík [33] gave an $O(n^3)$ algorithm for finding a bipartite subgraph of the type guaranteed in the lemma; Ngoc and Tuza [30] improved upon this by giving an algorithm running in time $O(m)$. The proof that we give is similar to the proofs of Erdős, Gyárfás and Kohayakawa and of Ngoc and Tuza, but is slightly simpler and also yields an $O(m)$ algorithm.

**Lemma 2.** *For a connected graph $G$,*

$$f(G) \geq \frac{e(G)}{2} + \frac{|G|-1}{4}.$$

**Proof.** Given an ordering of the vertices of $G$, we can partition $V(G)$ by using the greedy algorithm: at each step a vertex is added to whichever class contains fewer of its predecessors (or to either class if both classes contain the same number). If we write $e(v)$ for the number of predecessors of $v$ that are adjacent to $v$ then $\sum_{v\in V(G)} e(v) = e(G)$, and the size of the bipartite graph between the two vertex classes is at least

$$\sum_{v\in V(G)} \lceil e(v)/2 \rceil = \frac{e(G)}{2} + \frac{k}{2}$$

where $k$ is the number of vertices with an odd number of predecessors.

It is therefore enough to find an ordering of $V(G)$ in which at least $(n-1)/2$ vertices have an odd number of predecessors. For $|G| \leq 1$ this is trivial. If $|G| = n > 1$, then we first find a set of vertices $S$ such that $G[S]$ is a star and $G \setminus S$ is connected. Let $T$ be a spanning tree of $G$. If two endvertices of $T$ are adjacent, say $v$ and $w$, then let $S = \{v, w\}$. Otherwise, let $T'$ be the tree obtained by removing all endvertices of $T$, let $v$ be any endvertex of $T'$ and let $S$ contain $v$, together with all endvertices of $T$ adjacent to $v$. In either case, $S$ contains a vertex $v_s$ together with an independent set of neighbours of $v_s$ and $G \setminus S$ is connected. We repeat the process with $G \setminus S$, continuing until at most one vertex remains.

We order the vertices of $G$ one star at a time. Given a star $S$, let $R$ be the set of vertices we have already ordered. Let $S^+$ be those vertices of $S \setminus \{v_s\}$ with an odd number of neighbours in $R$ and let $S^- = S \setminus \left(S^+ \cup \{v_s\}\right)$. After $R$, take $S^+$ (in any order), followed by $v_s$, followed by $S^-$ (in any order). Note that the vertices in $S^+ \cup S^-$ all have an odd number of predecessors in this ordering, and $|S^+ \cup S^-| \geq |S|/2$. Thus in the complete ordering, since the stars together contain at least $n-1$ vertices, at least $(n-1)/2$ vertices have an odd number of predecessors.

It is easy to see that this proof gives an algorithm that runs in time $O(m)$. Note that the tree $T$ can be updated efficiently between the removal of successive stars. ■

In the proof above, we can avoid the need to generate a star partition by constructing more directly an ordering of the vertices. Begin with any ordering $v_1 < \cdots < v_n$ of $V(G)$ in which every vertex except $v_1$ has at least one predecessor. For each vertex calculate the number of predecessors, and let $X$ be the set of vertices with an even number of predecessors. For each $x \in X \setminus v_1$ find the largest predecessor of $x$. This can clearly all be done in time $O(m)$. Now suppose that two vertices have the same largest predecessor, say $v$. Reorder $V(G)$ by moving $x$ and $y$ to immediately before $v$, leaving $x$ and $y$ in the same order. The parity of $v$ does not change, since it has gained two predecessors, while $x$ and $y$ now have an odd number of predecessors. No other vertex has changed its set of predecessors. However, some vertices that previously had $x$ or $y$ as largest predecessor may now have $v$: we can check this by examining the neighbours of $x$ and $y$. By a similar argument, if any $x \in X$ has largest predecessor $y \in X$, then moving $x$ to a position immediately before $y$ gives an ordering in which $x$ and $y$ have an odd number of predecessors and only neighbours of $x$ (in the original ordering) can have a new largest predecessor. Repeating this stage of the algorithm, we note that each vertex is moved at most once (when its parity changes), and that we reexamine the neighbours of a vertex only when we move it (note also that if we begin by moving the largest possible vertex in $X$ then it is straightforward to update the predecessor lists). Thus this part of the algorithm runs in time $O(m+n) = O(m)$. Finally, suppose that no two vertices in $X$ have the same largest predecessor and no vertex in $X$ has largest predecessor in $X$. Now if $x \in X$ has no predecessors then either $x = v_1$, or (since vertices are only ever moved downwards) $x$ must have been moved at some point in the algorithm, which implies $x \notin X$. Thus every $x \in X \setminus v_1$ has a largest predecessor in $V(G) \setminus X$. It follows that we have an injection $X \setminus v_1 \to V(G) \setminus X$, so $|X| \leq (n-1)/2 + 1$, and thus at least $n - |X| \geq (n-1)/2$ vertices have an odd number of predecessors. The algorithm is now completed greedily as before.

Note that Lemma 2 furnishes a quick proof of the Edwards formula (1). Indeed, given a graph $G$, we may assume $G$ is connected or else identify one vertex from each component. Let $n = |G|$, so $e(G) \leq \binom{n}{2}$: it follows from the lemma that $f(G) \geq e(G)/2 + (n-1)/4$, and (1) follows by a simple calculation.

Lemma 3 was noted by several authors (see [3], [27], [28], [1]). We prove it here for completeness.

**Lemma 3.** *For a nonempty graph $G$,*

$$f(G) \geq \left(\frac{1}{2} + \frac{1}{2\chi(G)}\right) e(G).$$

**Proof.** Fix a colouring of $G$ with $t = \chi(G)$ colours, and let the colour classes be $V_1, \ldots, V_t$. Let $S \cup T$ be a random partition of $[t]$ into a set of size $\lfloor t/2 \rfloor$ and a set of size $\lceil t/2 \rceil$. Then, writing $m = e(G)$, the expected number of edges between $\bigcup_{i \in S} V_i$ and $\bigcup_{i \in T} V_i$ is

$$\left(\left\lfloor \frac{t^2}{4} \right\rfloor \Big/ \binom{t}{2}\right) m \geq \left(\frac{t^2-1}{4} \Big/ \binom{t}{2}\right) m = \left(\frac{1}{2} + \frac{1}{2t}\right) m.$$

Therefore some partition satisfies this inequality. ■

Lemma 3 also gives a fast proof of the Edwards formula, as observed independently by Alon [1] and Hofmeister and Lefmann [20].

We will also need a lemma concerning partitions of graphs whose edges are weighted with (positive or negative) integers. Note that as a consequence of this lemma, for $m = \binom{n}{2}$ we obtain the extremal graphs for the formula of Edwards.

**Lemma 4.** *Let $H$ be a graph whose edges have integer weights. If $n$ is an integer with*

$$w(H) \geq \binom{n}{2}$$

*then there is a partition of $H$ into two sets such that the total weight of edges between the sets is at least $\lfloor n^2/4 \rfloor$. For $n \neq 4$, the unique extremal graph is $K_n$ with all edges of weight 1. For $n = 4$ the extremal graphs are $K_4$ with all edges of weight 1 and the graphs obtained by taking the edge sum of two copies of $K_3$ with all edges of weight 1.*

**Proof.** The proof that such a partition exists is straightforward, since we may consider $H$ as a weighted complete graph, by adding an edge of weight 0 between every pair of nonadjacent vertices. If $H$ contains an edge with weight at most 0 then contracting that edge does not decrease the weight of the graph. Therefore we may assume that $H$ is a complete graph and all edges have weight at least 1, so $|H| \le n$. A random partition of $V(H)$ into sets of size $\lfloor |H|/2 \rfloor$ and $\lceil |H|/2 \rceil$ yields a bipartite subgraph of expected weight at least $\lfloor n^2/4 \rfloor$.

To derive the extremal graphs, note first that we can assume that all edges have nonnegative weight, since contracting an edge with negative weight increases the total weight, and we can then do better than $\lfloor n^2/4 \rfloor$ in the argument above (this remark also applies if $w(H) > \binom{n}{2}$). If $H$ is not $K_n$ with all edges of weight 1, then we can contract to a complete graph with at least one edge of weight greater than 1. Thus $|H| < n$ and all edges have weight at least 1. Writing $h = |H|$, a random bipartition into sets of size $\lceil h/2 \rceil$ and $\lfloor h/2 \rfloor$ yields a bipartite subgraph of expected weight at least

$$\frac{\lfloor h^2/4 \rfloor}{\binom{h}{2}} w(H) = \left\lfloor \frac{h^2}{4} \right\rfloor \binom{n}{2} \Big/ \binom{h}{2}. \tag{8}$$

If $h < n-1$ or $h = n-1$ and $n$ is odd then (8) is strictly larger than $\lfloor n^2/4 \rfloor$. Otherwise, $h = n-1$ is odd and, since all edges have weight at least 1, $H$ is the edge sum of $K_{n-1}$ with all weights 1 and a graph $H^*$ with $V(H^*) = V(H)$ and weight $\binom{n}{2} - \binom{n-1}{2} = n-1$. Furthermore, all bipartitions of $V(H)$ into two sets of size $\lfloor h/2 \rfloor$ and $\lceil h/2 \rceil$ yield a bipartite subgraph of size exactly $\lfloor n^2/4 \rfloor$, since otherwise some partition would exceed the expectation (8). Since every bipartition of $K_{n-1}$ into two sets of size $\lfloor (n-1)/2 \rfloor$ and $\lceil (n-1)/2 \rceil$ gives a bipartite subgraph of size $\lfloor (n-1)^2/4 \rfloor = n^2/4 - n/2$, it follows that every bipartition of $H^*$ into sets of size $\lceil (n-1)/2 \rceil$ and $\lfloor (n-1)/2 \rfloor$ gives a bipartite subgraph of size $n/2$.

If $H^*$ is not complete, let $v$ be a vertex that is not adjacent to every other vertex of $H^*$. We can partition $V(H^*) \setminus v$ into two sets $V_1$ and $V_2$

with $|V_1| = |V_2| = (n-2)/2$, such that $|\Gamma(v) \cap V_1| > |\Gamma(v) \cap V_2|$. Then adding $v$ to $V_1$ or $V_2$ gives two partitions into sets of size $\lfloor (n-1)/2 \rfloor$ and $\lceil (n-1)/2 \rceil$ that yield bipartite subgraphs with different sizes. It follows that $H^*$ is complete: the only possibility is $n = 4$, $H^* = K_3$ with all edges of weight 1. Thus for $n > 4$, $H$ must be $K_n$, with all edges of weight 1. For $n = 4$, a simple case check shows that $H$ can also be any graph obtained by taking the edge sum of two copies of $K_3$ with all edges of weight 1. ∎

Another bound on $f(G)$ was given by Poljak and Turzik [31], who showed that every connected graph $G$ with edge-weighting $w$ has a bipartite subgraph of weight at least

$$\frac{1}{2}w(G) + \frac{1}{4}\min_T w(T)$$

where the minimum is taken over spanning trees $T$ of $G$. Poljak and Turzik show that there is an algorithm running in time $O(n^3)$ that finds the required subgraph; Poljak and Tuza [32] show that the algorithm runs in time $O(mn)$. We show that there is an $O(m)$ algorithm: note that for unweighted graphs, we obtain a cut of weight at least $e(G)/2 + (|G| - 1)/4$, thus giving another proof of Lemma 2.

**Theorem 5.** *There is an algorithm running in time $O(m)$ that finds in every connected graph $G$ with $m$ edges and edge-weighting $w$ a cut of weight at least*

$$\frac{1}{2}w(G) + \frac{1}{4}\min_T w(T).$$

Before proving this theorem we need a lemma. We say that a collection of induced stars or single vertices $S_1, \ldots, S_t$ in a graph $G$ is a *tree-like star-covering* if every vertex in $G$ belongs to some $S_i$ and the graph with vertices $S_1, \ldots, S_t$ and edges between $S_i$ and $S_j$ iff $S_i \cap S_j \neq \emptyset$ is a tree.

**Lemma 6.** *There is an algorithm running in time $O(m)$ that finds a tree-like star-covering of any connected graph $G$ with $m$ edges.*

**Proof.** Recall that a rooted spanning tree $T$ of a graph $G$ is a *depth-first search (DFS) tree* if, for every $uv \in E(G)$, either the path from the root $r$

to $u$ in $T$ contains $v$ or the path from $r$ to $v$ contains $u$. In other words, if we delete $v \in V(T)$, then the components of $T \setminus v$ containing each of the children of $v$ are not joined by any edge. A DFS tree can be found in time $O(m)$ (see, for instance, [30]).

Let $T$ be a DFS tree in $G$ with root $x$. For every $v \in V(T)$ that is not an endvertex of $T$ let $T_v$ be the induced star containing $v$ and its children. If $v$ is an endvertex of $T$ then let $T_v = \{v\}$. Then $\{T_v : v \in V(T)\}$ is a tree-like star covering of $G$. ■

**Proof of Theorem 5.** Keeping the notation of the proof of Lemma 6, let

$$\mathcal{T}_0 = \{T_v : d_T(x, v) \equiv 0 \bmod 2\}$$

and

$$\mathcal{T}_1 = \{T_v : d_T(x, v) \equiv 1 \bmod 2\}.$$

Each of $\mathcal{T}_0$ and $\mathcal{T}_1$ is a collection of disjoint induced stars and single vertices, and every edge of $T$ is contained in some member of $\mathcal{T}_0$ or $\mathcal{T}_1$. We partition whichever of $\mathcal{T}_0$ and $\mathcal{T}_1$ has the greater weight, say $\mathcal{T}_i$ one star at a time. Suppose we have a partial partition $V_1 \cup V_2$ and wish to partition a star $T_v$. We greedily assign $v$ to one class and $T_v \setminus v$ to the other so that the weight of the partial partition is increased by at least $w(T_v, V_1 \cup V_2)/2 + w(T_v)/2$. Repeating for all stars, we obtain a cut of weight at least

$$\frac{1}{2}w(G) + \frac{1}{2}w(\mathcal{T}_i) \geq \frac{1}{2}w(G) + \frac{1}{4}w(T).$$

The algorithm clearly runs in time $O(m)$. ■

Finally, it will be useful to have the following remark.

**Lemma 7.** *If $W \subset V(G)$ and $H = G[W]$ then*

$$f(G) \geq f(H) + \frac{1}{2}\big(e(G) - e(H)\big).$$

**Proof.** Partition $H$, then add the remaining vertices from $G$ one at a time to whichever class has fewer neighbours. The resulting partition clearly satsfies the inequality. ■

With these lemmas in hand, we turn to the proof of Theorem 1.

**Proof of Theorem 1.** Let $n_0 = 5 \times 10^8$. Suppose that $n > n_0$ and $G$ is a graph with

$$m = \binom{n}{2} + \binom{k}{2} < \binom{n+1}{2} \tag{9}$$

edges and

$$f(G) \leq \min\left\{\left\lfloor\frac{n^2}{4}\right\rfloor + \left\lfloor\frac{k^2}{4}\right\rfloor, \left\lfloor\frac{(n+1)^2}{4}\right\rfloor\right\}. \tag{10}$$

We shall prove that $G$ is one of the extremal graphs given in the statement of the theorem. Note that we may assume that $G$ is connected by identifying one vertex from each component.

We begin by showing that $G$ consists of a very large complete graph on $n - O(\sqrt{n})$ vertices, together with $O(\sqrt{n})$ 'exceptional vertices'.

Clearly $|G| \geq n$ and $k < \sqrt{2n} + \frac{1}{2}$. Note also that $n > 140(n^{3/4} + k + 1)$. If $k < 2$, we are done by Lemma 4; so we may assume $k \geq 2$. Let $\chi = \chi(G)$. Now by (9),

$$\lfloor n^2/4 \rfloor + \lfloor k^2/4 \rfloor \leq \frac{m}{2} + \frac{n+k}{4}, \tag{11}$$

and so it follows from (10) that

$$f(G) \leq \frac{m}{2} + \frac{n+k}{4}. \tag{12}$$

Therefore, by Lemma 2,

$$|G| \leq n + k + 1. \tag{13}$$

Furthermore, Lemma 3 and (12) imply that

$$\frac{m}{2\chi} \le \frac{n+k}{4}$$

and so by (9)

$$\chi \ge \frac{2m}{n+k} > n - k - 1.$$

Now if $\overline{G}$ contains $2k+2$ independent edges, we can cover $\overline{G}$ by $|\overline{G}| - (2k+2) \le n-k-1$ edges and vertices, which is equivalent to colouring $G$ with at most $n-k-1$ colours. It follows that $\overline{G}$ contains at most $2k+1$ independent edges and therefore that some set $Y$ of at most $4k+2$ vertices meets all edges of $\overline{G}$. Let $X = V(G) \setminus Y$; then $G[X]$ is complete, and

$$|X| = |G| - |Y| \ge n - 4k - 2. \tag{14}$$

We have partitioned $G$ into a large complete graph $G[X]$ and a small set $Y$ which we shall regard as a set of exceptional vertices. Note that it follows from (14) that any partition of $X$ into two sets of equal size (or sizes differing by 1) corresponds to a bipartite subgraph of $G[X]$ of size

$$\left\lfloor \frac{|X|^2}{4} \right\rfloor \ge \frac{e(X)}{2} + \frac{|X|-1}{4} \ge \frac{e(X)}{2} + \frac{n-4k-3}{4} \tag{15}$$

Throughout the proof, we shall consider partial partitions of $V(G)$ in which we partition $Y$ and some vertices from $X$, and then extend these to partitions of $V(G)$ in which $X$ is split as evenly as possible.

We now show that every vertex in $Y$ has either very many or very few neighbours in $X$. Indeed, suppose some $v \in Y$ has $\left|\Gamma(v) \cap X\right| \ge 5k/2 + 2$ and $\left|X \setminus \Gamma(v)\right| \ge 5k/2 + 2$. We partition $G$ as follows. Since $n > n_0$ we have $n > 5k + 4$, so we can find a partition $X = X_1 \cup X_2$ with $|X_1| \le |X_2| \le |X_1| + 1$ and

$$\big|\left|\Gamma(v) \cap X_1\right| - \left|\Gamma(v) \cap X_2\right|\big| \ge \frac{5k}{2} + 2. \tag{16}$$

Adding $v$ to whichever of $X_1$ and $X_2$ contains fewer of its neighbours, it follows from (14) that we obtain a bipartition of $H = G\big[X \cup \{v\}\big]$ with at least

$$\frac{e(H)}{2} + \frac{|X|-1}{4} + \frac{5k}{4} + 1 > \frac{e(H)}{2} + \frac{n+k}{4}$$

edges between the two classes. Thus, by Lemma 7,

$$f(G) \geq f(H) + \frac{1}{2}\big(e(G) - e(H)\big) > \frac{m}{2} + \frac{n+k}{4}, \tag{17}$$

which contradicts (12). We may therefore assume that for every $v \in Y$, either $\big|\Gamma(v) \cap X\big| < 5k/2+2$ or $\big|X \setminus \Gamma(v)\big| < 5k/2+2$. Let

$$Y^+ = \big\{v \in Y \,:\, \big|\Gamma(v) \cap X\big| > |X| - 5k/2 - 2\big\}$$

$$Y^- = \big\{v \in Y \,:\, \big|\Gamma(v) \cap X\big| < 5k/2 + 2\big\}.$$

Then $Y^+ \cup Y^-$ is a partition of $Y$.

Next we show that the subgraph induced by $X \cup Y^+$ is nearly complete. Indeed, we claim that $e\big(\overline{G}[X \cup Y^+]\big) < 5k/2+2$. If not, then let $W \subset Y^+$ be minimal such that

$$e\big(\overline{G}[X \cup W]\big) \geq \frac{5k}{2} + 2. \tag{18}$$

Since $\big|X \setminus \Gamma(v)\big| < 5k/2+2$ for every $v \in Y^+$, we have

$$5k/2 + 2 \leq e\big(\overline{G}[X \cup W]\big) \leq 5k + 4.$$

Since $n > n_0$, it follows that $n > 4(5k+4)$, so we can find a partition $V_1 \cup V_2$ of $X \cup W$ such that $|V_1| \leq |V_2| \leq |V_1|+1$ and all the edges of $\overline{G}[X \cup W]$ are contained in $V_1$. Then since $\binom{|X \cup W|}{2} \geq e(X \cup W) + (5k/2) + 2$, it follows from (18) and (14) that

$$\begin{aligned} f\big(G[X \cup W]\big) &\geq |V_1|\,|V_2| \\ &\geq \left\lfloor \frac{|X \cup W|^2}{4} \right\rfloor \\ &\geq \frac{1}{2}\big(e(X \cup W) + 5k/2 + 2\big) + \frac{|X| + |W| - 1}{4} \\ &> \frac{1}{2}e(X \cup W) + \frac{n+k}{4}, \end{aligned}$$

and we are done, as in (17). Thus we may assume that

$$e\big(G[X \cup Y^+]\big) < \frac{5k}{2} + 2, \tag{19}$$

so $G[X \cup Y^+]$ is nearly a complete graph. In particular,

$$f\big(G[X \cup Y^+]\big) \geq \left\lfloor \frac{|X \cup Y^+|^2}{4} \right\rfloor \tag{20}$$

Now we show that there are not too many edges between $Y^-$ and $X \cup Y^+$. Note first that every vertex $v \in Y^-$ has fewer than $5k/2 + 2$ neighbours in $X \cup Y^+$: otherwise, since $e\big(\overline{G}[X \cup Y^+]\big) < 5k/2 + 2$ and $\big|\Gamma(v) \cap X\big| < 5k/2 + 2$ (and since $|X| > 6(5k/2 + 2) = 15k + 12$ which, as $n > n_0$, follows from (14)) we can find a partition of $X \cup Y^+$ into sets $W_1$ and $W_2$ with $|W_1| \leq |W_2| \leq |W_1| + 1$ such that all edges of $\overline{G}[X \cup Y^+]$ are contained in $W_1$ and $\big|\,|\Gamma(v) \cap W_1| - |\Gamma(v) \cap W_2|\,\big| \geq 5k/2 + 2$. Arguing in the same way as from (16) we arrive at a contradiction. Thus

$$\big|\Gamma(v) \cap (X \cup Y^+)\big| < \frac{5k+4}{2} \tag{21}$$

for every $v \in Y^-$.

Now suppose that

$$e(Y^-, X \cup Y^+) > 70n^{3/4}.$$

Let $Y_0 \subset Y^-$ be minimal such that $e(Y_0, X \cup Y^+) > 70n^{3/4}$, and let $U = \Gamma(Y_0) \cap (X \cup Y^+)$. Note that the minimality of $Y_0$ and (21) imply that $|U| < 70n^{3/4} + 5k/2 + 2$. Since $n > n_0$ it follows that $|U| < |X|/2$. Let $Y_0 = Y_1 \cup Y_2$ be a random partition, where each vertex of $Y_0$ is in $Y_1$ or $Y_2$ independently with probability 1/2. Let

$$U_1 = \left\{ u \in U \, : \, \left|\Gamma(u) \cap Y_1\right| > \left|\Gamma(u) \cap Y_2\right| \right\}$$

and let $U_2 = U \setminus U_1$. For $u \in U$, let $d_u = \left|\Gamma(u) \cap Y_0\right|$ and define $\Delta(u) = \left|\left|\Gamma(u) \cap Y_1\right| - \left|\Gamma(u) \cap Y_2\right|\right|$. Then

$$\mathbb{E}(\,\Delta(u)) = \mathbb{E}\big(\left|\,S(d_u)\right|\big),$$

where we write $S(d_u)$ for the position after $d_u$ steps of a simple symmetric random walk on $\mathbb{Z}$ starting from 0. It is easily checked that $\mathbb{E}\left|\,S(d)\right| \geq \sqrt{d}/2$, and so

$$\begin{aligned}
\mathbb{E}(\,e(Y_1, U_2) + e(Y_2, U_1)) &= \frac{1}{2}e(U, Y_0) + \frac{1}{2}\sum_{u \in U} \mathbb{E}(\,\Delta(u)) \\
&\geq \frac{1}{2}e(U, Y_0) + \frac{1}{2}\sum_{u \in U} d_u^{1/2}/2 \\
&\geq \frac{1}{2}e(U, Y_0) + e(U, Y_0)/12n^{1/4},
\end{aligned}$$

since $d_u \leq |Y_0| \leq |Y| \leq 4k + 2 < 9\sqrt{n}$ and so $d_u^{1/2} > d_u/3n^{1/4}$; also

$$\mathbb{E}(\,e(Y_1, Y_2)) = \frac{1}{2}e(Y_0).$$

Since $|U| < |X|/2$, we can extend the partition $U_1 \cup U_2$ of $U$ to a partition $T_1 \cup T_2$ of $X \cup Y^+$ with $|T_1| \le |T_2| \le |T_1| + 1$. Then, by (19) and (14),

$$\begin{aligned} e(T_1, T_2) &\ge \left\lfloor \frac{|X \cup Y^+|^2}{4} \right\rfloor - e(\overline{G}[X \cup Y^+]) \\ &\ge \frac{1}{2} e(X \cup Y^+) + \frac{|X \cup Y^+| - 1}{4} - \frac{5}{2}k - 2 \\ &> \frac{1}{2} e(X \cup Y^+) + \frac{n - 14k - 11}{4}. \end{aligned}$$

Thus, partitioning $X \cup Y^+ \cup Y_0$ into $T_1 \cup Y_2$ and $T_2 \cup Y_1$, we see that

$$\begin{aligned} f\big(G[X \cup Y^+ \cup Y_0]\big) &\ge \mathbb{E}\big( e(T_1 \cup Y_2, T_2 \cup Y_1)\big) \\ &= \mathbb{E}\big( e(T_1, T_2) + e(Y_1, Y_2) + e(Y_1, U_2) + e(Y_2, U_1)\big) \\ &> \frac{1}{2} e(X \cup Y^+ \cup Y_0) + \frac{n - 14k - 11}{4} + \frac{e(U, Y_0)}{12 n^{1/4}} \\ &> \frac{1}{2} e(X \cup Y^+ \cup Y_0) + \frac{n + k}{4}, \end{aligned}$$

provided $e(U, Y_0) > (45k + 33) n^{1/4}$, which follows from $e(U, Y_0) > 70 n^{3/4}$, since $n > n_0$. Thus it follows from Lemma 7 and (12) that we may assume

$$e(Y^-, X \cup Y^+) \le 70 n^{3/4}. \tag{22}$$

Next we prove that $|X \cup Y^+| = n$ or $n+1$. Now since $|Y^-| \le |Y| \le 4k+2$, we have

$$e(Y^-) \le \binom{4k+2}{2} = 8k^2 + 6k + 1.$$

So, since $n > n_0$, it follows that

$$\begin{aligned} e(X \cup Y^+) &\geq e(G) - e(Y^-) - e(Y^-, X \cup Y^+) \\ &\geq \binom{n}{2} + \binom{k}{2} - (8k^2 + 6k + 1) - 70n^{3/4} \\ &\geq \binom{n}{2} - \frac{15}{2}k^2 - \frac{13}{2}k - 1 - 70n^{3/4} \\ &> \binom{n}{2} - 16n. \end{aligned}$$

So in order to have enough vertices for the edges, we must have $|X \cup Y^+| \geq n - 16$, and thus by (12),

$$|Y^-| \leq k + 17. \tag{23}$$

Now if $|X \cup Y^+| \leq n - 1$, then $e(X \cup Y^+) \leq \binom{n-1}{2}$ and so, since $n > n_0$,

$$\begin{aligned} e(Y^-) &\geq \binom{n}{2} + \binom{k}{2} - e(X \cup Y^+) - e(Y^-, X \cup Y^+) \\ &\geq \binom{k}{2} + n - 1 - 70n^{3/4} \\ &> \binom{k}{2} + \frac{n}{2}, \end{aligned}$$

which contradicts (23). So $|X \cup Y^+| \geq n$. Similarly, we have $|X \cup Y^+| \leq n+1$, since, by (19),

$$\begin{aligned} \binom{n+2}{2} - e(\overline{G}[X \cup Y^+]) &> \binom{n+2}{2} - \frac{5k}{2} - 2 \\ &> \binom{n}{2} + \binom{k}{2}. \end{aligned}$$

We have shown that $|X \cup Y^+| = n$ or $|X \cup Y^+| = n+1$. If $|X \cup Y^+| = n+1$ then, by (20),

$$f(G) \geq \left\lfloor \frac{(n+1)^2}{4} \right\rfloor$$

with equality iff $Y^- = \emptyset$, in which case $G$ consists of $K_{n+1}$ with $\binom{n+1}{2} - m$ edges deleted.

Otherwise $|X \cup Y^+| = n$. Now let $H$ be the weighted graph consisting of all edges of $E(Y^-) \cup E(Y^-, X \cup Y^+)$ with weight $+1$ and all edges of $E\big(\overline{G}[X \cup Y^+]\big)$ with weight $-1$, so $H$ has total weight $\binom{k}{2}$. It follows from Lemma 4 that $H$ has a cut of weight at least $\lfloor k^2/4 \rfloor$.

Note that since $n > n_0$ it follows from (14), (19) and (22) that $|H| \leq |Y| + e(Y^-, X \cup Y^+) \leq 4k + 2 + 70n^{3/4} + 5k/2 + 2 < |X|/2$. We can therefore extend a partition of $H$ to a partition of $G$ in which $X \cup Y^+$ is evenly partitioned, so

$$f(G) \geq \left\lfloor \frac{n^2}{4} \right\rfloor + f(H) \geq \left\lfloor \frac{n^2}{4} \right\rfloor + \left\lfloor \frac{k^2}{4} \right\rfloor$$

with equality iff $H \cong K_k$, with all edges of weight 1 (or $H \cong K_4$ or $H \cong 2K_3$ when $k = 4$). It follows immediately that $G[X \cup Y^+]$ is complete, and the extremal graphs are as described in the statement of the theorem. ∎

What prevents us from extending the argument in the proof of Theorem 1 to graphs with $\binom{n}{2} + \binom{k}{2} + \binom{l}{2}$ edges? The problem is that when we remove the copy of $K_n$ in the argument above, we are left with a graph with weighted edges. If we have $\binom{k}{2}$ edges then Lemma 4 gives us the unique extremal graph, whereas to deal with $\binom{k}{2} + \binom{l}{2}$ edges we would need a version of Theorem 1 for weighted graphs. Our aim in the next section is to prove such a theorem. In particular, it will enable us to determine $f(m)$ exactly for a much wider range of $m$, and to within an additive constant for every value of $m$. It will not, however, yield all the extremal graphs.

## 3. Max cut for multigraphs

Our aim in this section is to determine $f(m)$ to within an additive constant for every integer $m$, and to determine $f(m)$ exactly for a larger range of values of $m$. In order to do this, we consider graphs with integer edge-weightings: note that, if all weights are positive, we can think of these as multigraphs, where the weight of an edge indicates its multiplicity. As in the unweighted case, for a graph $G$ with edge-weighting $w$, we write $f(G)$ for the maximal weight of a cut of $G$. We define $f_w(m)$ to be the minimum of $f(G)$ over graphs $G$ with $w(G) = \sum_{e\in E(G)} w(e) = m$ and all weights non-negative integers.

Note that the restriction to positive integers means that $f_w(m)$ is the minimum of $f(G)$ over a finite set of multigraphs. Thus there is a (very slow) algorithm to determine $f(m)$. However, we do not lose anything by allowing negative weights: if $w(G) = m$ and $G$ has an edge $xy$ with negative weight, then consider the graph $H = G/xy$ obtained by contracting the edge $xy$ to a single vertex $z$ and defining $w(vz) = w(vx) + w(vy)$ for $v \neq x, y$. Repeating the process until we obtain a graph $H$ with no edges of negative weight, it is clear that $w(H) \geq w(G)$. Since $f_w(m)$ is monotone increasing, it follows that $f(G) \geq f(H) \geq f_w(m)$.

What can we say about $f_w(m)$? Clearly it is subadditive: since $f(G \cup H) = f(G) + f(H)$ for any graphs $G$ and $H$, it follows that $f_w(m + r) \leq f_w(m) + f_w(r)$ and, similarly, $f(m + r) \leq f(m) + f(r)$. Furthermore, it follows from Lemma 4 that, for $m = \binom{n}{2}$, we have $f_w(m) = f(m) = \lfloor n^2/4 \rfloor$. All the work in this section will go into proving a lower bound for $f_w(m)$ for other values of $m$.

We begin with Theorem 8, which provides a recursive lower bound on $f_w(m)$ and hence $f(m)$ (as noted in the introduction, this was proved independently by Alon and Halperin [2]). The approach we use in proving the theorem is similar to that used in the proof of Theorem 1. However since we are dealing with weighted graphs the details are rather different.

**Theorem 8.** *Let $G$ be a graph with integer-valued edge-weighting $w$. Suppose $w(G) = m$. Then, provided $m > m_0$,*

$$f(G) \geq \min_{n\geq 1} \left\{ \left\lfloor \frac{n^2}{4} \right\rfloor + f_w\left( m - \binom{n}{2} \right) \right\}, \tag{24}$$

*where we define $f_w(r) = 0$ for $r < 0$.*

**Proof.** Let $G$ be a graph with $w(G) = m$ and $f(G) = f_w(m)$ that does not satisfy (24). We can consider $G$ as a weighted complete graph by defining $w(xy) = 0$ if $x$ and $y$ are nonadjacent. If $G$ contains an edge $xy$ with nonpositive weight then replace $G$ with $G/xy$. Clearly $w(G/xy) \geq w(G)$ and $f(G/xy) \leq f(G)$. Repeating this process, we may assume that $G$ is a weighted complete graph with all edges of positive weight. Let $m = w(G)$ and define the integer $n$ by

$$\binom{n}{2} \leq m < \binom{n+1}{2}. \tag{25}$$

Since $G$ is complete and every edge has weight at least 1, we have $|G| \leq n$. We will use the fact that, as in (6), for some $c, c_0 > 0$ and every integer $m$,

$$f(m) \leq \frac{m}{2} + \sqrt{\frac{m}{8}} + cm^{1/4} < \frac{m}{2} + \frac{n}{4} + c_0\sqrt{n}. \tag{26}$$

We will use $c_1, c_2, \ldots$ to refer to constants in the proof below; suitable constants can easily be determined. In several places, we shall assume that $n$ is larger than some fixed constant.

Note that the proof of Lemma 3 carries over straightforwardly to the weighted case (see also Section 5). In particular, since $G$ is complete, it follows from the weighted version of Lemma 3 that

$$f(G) \geq \left(\frac{1}{2} + \frac{1}{2|G|}\right) m$$

and so by (26),

$$\frac{m}{2|G|} \leq \frac{n}{4} + c_0\sqrt{n}.$$

Since $m \geq \binom{n}{2}$, we have

$$|G| \geq \frac{n(n-1)}{n + 4c_0\sqrt{n}} \tag{27}$$

$$= \frac{n-1}{1+4c_0/\sqrt{n}}$$

$$> n - c_1\sqrt{n}.$$

We now find a small set of vertices that meets all edges with weight greater than 1. Let $M$ be a matching of maximal weight in $G$. Consider the random partition $V(G) = V_1 \cup V_2$, where for each edge $xy \in M$ we independently assign $x \in V_1$ and $y \in V_2$ or $x \in V_2$ and $y \in V_1$ with equal probability; if there is a vertex not covered by $M$, we assign it to $V_1$ or $V_2$ with equal probability. The expected weight of edges between $V_1$ and $V_2$ is

$$w(M) + \frac{1}{2}\big(w(G) - w(M)\big) = \frac{1}{2}w(G) + \frac{1}{2}w(M).$$

It follows from (26) that any matching in $G$ has weight at most

$$\frac{n}{2} + 2c_0\sqrt{n}. \tag{28}$$

Now let $e_1, \ldots, e_k$ be a maximal set of independent edges of $G$ with $w(e_i) > 1$ for $i = 1, \ldots, k$. Extend this arbitrarily to a maximal matching $M$. Then $|M| \geq \big(|G| - 1\big)/2$, so by (27),

$$w(M) \geq \frac{|G|-1}{2} + k > \frac{n - c_1\sqrt{n} - 1}{2} + k$$

and so by (28) we have $k \leq \frac{1}{2}\big(c_1\sqrt{n} + 1\big) + 2c_0\sqrt{n} \leq c_2\sqrt{n}$. Let $Y$ be the set of vertices spanned by $e_1, \ldots, e_k$. Then

$$|Y| = 2k \leq 2c_2\sqrt{n} \tag{29}$$

and any edge of weight greater than 1 is incident with $Y$. Let $X = V(G)\backslash Y$: then by (27) and (29)

$$|X| \geq |G| - |Y| \geq n - (c_1 + 2c_2)\sqrt{n}. \tag{30}$$

Note that $X$ induces a complete graph with all edges of weight 1.

Now for $y \in Y$, consider the edges between $y$ and $X$, and order them in increasing order of weight (order edges of the same weight arbitrarily). Let $Z_1$ be the vertices in $X$ incident with the first $\lfloor |X|/2 \rfloor$ edges and let $Z_2 = X \setminus Z_1$. Consider the partition of $X \cup \{y\}$ into $Z_1 \cup \{y\}$ and $Z_2$: since this partitions $X$ into sets of size $\lfloor |X|/2 \rfloor$ and $\lceil |X|/2 \rceil$, we see that

$$f(X \cup \{y\}) \geq \left\lfloor \frac{|X|^2}{4} \right\rfloor + w(y, Z_2)$$

$$\geq \frac{1}{2}w(X) + \frac{|X|-1}{4} + \frac{1}{2}w(y, X) + \frac{1}{2}\left(w(y, Z_2) - w(y, Z_1)\right)$$

$$= \frac{1}{2}w\big(X \cup \{y\}\big) + \frac{|X|-1}{4} + \frac{w(y, Z_2) - w(y, Z_2)}{2}.$$

Hence, by Lemma 7 and (30),

$$f(G) \geq \frac{1}{2}w(G) + \frac{n - (c_1 + 2c_2)\sqrt{n} - 1}{4} + \frac{1}{2}\big(w(y, Z_2) - w(y, Z_1)\big)$$

and so, by (26),

$$(31) \qquad w(y, Z_2) - w(y, Z_1) \leq (2c_0 + c_2 + c_1/2)\sqrt{n} + 1 < c_3\sqrt{n}.$$

It follows, in particular, that for each $y \in Y$ there is an integer $t(y)$ such that all but at most $c_3\sqrt{n}$ of the edges between $y$ and $X$ have the same weight $t(y)$.

We have defined $t(v)$ for $v \in Y$; set $t(v) = 1$ for $v \in X$. Then $t(v)$ denotes the "typical" weight of edges incident with a vertex $v$. We could obtain a graph $H$ with the same vertex weights as $G$ from a complete graph on $\sum_{v \in V(G)} t(v)$ vertices by partitioning its vertices into sets $\{T_v : v \in V(G)\}$, where $|T_v| = t(v)$: contracting each set $T_v$ to a single vertex $v^*$ gives a graph in which all but $O(\sqrt{n})$ edges from each vertex $v^*$ have weight $t(v)$ (since all but $O(\sqrt{n})$ vertices $v^*$ in $H$ correspond to vertices $v$ in $X$ for which $t(v) = 1$). We shall show that in fact $G$ is not too far from $H$. Note that the edge in $H$ between vertices $v^*$ and $w^*$ has weight $t(v)t(w)$. With this in mind, for an edge $xy$ in $G$ we define

$$u(xy) = w(xy) - t(x)t(y).$$

Thus $u$ denotes the weight we have to add to each edge of $H$ in order to obtain the graph $G$.

We know that $u(e) = 0$ for every edge $e$ in $G[X]$. Suppose that for some $y \in Y$ we have

$$\sum_{x \in X} \left| u(xy) \right| \geq c_4 \sqrt{n}$$

where we define $c_4 = 2c_0 + c_2 + c_1/2 + 1$. Then since $w(xy) = t(y)$ for all but at most $c_3\sqrt{n}$ edges between $y$ and $X$, and $w(y, X) = \sum_{x \in X} \big( t(y) + u(xy) \big)$, we can partition $X$ into $Z_1 \cup Z_2$ as before (except that we order edges between $y$ and $X$ with increasing $u$-weight). The total weight (with weighting $w$) of edges between $Z_1 \cup \{y\}$ and $Z_2$ is then at least

$$\left\lfloor \frac{|X|^2}{4} \right\rfloor + \frac{w(y, X)}{2} + \frac{1}{2} \sum_{x \in X} \left| u(yx) \right| > \frac{w\big(X \cup \{y\}\big)}{2} + \frac{|X| - 1}{4} + \frac{c_4 n^{1/4}}{2}, \tag{32}$$

and hence by Lemma 7 and (30)

$$\begin{aligned} f(G) &\geq \frac{1}{2} w(G) + \frac{1}{4}\Big(n - (c_1 + 2c_2)\sqrt{n} - 1\Big) + \frac{1}{2} c_4 \sqrt{n} \\ &> \frac{m}{2} + \frac{n}{4} + c_0 \sqrt{n}, \end{aligned}$$

which contradicts (26).

Thus we may assume that, for every $y \in Y$,

$$\sum_{x \in X} \left| u(xy) \right| < c_4 \sqrt{n}. \tag{33}$$

Suppose that

$$\sum_{x\in X}\sum_{y\in Y}\left|u(xy)\right| > c_5 n^{3/4}$$

where $c_5 = 4c_4^{3/4}$. It follows from (33) that we can pick a subset $Y'$ of $Y$ such that

$$c_5 n^{3/4} < \sum_{x\in X}\sum_{y\in Y'}\left|u(xy)\right| < c_5 n^{3/4} + c_4\sqrt{n}. \tag{34}$$

Then the number of vertices $x \in X$ such that $u(xy) \neq 0$ for some $y \in Y'$ is at most $c_5 n^{3/4} + c_4\sqrt{n} < n/4$, provided $n$ is sufficiently large. We construct a partition of $Y' \cup X$ as follows. Let $Y' = Y_1 \cup Y_2$ be a random partition of $Y'$, where each $y \in Y'$ is in $Y_1$ or $Y_2$ independently with probability $1/2$. Let $X'$ be the set of vertices $x \in X$ such that $u(xy) \neq 0$ for some $y \in Y'$. Let

$$Z_1 = Y_1 \cup \left\{x \in X' \,:\, u(x, Y_1) < u(x, Y_2)\right\}$$

and

$$Z_2 = Y_2 \cup \left\{x \in X' \,:\, u(x, Y_2) \le u(x, Y_1)\right\}.$$

Finally, extend the partial partition $Z_1 \cup Z_2$ to a partition of $X \cup Y'$ by adding the remaining members of $X$ arbitrarily so that the final partition $W_1 \cup W_2$ satisfies

$$\sum_{v\in W_1} t(v) \le \sum_{v\in W_2} t(v) \le \sum_{v\in W_1} t(v) + 1. \tag{35}$$

(Note that this is possible since $Z_1 \cup Z_2$ contains at most $n/2$ elements of $X$, provided $n$ is sufficiently large.) Now by Lemma 9 below, for $x \in X'$ we have $\mathbb{E}\left|\, u(x, Y_1) - u(x, Y_2)\right| \ge \mathbb{E}\left(\sum_{i=1}^{U} \pm 1\right) \ge \frac{1}{2}\sqrt{U}$, where $U = \sum_{y\in Y'}\left|u(xy)\right|$.

So by (33), (34) and (35), the expected weight of edges joining $W_1$ and $W_2$ is at least

$$\mathbb{E}\Big( \big\lfloor t(X \cup Y')^2/4 \big\rfloor + u(Y')/2 + \sum_{x \in X} \max\big(u(x,Y_1), u(x,Y_2)\big) \Big)$$

$$\geq \frac{1}{2} w(X \cup Y') + \frac{|X|-1}{4} + \mathbb{E} \sum_{x \in X} \frac{\big|u(x,Y_1) - u(x,Y_2)\big|}{2}$$

$$\geq \frac{1}{2} w(X \cup Y') + \frac{|X|-1}{4} + \frac{1}{4} \sum_{x \in X} \Big( \sum_{y \in Y'} \big|u(xy)\big| \Big)^{1/2}$$

$$\geq \frac{1}{2} w(X \cup Y') + \frac{|X|-1}{4} + \frac{c_5 n^{3/4}}{4\sqrt{c_4} n^{1/4}}$$

$$\geq \frac{1}{2} w(X \cup Y') + \frac{|X|-1}{4} + c_4 \sqrt{n}.$$

As in (32), this yields a contradiction, so we may assume that

$$\sum_{x \in X} \sum_{y \in Y} \big|u(xy)\big| < c_5 n^{3/4}. \tag{36}$$

It follows that there are at most $c_5 n^{3/4}$ vertices of $X$ which are incident to an edge $e$ with $u(e) \neq 0$.

Let $G'$ be the graph with edges $\{e \in E(G) : u(e) \neq 0\}$ with edge-weighting $u$ and vertices $\{v \in V(G) : u(vw) \neq 0 \text{ for some } w \in V(G)\}$. Then, by (29) and (36),

$$|G'| \leq c_5 n^{3/4} + |Y| \leq c_6 n^{3/4}$$

which by (27) is smaller than $n/4$ for sufficiently large $n$.

Finally, let $W_1 \cup W_2$ be a partition of $G'$ such that the total weight of edges between $W_1$ and $W_2$ is at least $f(G')$. Since $|G'| < n/4$ provided $n$

is sufficiently large, it follows from (30) that we can extend $W_1 \cup W_2$ to a partition $V_1 \cup V_2$ of $V(G)$ such that

$$\left|t(V_1) - t(V_2)\right| \leq 1.$$

Let $t = \sum_{v \in V(G)} t(v)$. Then

$$\begin{aligned} w(G) &= \sum_{vw \in E(G)} t(v)t(w) + \sum_{vw \in E(G)} u(vw) \\ &\leq \binom{t}{2} + u(G'), \end{aligned}$$

while the weight of edges between $V_1$ and $V_2$ is at least

$$t(V_1)t(V_2) + u(V_1, V_2) \geq \left\lfloor \frac{t^2}{4} \right\rfloor + f(G').$$

Therefore

$$\begin{aligned} f(G) &\geq \left\lfloor \frac{t^2}{4} \right\rfloor + f(G') \\ &\geq \left\lfloor \frac{t^2}{4} \right\rfloor + f_w\left(w(G) - \binom{t}{2}\right). \qquad \blacksquare \end{aligned}$$

We have used an estimate in the proof above that is an immediate consequence of the following trivial lemma. We are interested in random sums $\sum \varepsilon_i a_i$, where the $a_i$ are independent Bernoulli random variables taking values $+1$ and $-1$ with probability $1/2$. We shall write $\pm$ instead of $\varepsilon_i$.

**Lemma 9.** *Let $s_1 + \cdots + s_k$ be a partition of $n$ and $t_1 + \cdots + t_l$ a refinement of $s_1 + \cdots + s_k$. Then*

$$\mathbb{E}\left|\sum_{i=1}^{k} \pm s_i\right| \geq \mathbb{E}\left|\sum_{i=1}^{l} \pm t_i\right|.$$

**Proof.** It is enough to consider the simple refinement when $l = k+1$, $s_i = t_i$ for $i < k$ and $s_k = t_k + t_{k+1}$. We may couple the sums $\sum \pm s_i$ and $\sum \pm t_i$ so that $\pm s_i$ and $\pm t_i$ have the same sign for $i < k$, while $\pm s_k$, $\pm t_k$ and $\pm t_{k+1}$ are independent. Let $S = \sum_{i=1}^{k-1} \pm s_i = \sum_{i=1}^{k-1} \pm t_i$. We must show

$$\mathbb{E}|S \pm s_k| \geq \mathbb{E}|S \pm t_k \pm t_{k+1}|.$$

Now for real numbers $\alpha \geq \beta \geq 0$ and $L \geq 0$,

$$\begin{aligned} |L+\alpha| + |L-\alpha| &\geq L + \alpha + |L-\alpha| \\ &\geq L + \alpha + |L-\beta| + \beta - \alpha \\ &= |L+\beta| + |L-\beta|, \end{aligned}$$

so in general for $|\alpha| \geq |\beta|$ and any $L$,

$$|L+\alpha| + |L-\alpha| \geq |L+\beta| + |L-\beta|.$$

Conditioning on the value of $S$, we see that since $|t_k - t_{k+1}| \leq |t_k + t_{k+1}|$, we have

$$\big|S + (t_k - t_{k+1})\big| + \big|S - (t_k - t_{k+1})\big| \leq |S + t_k + t_{k+1}| + |S - t_k - t_{k+1}|$$

and so

$$\mathbb{E}|S \pm t_k \pm t_{k+1}| \leq \frac{1}{2}|S + t_k + t_{k+1}| + \frac{1}{2}|S - t_k - t_{k+1}| = \mathbb{E}|S \pm s_k|.$$

The result follows immediately. ■

For what value of $n$ is the quantity in Theorem 8 minimized? Suppose $\binom{n_0}{2} \leq m < \binom{n_0+1}{2}$, say $m = \binom{n_0}{2} + r$. Clearly we must have $n \leq n_0 + 1$, and it follows from (30) that we may assume $n > n_0 - c\sqrt{n_0}$. We claim that (24) is minimized with $n = n_0$ or $n = n_0 + 1$. Now since the argument of Lemma 3 applies to multigraphs as well as graphs, we can deduce the Edwards formula for multigraphs:

$$f_w(m) \geq \frac{m}{2} + \sqrt{\frac{m}{8}} + O(1).$$

Since $f_w(m) \leq f(m)$, it follows from (6) that

$$f_w(m) = \frac{m}{2} + \sqrt{\frac{m}{8}} + O(m^{1/4}). \tag{37}$$

If $n = n_0 - t$, with $0 \leq t \leq c\sqrt{n}$, then

$$m - \binom{n}{2} = t(n_0 - t) + \binom{t}{2} + r$$

so by (37),

$$\left\lfloor \frac{n^2}{4} \right\rfloor + f_w\left(m - \binom{n}{2}\right)$$

$$= \frac{m}{2} + \frac{n}{4} + \left(\frac{t(n_0 - t) + \binom{t}{2} + r}{8}\right)^{1/2} + O(n_0 t + r + 1)^{1/4}.$$

Provided $n$ is sufficiently large, and $0 \leq t < c\sqrt{n_0}$, this is minimal when $t = 0$. We conclude the following.

**Theorem 10.** *For every sufficiently large positive integer $m$,*

$$f_w(m) = \min\left\{ \left\lfloor \frac{(n+1)^2}{4} \right\rfloor, \left\lfloor \frac{n^2}{4} \right\rfloor + f_w\left(m - \binom{n}{2}\right) \right\}, \tag{38}$$

*where $n$ is defined by*

$$\binom{n}{2} \leq m < \binom{n+1}{2}.$$

As remarked in the introduction, probably $f_w(m) = f(m)$ for every $m$. Even if this is not true, it seems likely that (38) holds with $f(m)$ in place of $f_w(m)$ when $m$ is sufficiently large.

## 4. Extremal graphs for Max Cut

We can apply Theorem 10 to obtain extremal graphs and multigraphs in more cases than Theorem 1 and Lemma 4. Let us note first that any integer $m$ can be written in the form

$$m = \binom{n_1}{2} + \binom{n_2}{2} + \cdots + \binom{n_k}{2}$$

for some $k > 0$, where $n_1 > \cdots > n_k \geq 2$ and each $n_i$ in turn is chosen to be as large as possible. For $1 \leq i < k$, let

$$M_i = \left\lfloor \frac{n_1^2}{4} \right\rfloor + \cdots + \left\lfloor \frac{n_{i-1}^2}{4} \right\rfloor + \left\lfloor \frac{(n_i+1)^2}{4} \right\rfloor$$

and define

$$M = \left\lfloor \frac{n_1^2}{4} \right\rfloor + \cdots + \left\lfloor \frac{n_k^2}{4} \right\rfloor$$

then it follows by repeated application of Theorem 10 that, provided $n_{k-1}$ is sufficiently large,

$$f_w(m) = \min\{M_1, \ldots, M_{k-1}, M\}.$$

For $1 \leq i < k$, we can obtain a graph $G$ with $m$ edges and $f(G) = M_i$ by deleting $\binom{n_i+1}{2} - \binom{n_i}{2} - \cdots - \binom{n_k}{2}$ edges from the graph

$$K_{n_1} \cup \cdots \cup K_{m_{i-1}} \cup K_{n_i+1}, \tag{39}$$

while the graph

$$K_{n_1} \cup \cdots \cup K_{n_k} \tag{40}$$

has $m$ edges and no bipartite subgraph with more than $M$ edges. Note that in both (39) and (40), we could instead take any edge-disjoint union of the complete graphs. Thus there may be many possible extremal graphs.

Recall that in the case $k = 2$, Theorem 1 asserts that for sufficiently large $m$, the extremal graphs are precisely the graphs (39) and (40) and their variants obtained by taking different edge-disjoint unions (note that the case $n_k = 4$ is special, since we can take two copies of $K_3$ instead of $K_4$).

Keeping the notation of the last few paragraphs, we can extend Theorem 10 for graphs as follows.

**Theorem 11.** *Let $m$ be a positive integer and define $k$, $n_1, \ldots, n_k$, and $M_1, \ldots, M_{k-1}, M$ as above. Suppose that*

$$M < \min\{M_1, \ldots, M_{k-1}\}. \tag{41}$$

*Then, provided $n_{k-1}$ is sufficiently large,*

$$f(m) = M$$

*and the extremal graphs are obtained by taking an edge-disjoint union of $K_{n_1}, \ldots, K_{n_k}$, unless $n_k = 4$, in which case there is an additional set of extremal graphs obtained by taking an edge-disjoint union of $K_{n_1}, \ldots, K_{n_{k-1}}, K_3, K_3$.*

**Proof.** We argue by induction on $k$. For $k = 1$, the result follows immediately from Lemma 4. For $k \geq 2$, we know from Theorem 10 and example (40) that $f(m) = f_w(m) = M$. Let $G$ be a graph with $m$ edges and $f(G) = f(m)$. As in the proof of Theorem 1, we can decompose $G$ as the edge-sum of $K_n$ and $H$, where $H$ is a weighted graph in which all edges are weighted $\pm 1$. Furthermore, any partition of $H$ can be extended to an optimal partition of $K_n$, so $f(G) = f(K_n) + f(H)$ and we must therefore have $n = n_1$ and so $w(H) = \binom{n_2}{2} + \cdots + \binom{n_k}{2}$. If $H$ has an edge $xy$ with negative weight, then contracting $xy$ gives a weighted graph $H'$ with

$$w(H') > \binom{n_2}{2} + \cdots + \binom{n_k}{2}.$$

It follows from Theorem 10 and the inductive hypothesis that

$$f(H') \geq \left\lfloor \frac{n_1^2}{4} \right\rfloor + \cdots + \left\lfloor \frac{n_k^2}{4} \right\rfloor + 1$$

and so, since $f(H) \geq f(H')$, we obtain $f(G) \geq f(K_{n_1}) + f(H) > M$, which is a contradiction.

Thus all edges of $H$ must have weight $+1$, so we can consider $H$ as an unweighted graph, with $\binom{n_2}{2} + \cdots + \binom{n_k}{2}$ edges. For $2 \leq i < k$, let

$$M_i' = \left\lfloor \frac{n_2^2}{4} \right\rfloor + \cdots + \left\lfloor \frac{n_{i-1}^2}{4} \right\rfloor + \left\lfloor \frac{(n_i+1)^2}{4} \right\rfloor$$

and let

$$M' = \left\lfloor \frac{n_2^2}{4} \right\rfloor + \cdots + \left\lfloor \frac{n_k^2}{4} \right\rfloor$$

it follows from (41) that

$$M' < \min\{M_2', \ldots, M_{k-1}'\}.$$

Thus we may apply the inductive hypothesis to $H$: the result follows immediately. ■

A similar argument gives the following result for weighted graphs.

**Theorem 12.** *Under the conditions of Theorem* 11,

$$f_w(m) = M$$

*and the extremal weighted graphs are the edge sums of $K_{n_1}, \ldots, K_{n_k}$, unless $n_k = 4$, in which case the edge sums of $K_{n_1}, \ldots, K_{n_{k-1}}, K_3, K_3$ are also extremal.*

**Proof.** We argue as in the proof of Theorem 11, except that we use the decomposition of Theorem 21 below. Note that in the decomposition $K_t^* \oplus H$ if $f(K_t^*) = \lfloor t^2/4 \rfloor$ then $K_t^*$ must not have been contracted ■

What happens when $M \geq \min\{M_1, \ldots, M_k\}$? We conjecture that the natural extension of Theorem 1 should hold: the extremal grahs are obtained by deleting edges in (39). The weighted case seems more complicated.

## Part II: Algorithms for Max Cut

### 5. An extremal algorithm for Max Cut

In this second part of the paper, we turn from extremal questions to the problem of finding polynomial time algorithms that give large cuts of a graph or edge-weighted graph. In this section we describe a linear time algorithm that, given a graph with total edge weight $m$, gives a bipartite subgraph of weight at least $f(m)$. In subsequent sections, we give a linear time algorithm that, for graphs $G$ of weight $m$, finds a cut of weight at least $m/2+\sqrt{m/8}+k$ if such a cut exists and otherwise finds an optimal cut, and an algorithm that approximates the order of magnitude of $f(G) - m/2 - \sqrt{m/8}$.

We remark that it often appears to be easier to find efficient algorithms for partitioning unweighted graphs than it is for partitioning weighted graphs. We shall assume below that we are dealing with graphs that have integer edge-weightings, where we allow both positive and negative weights. We may also assume that our graphs are connected: given a graph $G$ with $n$ vertices and $e$ edges, we can identify a vertex from each component in time $O(e + n)$ to obtain a graph $H$ with $f(G) = f(H)$; any bipartition of $H$ yields an equivalent bipartition of $G$ in time $O(n)$. (We should also note that we have assumed that all arithmetical operations can be performed in unit time, regardless of the magnitude of edge-weights.)

The main result of this section is the following linear time algorithm.

**Theorem 13.** *There is an algorithm that, given a graph $G$ with $e$ edges, integer-valued edge-weighting $w$ and total weight $m$ finds a cut of weight at least $f_w(m)$ in time $O\big(e+|G|\big)$.*

By taking $w \equiv 1$, we obtain the following immediate corollary for unweighted graphs.

**Corollary 14.** *There is an algorithm that, given a multigraph $G$ with $m$ edges and $n$ vertices, finds in time $O(m+n)$ a bipartite subgraph of $G$ with at least $f_w(m)$ edges.*

Many of the results from Sections 2 and 3 have efficient corresponding algorithms. Let us note first that Theorem 5 has the following immediate corollary.

**Lemma 15.** *There is an algorithm that, given a connected graph $G$ with $n$ vertices and $e$ edges, and an edge-weighting $w$ with positive integers and total weight $m$, finds in time $O(e+n)$ a bipartite subgraph of $G$ with weight at least*

$$\frac{w(G)}{2}+\frac{|G|-1}{4}.$$

We shall find it useful to have the following lemma.

**Lemma 16.** *There is an algorithm that, given a graph $G$ with $e$ edges and edge-weighting $w$, finds in time $O(e+n)$ a contraction of $G$ to a complete graph in which all edges have positive weight.*

**Proof.** Begin by deleting all edges with weight 0. We then take a greedy colouring of $G$ (which takes time $O(n+e)$) and contract each colour class to a single vertex to obtain a weighted complete graph $H$ with $|H| = O\big(\sqrt{e}\,\big)$. We now repeatedly contract edges of nonpositive weight until we obtain a graph with all edge weights positive. Since each contraction takes time $O\big(\sqrt{e}\,\big)$ and there are at most $O\big(\sqrt{e}\,\big)$ contractions, the algorithm terminates in time $O(e)$ (note that we can deal with all edges with nonpositive weight in time $O(e)$ by processing one vertex at a time). ■

We shall also need algorithmic versions of Lemmas 3, 4 and 7, and a result concerning weighted matchings. Most of the following lemma can be found in Hofmeister and Lefmann [19].

**Lemma 17.** *We consider graphs $G$ with $n$ vertices, $e$ edges and edge-weighting $w$.*

(i) *There is an algorithm running in time $O(e+n)$ that, given a matching $M$ in $G$, finds a cut with weight at least*

$$\frac{1}{2}w(G) + \frac{1}{2}w(M).$$

(ii) *There is an algorithm running in time $O(e+n)$ that, given a proper $k$-colouring of $G$, finds a cut with weight at least*

$$\left(\frac{1}{2} + \frac{1}{2k}\right) w(G).$$

*In particular, there is an $O(e+n)$ algorithm that finds a cut with weight at least*

$$\left(\frac{1}{2} + \frac{1}{2|G|}\right) w(G).$$

(iii) *Given a weighted graph $G$ and a partial partition $V_1 \cup V_2$ of $V(G)$, we can find in time $O(e+n)$ a cut of weight at least*

$$w(V_1, V_2) + \frac{1}{2}\big(w(G) - w(V_1 \cup V_2)\big).$$

(iv) *There is an algorithm running in time $O(e+n)$ that finds a matching of weight at least $w(G)/n$.*

**Proof.** Parts (i), (ii) and (iv) are obtained by applying algorithms from [19]. Note that if $|G|$ is odd, we may add in an isolated vertex.

Part (iii) follows by using the greedy algorithm: add each vertex of $V(G) \setminus (V_1 \cup V_2)$ in turn to whichever side of the partition gives the heavier cut. ■

We now prove the main result of this section.

**Proof of Theorem 13.** Let $m_0$ be large enough for (38) to apply: that is, for $m \geq m_0$, and $\binom{n}{2} \leq m < \binom{n+1}{2}$, $f_w(m) = \min\left\{ \lfloor (n+1)^2/4 \rfloor, \lfloor n^2/4 \rfloor + f_w\left(m - \binom{m}{2}\right) \right\}$. By (1) (for weighted graphs) and (6), there is $m_1$ such that, provided $m \geq m_1$,

$$\frac{m}{2} + \sqrt{\frac{m}{8}} - 1 \leq f_w(m) \leq \frac{m}{2} + \sqrt{\frac{m}{8}} + 2m^{1/4}. \tag{42}$$

The main part of our algorithm will apply to edge-weighted graphs with weight at least $M = \max\{m_0, m_1, K\}$, where $K$ is a large fixed constant; we deal separately with graphs of smaller weight.

We begin by contracting $G$ to a complete weighted graph with the algorithm from Lemma 16. We may thus assume that $G$ is a complete weighted graph with $e$ edges, all of positive weight, and that $w(G) = m$. We shall show that we can find a cut of weight at least $f_w(m)$ in time $O(e)$.

If $w(H) \leq M$, then since there are only finitely many graphs with positive edge weights and total weight at most $M$ (that is, multigraphs with at most $M$ edges), we can examine all partitions of $H$ in fixed time, or else store all optimal partitions as a look-up table. Note that this may introduce a large constant into the time or space complexity of the algorithm: we return to this point after the proof.

We may therefore assume that $G$ has weight $m \geq M$. Our algorithm follows parallel to the proof of Theorem 8. Note that if at any time we find a cut with weight at least $m/2 + \sqrt{m/8} + 2m^{1/4}$ then we can halt the algorithm immediately.

Define the integer $n$ by

$$\binom{n}{2} \leq m < \binom{n+1}{2}. \tag{43}$$

Then, by (42),

$$f_w(m) \leq \frac{m}{2} + \frac{n}{4} + 3\sqrt{n}. \tag{44}$$

Since $G$ is complete we have

$$|G| \leq n.$$

On the other hand, by Lemma 17(ii) we can find in time $O\big(e + |G|\big)$ a bipartite subgraph of $G$ with weight at least $\left(\frac{1}{2} + \frac{1}{2|G|}\right) m$. Thus we can halt the algorithm if

$$\text{(45)} \qquad \left(\frac{1}{2} + \frac{1}{2|G|}\right) m > \frac{m}{2} + \sqrt{\frac{m}{8}} + 2m^{1/4},$$

which is true for sufficiently large $m$ unless

$$\text{(46)} \qquad |G| > n - 8\sqrt{n}$$

Note that this implies $m$ is $O(e)$.

Now we find a small set of edges that meets all edges in $G$ of weight more than 1. We can find a maximal matching $M$ in $G$ in time $O(e)$, by choosing greedily edges of weight more than 1 and then filling out with edges of weight 1. By Lemma 17(i) we can find in time $O(e)$ a bipartite subgraph of $G$ with weight at least $\big(m + w(M)\big)/2$. Thus we are done if $w(M) > \sqrt{m/2} + 4m^{1/4}$. Otherwise, provided $m$ is sufficiently large, using (43) we see that $M$ contains at most

$$\text{(47)} \qquad \sqrt{\frac{m}{2}} + 4m^{1/4} - \left\lfloor \frac{|G|}{2} \right\rfloor < 7\sqrt{n}$$

edges of weight greater than 1. We obtain either a bipartite subgraph with weight at least $f_w(m)$, in which case we halt the algorithm, or else a set of at most $7\sqrt{n}$ edges, and hence a set $Y$ of at most $14\sqrt{n}$ vertices of $G$, meeting all edges with weight more than 1.

Let $X = V(G) \setminus Y$, so by (46) and (47),

$$|X| = |G| - |Y| \geq n - 22\sqrt{n}.$$

Note that $G[X]$ is a complete graph in which all edges have weight 1. For $y \in Y$, let us consider the edges between $y$ and $X$. As in the proof of

Theorem 8, we order the edges into increasing order of weight, which takes time $O(n \log n)$, which is $O(e)$, and partition $X \cup \{y\}$ into $Z_1 \cup \{y\}$ and $Z_2$, where $\{yz : z \in Z_1\}$ are the lightest $\lfloor |X|/2 \rfloor$ edges and $\{yz : z \in Z_2\}$ are the heaviest $\lceil |X|/2 \rceil$ edges. Extending to a partition of $V(G)$ with the algorithm of Lemma 17(iii), we obtain a bipartite subgraph of weight at least $w(G)/2 + \left(|X|-1\right)/4 + \left(w(y,Z_2) - w(y,Z_1)\right)/2$, which is at least

$$\frac{w(G)}{2} + \frac{n - 15\sqrt{n} - 1}{4} + \frac{w(y,Z_2) - w(y,Z_1)}{2}. \tag{48}$$

Either this is a bipartite subgraph of weight at least $f(m)$ or, by (44),

$$w(y,Z_2) - w(y,Z_1) < 15\sqrt{n}. \tag{49}$$

It follows that all but $15\sqrt{n}$ of the edges between $y$ and $X$ have the same weight, say $t(y)$. Note that it is easy to check that (49) is satisfied and to determine $t(y)$ for each $y \in Y$ in total time $O\left(|Y| n \log n\right)$ which is $O(e)$.

Setting $t(v) = 1$ for $v \in X$, we may assume that $t(v)$ has been defined for all $v \in V(G)$. For $x, y \in V(G)$, define $u(xy) = w(xy) - t(x)t(y)$. It follows from (49) that, for $y \in Y$,

$$\sum_{x \in X} \left| u(xy) \right| < 15\sqrt{n}.$$

Now it is straightforward to calculate $U := \sum_{x \in X, y \in Y} \left| u(x,y) \right|$ in time $O(e)$. If $U > 30n^{3/4}$ then we can find a set $Y' \subset Y$ with

$$30n^{3/4} \le \sum_{x \in X, y \in Y} \left| u(xy) \right| < 30n^{3/4} + 15\sqrt{n}$$

by choosing vertices one at a time from $Y$ until both inequalities are satisfied. Let $Z = \left\{ x \in X : u(xy) \neq 0 \text{ for some } y \in Y' \right\}$. Then $|Z| \le 30n^{3/4} + 15\sqrt{n}$. In the proof of Theorem 10, we used a random bipartition $Y' = Y_1 \cup Y_2$, and defined $Z_1 = \left\{ x \in Z : u(x, Y_1) \ge u(x, Y_2) \right\}$ and $Z_2 = Z \setminus Z_1$. Extending such a partition here to a partition $W_1 \cup W_2$ of $X \cup Y'$ satisfying (35), we get

$$(50)\quad \mathbb{E}\big( w(W_1, W_2)\big) \geq \frac{1}{2} w(X \cup Y') + \frac{|X|-1}{4} + \sum_{x \in X, y \in Y'} \big| u(x,y)\big| / 3n^{1/4}$$

$$> \frac{1}{2} w(X \cup Y') + \frac{|X|}{4} + 10\sqrt{n}.$$

It follows that as in the proof of Theorem 8 after (34) we can extend the partition as in (35) and obtain a cut of $G$ with weight at least

$$(51)\qquad \frac{1}{2} w(G) + 10\sqrt{n} + \frac{1}{4}|X| > \frac{1}{2} m + \frac{n}{4} + 3\sqrt{n}$$

which suffices by (44). Thus it is enough to find a bipartition of $X \cup Y'$ that does at least as well as (50). It follows from Lemma 18 below that we can do this in time $O\big(|Y'|^2 |U| + e\big)$ which by assumption is at most $c\big(\sqrt{n}\big)^2 n^{3/4} + e$, which is $O(e)$. (Note that we apply Lemma 18 with vertex sets $V_1 = Y'$, $V_2 = X$ and edge set $E(Y') \cup E(Y, X)$.)

Otherwise, $U \leq 30n^{3/4}$, so $\big\{ x \in X : u(xy) = 0 \text{ for all } y \in Y \big\}$ contains at least $n/2$ vertices, provided $U$ is sufficiently large. Thus we may consider $G$ as the edge sum of a graph $H_1$ with edge weights given by $u$ and a graph $H_2$ with edge weights given by $t$. Since $\big|X \setminus V(H_1)\big| > n/2$, any partition of $H_1$ can be extended to a partition of $V(G)$ in which the two halves have $t$-weight differing by at most 1. This corresponds to a bipartite subgraph of $H_2$ with weight at least $\big\lfloor t(G)^2/4 \big\rfloor$.

It is therefore enough to find a bipartite subgraph of $H_1$ with $u$-weight at least $f\big(u(H_1)\big)$, which we do recursively with the algorithm above. ■

In the proof above, we have used a lemma which we now give.

**Lemma 18.** *There is a polynomial time algorithm that, given a graph $G$ with $e$ edges, integer-valued edge-weighting $w$ and total weight $m$, and* *a bipartition $V(G) = V_1 \cup V_2$, where $V_2$ is independent, finds a bipartite subgraph with weight at least*

$$\frac{1}{2}w(G) + \sum_{v \in V_2} d_v^{1/2}/2 \tag{52}$$

*where $d_v = \sum_{u \in V_1} \left|w(vu)\right|$. If all edges have nonzero weight and $G$ contains no isolated vertices then the algorithm runs in time $O(|V_1|^2 D + e)$, where $D = \sum_{v \in V_2} d_v$.*

**Proof.** We obtain a bipartition of $V(G)$ by first partitioning $V_1$ as $W_1 \cup W_2$ and then adding each vertex $v$ of $V_2$ to $W_1$ if $w(v, W_1) \leq w(v, W_2)$ and to $W_2$ otherwise. If we take a random partition of $V_1$, then as in the proof of Theorem 8, we see that the expected weight of edges in the bipartite subgraph given by the resulting partition of $V(G)$ is at least (52).

We derandomize this as follows. First delete all edges with weight 0: this takes time $O(e)$. Now partition $V_1$ one vertex at a time. Suppose we have a partition $V_1 = W_1' \cup W_2' \cup W_3$, where $W_3$ is the set of vertices we have not yet assigned, and we wish to add the vertex $v \in W_3$ to $W_1'$ or $W_2'$. Consider a random bipartition $W_1 \cup W_2$ of $V_1$, chosen uniformly among all partitions that extend the partial partition $W_1' \cup W_2'$: define a random variable $X = w(W_1, W_2)$ and, for each $x \in V_2$, define $Y_1(x) = w(x, W_1)$ and $Y_2(x) = w(x, W_2)$. Then the bipartite subgraph obtained by greedily adding vertices in $V_2$ to $W_1$ or $W_2$ has weight

$$Z = X + \sum_{x \in V_2} \max\left\{Y_1(x), Y_2(x)\right\}.$$

We compare $\mathbb{E}(Z \,|\, v \in W_1)$ and $\mathbb{E}(Z \,|\, v \in W_2)$ explicitly. First of all, $\mathbb{E}(X \,|\, v \in W_1) - \mathbb{E}(X \,|\, v \in W_2) = \frac{1}{2}\left(w(v, W_2') - w(v, W_1')\right)$, which is easily calculated in time $O(|V_1|)$. For $x \in V_2$, the distribution of $w(x, W_1)$ is the result of a random walk with initial value $w(x, W_1')$ and (independent, equiprobable) increments $\left\{0, w(xy)\right\}$ for each edge $xy$ with $y \in V_1 \setminus (W_1' \cup W_2')$. Since the walk has integer values in $[-d_x, d_x]$ and there are at most $|V_1|$ steps, we can update the distribution after each step in time $O(d_x)$ and thus obtain the distribution of $w(x, W_1)$ in time $O(|V_1| d_x)$. Since $w(x, W_1) + w(x, W_2) = w(x, V_1)$, we obtain the joint distribution of $Y_1(x)$ and $Y_2(x)$ and hence the distribution of $\max\left\{Y_1(x), Y_2(x)\right\}$ in time $O(|V_1| d_x)$. Similar comments apply to the distributions conditioned on $v \in W_1$ or $v \in W_2$, so we can determine the conditional distribution of $\sum_{x \in V_2} \max\{Y_1(x), Y_2(x)\}$ in time $O(|V_1|D)$. We can therefore decide to assign each vertex to $W_1$ or $W_2$ so as to maximize the expectation of $Z$ in time $O(|V_1|D)$ and hence obtain a partition satisfying (52) in time $O(|V_1|^2 D)$. ∎

In Theorem 13 we may have an extremely large look-up table. This can be avoided by slightly weakening the results we expect from our algorithm. For graphs with large total weight we run the algorithm as before. However, when we reach the point where we want to bipartition a small subgraph we use an approximation algorithm instead. Thus we lose at most some fixed additive constant on the bound given in the theorem.

## 6. An algorithm for graphs without large cuts

A parametrized problem with parameter $k$ is said to be *fixed parameter tractable* (see [8]) if there is an algorithm running in time $O(f(k)N^t)$, where $N$ is the input size, $t$ is a constant and $f$ is any (not necessarily polynomial) function. Mahajan and Raman [29] show that the problem of finding a cut of size at least $\lceil m/2 \rceil + k$ if one exists in a graph with $m$ edges and $n$ vertices is fixed parameter tractable and give algorithms running in time $O(m2^{4k} + n^3)$ and $O(2^{ck^2} + m + n)$, and deduce that if $k \in O(\log mn)$ then there is a polynomial time algorithm. We remark that the first algorithm relies on finding a cut of weight at least $m/2 + (n-1)/4$; since Ngoc and Tuza [30] (see also Lemma 2 above) have found an $O(m+n)$ algorithm, the first algorithm of Mahajan and Raman can be improved to $O(m2^{4k} + n)$.

Note that by the bound (1) of Edwards, every graph of weight $m$ satisfies $f(G) \geq m/2 + \sqrt{m/8} + O(1)$, while we know that $f_w(m) \leq m/2 + \sqrt{m/8} + O(m^{1/4})$. The algorithms of Mahajan and Raman for finding cuts of size at least $\lceil m/2 \rceil + k$ exploit the fact that $f(m) - \lceil m/2 \rceil$ grows as fast as $c\sqrt{m}$: for instance, if $k < (n-1)/4 \leq \sqrt{m/8} + O(1)$ we can use the Edwards result (1), whereas if $k \geq (n-1)/4$ we can examine all partitions explicitly. Our aim in this section is to show that the problem of finding a cut of size at least $m/2 + \sqrt{m/8} + k$, if such a cut exists, is fixed-parameter tractable, and give an algorithm runing in time $O(2^{ck^4} + e + n)$.

We begin with an extension of Lemma 15 for graphs with arbitrary non-zero integer edge-weightings.

**Lemma 19.** *There is an algorithm running in time $O\big(e+|H|\big)$ that given a connected graph $H$ with $e$ edges and an edge-weighting $w$ with non-zero integers, finds a cut of weight at least*

$$\frac{w(H)}{2}+\frac{|H|-1}{4}.$$

**Proof.** As in the proof of Theorem 5 we can find a collection $\mathcal{S}$ of vertex-disjoint induced stars containing at least $\big(|H|-1\big)/2$ edges. We partition the edges of $\mathcal{S}$ into $\mathcal{S}^+$ and $\mathcal{S}^-$, the edges of positive and negative weight respectively. Contracting every edge in $\mathcal{S}^-$ gives a graph of weight $w(H)-w(\mathcal{S}^-)$, in which the edges of $\mathcal{S}^+$ form a collection of induced stars. We take each star $S$ in turn: if $|S|>2$ then we identify its endvertices to obtain an edge of weight $w(S)$. We end up with a matching of weight $w(\mathcal{S}^+)$, and the algorithm of Lemma 17(i) gives a cut of weight $\frac{1}{2}\big(w(H)-w(\mathcal{S}^-)\big)+\frac{1}{2}w(\mathcal{S}^+)\geq w(H)/2+\big(|H|-1\big)/4$. ■

We will also need an extension of the Edwards bound (1) to weighted graphs.

**Theorem 20.** *There is an algorithm running in time $O\big(e+|H|\big)$ that given a graph $H$ with $e$ edges and integer-valued edge-weighting $w$, finds a cut of weight at least*

$$\frac{w(H)}{2}+\sqrt{\frac{U}{8}+\frac{1}{64}}-\frac{1}{8}, \tag{53}$$

*where $U=\sum_{e\in E(H)}\big|w(e)\big|$.*

**Proof.** Let $h=\sqrt{U/8+1/64}-1/8$. We begin by deleting edges of weight 0. We may assume $H$ is connected or else identify a vertex from each component. If $|H|-1>4h$ then Lemma 19 provides a cut of weight at least $w(H)/2+\big(|H|-1\big)/4\geq w(H)/2+h$ as required. Otherwise, $|H|-1\leq 4h$. Let $u$ be the weighting on $E(H)$ defined by $u(e)=\big|w(e)\big|$, so $u(H)=U$. Then the algorithm of Lemma 17(iv) gives a matching $M$

with $u(M) \geq U/|H|$. Let $M^+$ be the edges of $M$ with positive weight and $M^-$ the edges with negative weight. As in the proof of Lemma 19, we contract the edges of $M^-$ and apply the algorithm of Lemma 17(i) to $M^+$ (with edge-weighting $w$) to obtain a cut of weight at least $\frac{1}{2}\big(w(H) - w(M^-)\big) + \frac{1}{2}w(M^+) \geq w(H)/2 + U/2|H| \geq w(H)/2 + U/(8h+2)$, and a simple calculation shows $U/(8h+2) = h$. ∎

Our main tool will be the following decomposition result.

**Theorem 21.** *Let $c > 0$ be a constant. There is an algorithm running in time $O\big(e + |G|\big)$ that takes as input a graph $G$ with $e$ edges, integer-valued edge-weighting $w$ and total weight $m$ and gives as output either a cut of weight at least $m/2 + \sqrt{m/8} + cm^{1/4}$ or a decomposition of $G$ as an edge sum*

$$G \equiv K_t^* \oplus H$$

*where $K_t^*$ is obtained from a complete graph $K_t$ of order $t = n + O(1)$ with all edges of weight 1 by contracting at most $O(m^{1/4})$ edges and $H$ is a graph such that $\big|V(H) \cap V(K_t^*)\big| = O(n^{3/4})$. Here the integer $n$ is defined by $\binom{n}{2} \leq m < \binom{n+1}{2}$.*

**Proof.** The algorithm in the proof of Theorem 13 began by contracting $G$ to a complete graph with positive edge-weights. This worked since the total weight was not reduced, and we were looking for a cut of weight at least $f_w(m)$. Here, however, we must be more careful: we may be looking for a cut of weight more than $f_w(m)$, and such a cut could be destroyed by any compression of $G$. For instance, the path $P_4$ with four vertices and all edges with weight 1 has $f(P_4) = 3$, while its contraction to $C_3$, all edges with weight 1, has $f(C_3) = 2$.

We may clearly delete all edges of weight 0. Identifying one vertex from each component, we may assume $G$ is connected. Define, as usual, the integer $n$ by $\binom{n}{2} \leq m < \binom{n+1}{2}$. Let $M$ be large enough so that, if $m \geq M$, (38) holds, and

$$\frac{m}{2} + \sqrt{\frac{m}{8}} - 1 \leq f_w(m) \leq \frac{m}{2} + \sqrt{\frac{m}{8}} + 2m^{1/4} \leq \frac{m}{2} + \frac{n}{4} + 3\sqrt{n}. \tag{54}$$

Suppose first that $m \le M$. Let $U = \sum_{e\in E(G)} \left|w(e)\right|$. If $U > 4(8c^2 + 1)M$ then Theorem 21 provides the required partition. Thus we may assume that $U \le 4(8c^2+1)M$, and since $G$ has at most $U$ edges we can examine all partitions explicitly in constant time.

Otherwise, we have $w(G) = m > M$. Note first that we may assume

$$|G| > n - 12c\sqrt{n}. \tag{55}$$

Otherwise, applying the algorithm of Lemma 17(ii), as in (45) we obtain a cut of weight at least $m/2 + \sqrt{m/8} + cm^{1/4}$, provided $m$ is sufficiently large. For a similar reason, we may assume $m$ is $O(e)$: otherwise, we may apply Lemma 16 to obtain a contraction of $G$ that fails (55).

On the other hand, the algorithm of Lemma 17(i) gives a bipartite subgraph of weight at least $w(G)/2 + \big(|G|-1\big)/4$ and so we can halt the algorithm unless

$$|G| < n + 14c\sqrt{n}. \tag{56}$$

Now we show that we may assume that $G$ is nearly complete. We run the algorithm in Lemma 16 on $G$ with edge-weighting $u$ defined by $u(e) = \max\big\{\, w(e), 0\big\}$ for $e \in E(G)$, and obtain a contraction to a weighted-complete graph $H$. If $H$ satisfies

$$|H| \le n - 12c\sqrt{n} \tag{57}$$

then applying the same contraction to $G$ with edge-weighting $w$ gives a graph with weight at least $w(G)$ satisfying (55), which we partition with the algorithm of Lemma 17. Otherwise let $W^-$ be the set of vertices of $G$ which are identified with some other vertex in the contraction from $G$ to $H$. Then $Y^-$ meets all edges of $G$ with weight at most 0, and

$$|Y^-| \le 2\big(|G| - |H|\big) \le 52c\sqrt{n}. \tag{58}$$

Let $Y^+ = V(G) \setminus Y^-$. Then $G[Y^+]$ is a complete graph in which all edges have positive weight, and it follows from (55) and (58) that

$$|Y^+| > n - 64c\sqrt{n}. \tag{59}$$

We claim that $G[Y^+]$ cannot contain too many edges with weight greater than 1. We choose greedily a maximal matching $e_1, \ldots, e_t$ in $G[Y^+]$, where at each stage we choose an edge of weight more than 1 if possible, and otherwise any edge. Using the algorithm of Lemma 17(i) we obtain in time $O(e)$ a cut of weight at least $\frac{1}{2}w(Y^+) + \frac{1}{2}\sum_{i=1}^{t} w(e_i)$. By Lemma 17(iii), we can extend this to a cut of $G$ of weight at least $\frac{1}{2}w(G) + \frac{1}{2}\sum_{i=1}^{t} w(e_i)$. It follows from (54) that either we can halt the algorithm or $\sum_{i=1}^{t} w(e_i) \le n/2 + 6\sqrt{n}$. By (59), $\{e_1, \ldots, e_t\}$ contains fewer than $40c\sqrt{n}$ edges of weight more than 1, and since they were chosen greedily these edges meet every edge with weight greater than 1. Therefore there is a set $Z$ of vertices that meets every edge in $G[Y^+]$ with weight greater than 1, where

$$|Z| \le 80c\sqrt{n}. \tag{60}$$

Let $Y = Y^- \cup Z$ and $X = V(G) \setminus Y$. Then by (58) and (60),

$$|Y| \le 144c\sqrt{n}$$

while by (55),

$$|X| \ge n - 156c\sqrt{n}. \tag{61}$$

The rest of the algorithm is broadly similar to the algorithm in the proof of Theorem 13. Note that $G[X]$ is a complete graph in which all edges have weight 1. As before, for each $y \in Y$ we partition $X$ into sets $Z_1$ and $Z_2$ according to the weight of edges from $Y$, and then extend to a partition of $G$. As in (48), we obtain a cut of weight at least

$$\frac{w(G)}{2} + \frac{|X|-1}{4} + \frac{w(y,Z_2) - w(y,Z_1)}{2}$$

so by (61) we are done unless

$$w(y,Z_2) - w(y,Z_1) < 85c\sqrt{n}.$$

It follows that all but at most $85c\sqrt{n}$ edges between $y$ and $X$ have the same weight $t(y)$. As before, let $t(x) = 1$ for $x \in X$, and for $x, y \in V(G)$ define

$$u(xy) = w(xy) - t(x)t(y).$$

Once again, we calculate $U = \sum_{x\in X, y\in Y} \left|u(xy)\right|$. If $U/cn^{3/4}$ is sufficiently large then as in (50) and (51) we can find a bipartite subgraph of weight at least $m/2 + \sqrt{m/8} + cm^{1/4}$.

Otherwise, $U$ is $O(n^{3/4})$, so provided $m$ is sufficiently large we have $U < n/4$ and we may decompose $G$ as the edge sum of a graph $H$ with edge weights given by $u$ and a graph $K_t^*$ with edge weights given by $t$. Finally, note that $|K_t^*| = n + O(\sqrt{n}) = n + O(m^{1/4})$, while a simple calculation shows that if $f(G) \le m/2 + \sqrt{m/8} + cm^{1/4}$ then $t = n + O(1)$; note that $K_t^*$ is obtained from $K_t$ by a sequence of at most $O(m^{1/4})$ contractions. ■

Note that in the decomposition $G = K_t^* \oplus H$ of Theorem 21 we have $f(G) = f(K_t^*) + f(H)$. Furthermore, it is straightforward to find an optimal partition of $K_t^*$ by splitting into two classes with $t$-weight as equal as possible.

We can now prove the main theorem of this section.

**Theorem 22.** *There is an algorithm running in time $O(2^{ck^4} + e + n)$ that, given a weighted graph $G$ with $e$ edges, edge-weighting $w$ and total weight $m$, and an integer $k$, finds a cut of $G$ with weight $f(G)$ if $f(G) \le m/2 + \sqrt{m/8} + k$ and otherwise a cut with weight at least $m/2 + \sqrt{m/8} + k$.*

**Proof.** If $\sum_{e\in E(G)} \left|w(e)\right| > 4(m + 8k^2)$ then we are done by Theorem 20. If $k > m^{1/4}$ then we examine all partitions explicitly. Otherwise, we apply Theorem 21 with $c = 1$: either we obtain a cut of weight at least $m/2 + \sqrt{m/8} + k$, or we obtain a decomposition $G = K_t^* \oplus H$ where $f(G) = f(K_t^*) + f(H)$ and $K_t^*$, $H$ have edge-weightings $w_1$, $w_2$ respectively. We can calculate $f(K_t^*)$ exactly; so if $w_2(H) > 0$, define $l$ by $m/2 + \sqrt{m/8} + k = f(K_t^*) + w_2(H)/2 + \sqrt{w_2(H)/8} + l$, and repeat the algorithm with $H$ and $l$ (note that $l \le k$). If $w_2(H) \le 0$ then define an edge-weighting $u$ by $u(e) = \left|w_2(e)\right|$ for $e \in E(H)$. If $\left|u(H)\right| > 8k^2$ then Theorem 20 gives a cut of weight at least $w_2(H)/2 + k$, which gives a cut of $G$ of weight at least $w_1(K_t^*)/2 + \sqrt{w_1(K_t^*)/8} + w_2(H)/2 + k \ge m/2 + \sqrt{m/8} + k$; otherwise, $\left|u(H)\right| < 8k^2$ and we can examine all partitions explicitly. ■

Theorem 22 gives a polynomial time algorithm for $k \le c(\log n)^{1/4}$. However, we cannot expect to go as high as $n^{\varepsilon}$ for any $\varepsilon > 0$, since then considering graphs of form $K_N \cup G$, where $N > |G|^{2/\varepsilon}$, would give a polynomial time algorithm for Max Cut. In particular, we cannot expect to find optimal cuts for all graphs with $f(G) \le m/2 + \sqrt{m/8} + cm^{1/4}$. On the other hand, we have given an algorithm that finds a cut of weight at least $f_w(m)$, which can be as large as $m/2 + \sqrt{m/8} + cm^{1/4}$. We are led naturally to the following problem.

**Problem 1.** *Let $k > 0$ be a fixed integer. Is there a polynomial time algorithm that given a graph $G$ of weight $m$ finds either a cut of weight at least $f_w(m) + k$ or else an optimal cut?*

Note that $k = 0$ is Theorem 13. However, if $m$ satisfies $m = \binom{n+1}{2} - n^{\varepsilon}$ and $f_w(m) = f(K_{n+1}) = \lfloor (n+1)^2/4 \rfloor$, for instance, then we may have $G = K_{n+1} \oplus H$, where $H$ has weight $-n^{\varepsilon}$. Theorem 21 will find this decomposition, and $H$ has a trivial cut of weight 0, but determining whether $f(H) \ge k$ is probably NP-hard.

## 7. A weak approximation algorithm for graphs that are close to extremal

In this section we concentrate on the existence of polynomial time algorithms for estimating the value of $f(G) - m/2 - \sqrt{m/8}$ for a graph of weight $m > 0$. Let us note first that we cannot expect an arbitrarily good approximation algorithm since this would immediately yield a good approximation algorithm for the Max Cut problem. For since $f(G) \ge m/2$, an algorithm approximating $f(G) - m/2 - \sqrt{m/8}$ within a factor $1 + \varepsilon$ gives $f(G)$ to within a factor $1 + \varepsilon/2$. Since Håstad [18] has shown that it is NP-hard to approximate $f(G)$ within any factor less than 17/16, it follows that it is NP-hard to approximate $f(G) - m/2 - \sqrt{m/8}$ to within any factor less than 9/8. However, we have the following weak approximation.

**Theorem 23.** *There is an algorithm running in time $O(e + n)$ that, given a graph $G$ with $e$ edges, edge-weighting $w$ and total weight $m$, either finds* *an optimal partition or gives a real number $\alpha$ such that*

$$\frac{m}{2} + \sqrt{\frac{m}{8}} + m^{\alpha} \leq f(G) \leq \frac{m}{2} + \sqrt{\frac{m}{8}} + m^{4\alpha}.$$

**Proof.** Let $U = \sum_{e \in E(G)} \left| w(e) \right|$. We may assume $U > U_0$, for some large constant $U_0$, or else examine all partitions explicitly. If $U > 4m$ then Theorem 20 gives a cut of weight at least $m/2 + \sqrt{U/8} - 1 > m/2 + \sqrt{m/8} + U^{1/4}$, so we can choose $\alpha = (\log_m U)/4$. Otherwise, applying the algorithm of Theorem 22 with $c = 4$ gives either a cut of weight at least $m/2 + \sqrt{m/8} + 4m^{1/4}$, so we can choose $\alpha$ such that $m^{\alpha} = 4m^{1/4}$ (note that $f(G) \leq U$), or else a decomposition of $G$ as $K_t^* \oplus H$, where $H$ has weighting $w'$.

If $w'(H) \leq 0$ then let $U' = \sum_{e \in E(H)} \left| w'(e) \right|$. If $U' < U_0$ we determine $f(H)$, and hence also $f(G) = f(K_t^*) + f(H)$, explicitly. Otherwise, since we have $f(G) \leq f(K_t^*) + U'$ and $f(G) = f(K_t^*) + f(H) \geq m/2 + \sqrt{m/8} + \sqrt{U'/8} - 2$, we may pick $\alpha$ such that $m/2 + \sqrt{m/8} + m^{\alpha} = f(K_t^*) + \sqrt{U'/8} - 2$.

Otherwise $w'(H) \geq 0$, so repeating the algorithm we can find $\beta$ such that

$$\frac{w'(H)}{2} + \sqrt{\frac{w'(H)}{8}} + w'(H)^{\beta} \leq f(H) \leq \frac{w'(H)}{2} + \sqrt{\frac{w'(H)}{8}} + w'(H)^{4\beta}.$$

Then we may pick $\alpha$ such that

$$\frac{m}{2} + \sqrt{\frac{m}{8}} + m^{\alpha} = f(K_t^*) + \frac{w'(H)}{2} + \sqrt{\frac{w'(H)}{8}} + w'(H)^{\beta}.$$

■

Compare the problem of approximating $f(G) - m/2 - \sqrt{m/8}$ with the problem of approximating $f(G)$. Since $f(G) \geq m/2$ for every graph $G$ with weight $m$, there is a "cushion" of weight $m/2$ in measuring the effectiveness of approximation algorithms for $f(G)$. For instance, the trivial decision algorithm that always returns the value $m/2$ and the greedy algorithm that achieves this bound manage to approximate within a factor 2 (improving beyond this is a different matter, however, and the 1.1383-approximation algorithm of Goemans and Williamson [17] is a tour de force). This contrasts with other approximation problems, such as Chromatic Number or Independent Set, where no such trivial approximation algorithm exists.

The problem of approximating $f(G) - m/2 - \sqrt{m/8}$ therefore seems more difficult than that of approximating $f(G)$.

**Problem 2.** *Is there a polynomial time algorithm that approximates* $f(G) - m/2 - \sqrt{m/8}$ *within a constant factor for graphs of weight* $m$*? What about* $f(G) - f_w(m)$*?*

Slightly easier, perhaps, is the following.

**Problem 3.** *Is there a polynomial time algorithm that approximates* $f(G) - m/2$ *within a constant factor for graphs of weight* $m$*?*

Note that the graphs that concern us here are graphs with $f(G) \leq 0.63w(G)$. For graphs with $f(G) > 0.63w(G)$, the algorithm of Goemans and Williamson gives an approximation algorithm for $f(G) - m/2 - \sqrt{m/8}$.

## Part III: Related problems

### 8. The Max $k$-Cut problem

In the first two parts of this paper we have concentrated entirely on the Max Cut problem. In this section, we give a brief account of related results for Max $k$-Cut. For an edge-weighted graph $G$ we write $f_k(G)$ for the maximal weight of a $k$-cut; we consider an unweighted graph as having weight 1 on every edge, so $f_k(G)$ is the maximal size of a $k$-partite subgraph. We define $f_k(m)$ to be the minimum of $f_k(G)$ over graphs with integer edge-weightings and total weight $m$.

We begin with an analogue of the Edwards bound (1) for $k$-partite graphs. Note first that if $n = rk + s$, where $0 \leq s < k$, then writing

$m = \binom{n}{2}$,

$$(62)\qquad f_k(K_n) = \binom{n}{2} - s\binom{r+1}{2} - (k-s)\binom{r}{2}$$

$$= \left(1 - \frac{1}{k}\right)\binom{n}{2} + \frac{r(k-1)}{2} + \frac{s(s-1)}{2k}$$

$$= \left(1 - \frac{1}{k}\right)\binom{n}{2} + \frac{k-1}{2k}n - \frac{s(k-s)}{2k}$$

$$\geq \left(1 - \frac{1}{k}\right)m + \frac{k-1}{2k}n - \frac{k}{8}$$

$$= \left(1 - \frac{1}{k}\right)m + \frac{k-1}{2k}\sqrt{2m + \frac{1}{4}} + \frac{k^2 - 2k + 2}{8},$$

with equality if and only if $k$ is even and $s = k/2$.

**Theorem 24.** *Let $G$ be a weighted graph with total weight $m$. Then*

$$(63)\qquad f_k(G) \geq \left(1 - \frac{1}{k}\right)m + \frac{k-1}{2k}\sqrt{2m + \frac{1}{4}} + \frac{k^2 - 2k + 2}{8}$$

*If $m \geq \binom{n}{2}$ then*

$$f_k(G) \geq f_k(K_n)$$

*where for $m > m_0$ we have equality if and only if $G$ is a copy of $K_n$, where all edges have weight 1.*

**Proof.** Note first that, writing $g(m)$ for the right hand side of (63), it is easily checked that $g(m)$ is monotonic increasing, while $g(m)/m$ is monotonic decreasing for $m > 0$. We consider $G$ as a weighted complete graph. Contracting any edge of weight at most 0 we obtain a weighted complete graph $H$ in which all edges have positive weight; furthermore, any partition of $H$ corresponds to a partition of $G$ with the same cut-weight. It is therefore enough to prove the theorem for weighted complete graphs in which all edges have positive weight.

Now let $r = |H|$ and $R = \binom{r}{2}$, so $w(H) \geq R$. A random partition of $H$ into $k$ pieces, of size as equal as possible, gives a subgraph with expected weight

$$\frac{f_k(K_r)}{R}w(H) \geq \frac{g(R)}{R}w(H) \geq g\big(w(H)\big)$$

since $g(R)/R \geq g\big(w(H)\big)/w(H)$. Thus there must be some $k$-partite subgraph with weight at least $g\big(w(H)\big) \geq g(m)$.

If $m \geq \binom{n}{2}$, $f_k(G) = f_k(K_n)$ and $G$ either has an edge of negative weight or is not isomorphic to $K_n$ with all edge-weights 1, then compressing all edges of negative weight (and possibly some further edges), we may assume that $|G|$ is a complete weighted graph on fewer than $n$ vertices, with all edges of positive weight and $w(G) \geq \binom{n}{2}$. Now since $f(K_n)/\binom{n}{2}$ is monotonic decreasing and $f(K_{n-k})/\binom{n-k}{2} > f(K_n)/\binom{n}{2}$, consider random partitions into $k$ sets of size $\lfloor |G|/k \rfloor$ and $\lceil |G|/k \rceil$. We see that $|G| > n-k$ and so $G$ can be decomposed as the edge sum of $K_{|G|}$ and a weighted graph with at most $(k-1)n$ edges. Provided $m$ is sufficiently large, it is now straightforward to construct two $k$-cuts (into sets of size $\lfloor |G|/k \rfloor$ and $\lceil |G|/k \rceil$ ) with different weights, and so some $k$-cut must have weight more than $w(G)f\left(K_{|G|}\right)/\binom{|G|}{2} \geq f(K_n)$. ∎

Ngoc and Tuza [30] prove that, for $k \geq 3$, every connected graph $G$ with order $n$, size $m$ and $\chi(G) > k$ satisfies

$$f_k(G) \geq \left(1 - \frac{1}{k}\right)m + \frac{1}{k}(n-1) + \frac{k-3}{2}. \tag{64}$$

Furthermore, they give an $O(m)$ algorithm that achieves this quickly. They note that every graph obtained by attaching trees to the vertices of a copy of $K_k$ is extremal.

Let us note that if $G$ has a vertex of degree at most $k-1$ then

$$f_k(G) = f_k(G \setminus v) + d_G(v)$$

since deleting $v$ from $G$ and partitioning the rest of the graph optimally, we can then add $v$ to a class in which it has no neighbours. Thus we can restrict our attention to graphs $G$ with $\delta(G) \geq k$. If we allow the minimal degree to grow, then we can improve on (64).

**Theorem 25.** *Let $G$ be a graph with $\delta(G) \geq \omega(n)$, where $\omega(n) \to \infty$ as $n \to \infty$. Then*

$$f_k(G) \geq \left(1 - \frac{1}{k}\right) m + \frac{k-1}{2k} n + o(n).$$

**Proof.** (Sketch) Take a random ordering and run the greedy algorithm. If $x$ has $rk+s$ predecessors, where $0 \leq s \leq k-1$, then the most even partition of earlier vertices has $s$ big classes and $k-s$ small classes. Thus we can pick a class in which $x$ has at most $rk$ neighbours, increasing the cutweight by at least $r(k-1)+s = \left(1-\frac{1}{k}\right)(rk+s) + s/k$. Since $s$ is asymptotically uniformly distributed on $\{0, \ldots, k-1\}$, the expectation of $s/k$ tends to $(k-1)/2$. Summing over all vertices gives the bound above. ■

The proof of Theorem 25 works in the same way as the proof of Lemma 2: we order the vertices and then partition greedily. One method for obtaining a good ordering was given immediately after Lemma 2, and it is natural to ask whether the same approach (taking an ordering in which every vertex except one has at least one predecessor and then successively modifying the order to obtain better orders) would work for larger $k$. In fact, the method does work for $k = 3$, but for $k > 3$ appears to give a poor bound. It would be interesting to know under what conditions the $o(n)$ term can be removed in Theorem 25. For instance, a connectivity condition may be enough.

**Problem 4.** *Does every $(k-1)$-connected graph $G$ satisfy*

$$f_k(G) \geq \frac{k-1}{k} m + \frac{k-1}{2k} n + O(1).$$

A best possible bound of this form would be $f_k(G) \geq \frac{k-1}{k} m + \frac{k-1}{2k} n - \frac{r(k-r)}{2k}$. A similar question arises if we assume $\delta(G) \geq k$. It would also be interesting to find efficient algorithms giving bounds of this form.

The results of earlier sections mostly carry over to the $k$-partite case.

**Theorem 26.** *Let $G$ be a graph with edge-weighting $w$ and total weight $w(G) = m$. Then provided $m$ is sufficiently large*

$$f_k(G) \geq \min_{n \geq 0} \left\{ f_k(K_n) + f_k\left(m - \binom{n}{2}\right)\right\}.$$

As in the proof of Theorem 1, we need several lemmas. An analogue to Lemma 2 is given by Theorem 25, while using a greedy algorithm as in the proof of Lemma 7 we see that, for any graph $G$ with edge-weighting $w$ and any set $W \subset V(G)$, writing $H = G[W]$,

$$f_k(G) \geq f_k(H) + \frac{k-1}{k}(w(G) - w(H)) \tag{65}$$

A version of Lemma 4 is given by Theorem 24; this leaves Lemma 3. We have the following straightforward result.

**Lemma 27.** *For a nonempty edge-weighted graph $G$ with total weight $m$,*

$$f_k(G) \geq \frac{k-1}{k}\left(1 + \frac{1}{\chi(G)}\right) m.$$

**Proof.** Fix a colouring of $G$ with $c = \chi(G)$ colours and, as in Lemma 3, take a random partition of the colour classes into $k$ sets of as equal size as possible. The expected size of the cut obtained is at least

$$\frac{f(K_c)}{\binom{c}{2}} m \geq \frac{k-1}{k}\left(1 + \frac{1}{c}\right) m.$$

■

Finally, we remark that as in the bipartite case there is $c_k > 0$ such that, for all $m$,

$$f_k(m) \leq \frac{k-1}{k} m + \frac{k-1}{2k}\sqrt{2m} + c_k m^{1/4}. \tag{66}$$

We can now proceed to the proof of the theorem.

**Proof of Theorem 26 (Sketch).** Let $G$ be a counterexample with smallest weight. As in the proof of Theorem 8, we may contract edges of weight at most 0, and assume that $G$ is a weighted complete graph with all edges of positive weight. Define $n$ as in (25), so $|G| \leq n$. Now it follows from Lemma 27 that

$$f_k(G) \geq \frac{k-1}{k}\left(1+\frac{1}{|G|}\right)m.$$

Therefore, by (66),

$$\frac{k-1}{k}\frac{m}{|G|} \leq \frac{k-1}{2k}\sqrt{2m} + c_k m^{1/4}$$

and so

$$|G| \geq n - c_1\sqrt{n}. \tag{67}$$

Now let $C_1, \ldots, C_r$ be any collection of pairwise vertex-disjoint complete subgraphs of $G$, each with $k$ vertices. We consider a random partition $V(G) = V_1 \cup \ldots \cup V_k$ as follows. Each $C_i$ is randomly partitioned with one vertex in each vertex class $V_j$, and the remaining vertices are independently assigned to each vertex class with equal probability. Then the expected weight of the resulting $k$-partite subgraph is

$$\sum_{i=1}^{r} w(C_i) + \frac{k-1}{k}\left(w(G) - \sum_{i=1}^{r} w(C_i)\right) = \frac{k-1}{k}w(G) + \frac{1}{k}\sum_{i=1}^{r} w(C_i).$$

Thus it follows from (66) that

$$\sum_{i=1}^{r} w(C_i) \leq \frac{1}{2}(k-1)n + c_2\sqrt{n}. \tag{68}$$

Now if we greedily choose $C_1, C_2, \ldots$ to be vertex disjoint complete subgraphs with $k$ vertices and maximal weight, it follows from (68) and (67) that at most $c_3\sqrt{n}$ of these subgraphs have weight more than $\binom{k}{2}$. Thus there is a set $Y$ of $c_3k\sqrt{n}$ vertices meeting all edges with weight more than 1. Setting $X = V(G) \setminus Y$, we may assume that

$$|Y| \leq c_4\sqrt{n}$$

and

$$|X| \geq n - c_4\sqrt{n}.$$

Note that $G[X]$ is a complete graph in which all edges have weight 1.

Now for $y \in Y$, we can order the edges between $y$ and $X$ in increasing order of weight. Partitioning $X$ into $k$ pieces and using (65) we see that, as in (31), all but at most $c_5\sqrt{n}$ of the edges between $y$ and $X$ have the same weight, say $t(y)$.

As before, we set $t(y) = 1$ for $y \in X$, and define

$$u(xy) = w(xy) - t(x)t(y).$$

the remainder of the proof of Theorem 8 goes through essentially unchanged, except with $k$ vertex classes instead of 2. ■

We note that it should be straightforward to generalize the algorithms of Sections 5, 6 and 7 to the $k$-partite case.

## 9. Maximum Cuts in directed graphs

In this final section, we make a few remarks concerning the analogue of Max Cut for directed graphs. For a directed graph $H$ with edge-weighting $w$ and a subset $S \subset V(H)$, we define

$$w(S, V \setminus S) = \sum_{x \in S} \sum_{y \in V \setminus S} w(xy)$$

where $w(xy) = 0$ if $xy \notin E(H)$. Then

$$g(H) = \max_{S \subset V(H)} w(S, V \setminus S).$$

For $m \geq 1$ we define $g(m)$ to be the maximum of $g(H)$ over all directed graphs whose edges are weighted with non-negative integers and have total weight $m$. It is easy to see that

$$g(m) \geq \lceil f(m)/2 \rceil .$$

Indeed, given a directed graph $H$, let $G$ be the underlying (undirected) graph, where we define

$$w_G(xy) = w(xy) + w(yx).$$

Then $G$ has total weight $m$, and therefore a cut of weight at least $f(m)$. Let $V(G) = V_1 \cup V_2$ be such a cut: then

$$w_H(V_1, V_2) + w_H(V_2, V_1) = w_G(V_1, V_2) \geq f(m)$$

so one of $w_H(V_1, V_2)$ and $w_H(V_2, V_1)$ is at least $\lceil f(m)/2 \rceil$.

**Lemma 28.** *If* $m = \binom{2n+1}{2}$ *then* $g(m) = \binom{n+1}{2} = f(m)/2$.

**Proof.** It is enough to construct a directed graph $H$ with $w(H) = m$ and $g(H) \leq f(m)/2$. We define $H$ by taking the vertices of $\mathbb{Z}_{2n+1}$, and adding edges from $i$ to $i + j$, for any $i$ and $1 \leq j \leq n$. Thus $H$ is a regular oriented tournament, every vertex has indegree and outdegree $n$. Consider an arbitrary subset $S \subset V(H)$ with $|S| = h$. Clearly

$$g\big(S, V(H) \setminus S\big) = \sum_{v \in S} d^+(v) - e(S) \tag{69}$$

$$= nk - \binom{k}{2}$$

This is maximal when $k = n$ or $k = n + 1$, when

$$g(S, V(H) \setminus S) = n^2 - \binom{n}{2} = \binom{n+1}{2}. \qquad \blacksquare$$

Let us determine the extremal graphs for this result. Suppose $H$ is a directed graph with $w(H) = m = \binom{2n+1}{2}$ and $g(H) = \binom{n+1}{2}$. Let $G$ be the underlying multigraph: then $f(G) \leq n(n+1) = f(m)$. So by Lemma 4, $G$ is a copy of $K_{2n+1}$ in which all edges have weight 1. Therefore $H$ is a directed tournament of order $2n + 1$. If $H$ is not regular, then pick $n$ vertices with as large an outdegree as possible. Since their average outdegree must be greater than $n$ it follows that we do better than (69). Since we must have equality in $n$, $H$ must be a regular tournament. However, it follows from (69) that any regular tournament is extremal. Similar results follow when $m$ is of form $\binom{2n+1}{2} + \binom{2k+1}{2}$, for $n > k$ sufficiently large, and so on, using Theorem 1 and Theorem 12.

What about if $m = \binom{2n}{2}$? The closest we have to a regular tournament in this case is a tournament with $2n$ vertices in which $n$ vertices have outdegree $n - 1$ and $n$ vertices have outdegree $n$. If we take a subset $S$ of $k \leq n$ vertices, then

$$w(S, V \setminus S) = \sum_{x \in S} d^+(x) - \binom{k}{2} \leq kn - \binom{k}{2}.$$

This can be attained by taking the vertices with outdegree $n$ and is then maximal when $k = n$, and

$$w(S, V \setminus S) = n^2 - \binom{n}{2} = \binom{n+1}{2}.$$

For $k \geq n + 1$, picking $k$ vertices with outdegrees as large as possible, we have

$$w(S, V \setminus S) = n^2 + (k - n)(n - 1) - \binom{k}{2}.$$

This is maximal when $k = n + 1$, and

$$w(S, V \setminus S) = n^2 + n - 1 - \binom{n+1}{2} = \binom{n+1}{2} - 1.$$

Thus if $H$ is a tournament of this form then

$$g(H) = \binom{n+1}{2}.$$

Clearly this is optimal among tournaments. However, if $m = \binom{2n}{2}$ then $f(m) = n^2$, so $g(H) = f(m)/2 + n/2$.

It would be interesting to determine the behaviour of $g(m) - f(m)/2$ for arbitrary $m$. It seems likely that, with more work, results similar to those proved above for Max Cut could be proved for the directed problem.

## References


[1] N. Alon, Bipartite Subgraphs, *Combinatorica,* **16** (1996), 301–311.

[2] N. Alon and E. Halperin, Bipartite subgraphs of integer-weighted graphs, *Discrete Math.,* **181** (1998), 19–29.

[3] L. D. Anderson, D. D. Grant and N. Linial, Extremal $k$-colourable subgraphs, *Ars Combinatoria,* **16** (1983), 259–270.

[4] S. Arora, D. Karger and M. Karpinski, Polynomial time approximation schemes for dense instances of NP-hard problems, in: *Proc. 27th ACM STOC* (1995), 284–293.

[5] H. L. Bodlaender and K. Jansen, On the complexity of the Maximum Cut problem, in: *G. Goos and J. Hartmanis, eds, 11th Annual Symposium on Theoretical Aspects of Computer Science, LNCS 775,* 769–780.

[6] B. Bollobás and A. D. Scott, Exact bounds for judicious partitions of graphs, *Combinatorica,* **19** (1999), 473–486.

[7] B. Bollobás and A. D. Scott, Problems on judicious partitions, to appear.

[8] R. G. Downey and M. R. Fellows, Fixed parameter tractability and completeness I: Basic theory, *SIAM J. Comput.,* **24** (1995), 873–921.

[9] C. S. Edwards, Some extremal properties of bipartite subgraphs, *Canadian J. Math.*, **25** (1973), 475–485.

[10] C. S. Edwards, An improved lower bound for the number of edges in a largest bipartite subgraph, Proc 2nd Czech. Symposium on Graph Theory, Prague (1975), 167–181.

[11] P. Erdős, Some recent problems in Combinatorics and Graph Theory, Proc. 26th Southeastern International Conference on Graph Theory, Combinatorics and Computing, Boca Raton, 1995, Congressus Numerantium, to appear.

[12] P. Erdős, A. Gyárfás and Y. Kohayakawa, The size of the largest bipartite subgraphs, *Discrete Maths.*, **177** (1997), 267–271.

[13] W. Fernandez-de-la-Vega and M. Karpinski, Polynomial time approximability of dense weighted instances of Max Cut, Research report no 85171-CS, University of Bonn, 1997.

[14] W. Fernandez de la Vega, Max-cut has a randomized approximation scheme in dense graphs, *Random Structures and Algorithms,* **8** (1996), 187–198.

[15] A. Frieze and M. Jerrum, Improved approximation algorithms for MAX $K$-CUT and MAX BISECTION, *Algorithmica,* **18** (1997), 67–81.

[16] M. R. Garey, D. S. Johnson and L. Stockmeyer, Some simplified NP-complete graph problems, *Theoretical Computer Science,* **1** (1976), 237–267.

[17] M. X. Goemans and D. P. Williamson, 0.878 approximation algorithms for MAX CUT and MAX 2-SAT, Proc. 26th ACM Symp. Theory of Comput., 1994, 422–431; see also Improved approximation algorithms for maximum cut and satisfiability problems using semidefinite programming, *J. ACM,* **42** (1995), 1115–1145.

[18] J. Håstad, Some optimal inapproximability results, in: *Proc. 29th ACM Symp. on Theory of Comp.* (1995), 1–10.

[19] T. Hofmeister and H. Lefmann, A combinatorial design approach to Max Cut, *Random Structures and algorithms,* **9** (1996), 163–175.

[20] T. Hofmeister and H. Lefmann, On $k$-partite subgraphs, *Ars Combinatoria,* **50** (1998), 303–308.

[21] D. S. Johnson, The NP-completeness column: an ongoing guide, *J. Algorithms,* **6** (1985), 434–451.

[22] Y. Kajitani, J. D. Cho and M. Sarrafzadeh, New approximation results on graph matching and realted problems, Proc. 20th Workshop on Graph Theoretic Concepts in Computer Science, LNCS 903, 1994, 343–358.

[23] V. Kann, S. Khanna, J. Lagergren, A. Panconesi, On the Hardness of Approximating Max k-Cut and Its Dual, 1997, Article 2, *Chicago J. Theoretical Comp. Sci.*

[24] R. M. Karp, Reducibility among combinatorial problems, in: *R. E. Miller and J. W. Thatcher, eds, Complexity of Computer Computations,* Putnam Press (New York, 1972), 85–103.

[25] M. Karpinski, Polynomial time approximation schemes for some dense instances of NP-hard optimization problems, in: *Randomization and approximation techniques in computer science* (Bologna 1997), Lecture Notes in Computer Science **1269**, 1–14.

[26] M. Laurent, Max-Cut problem, in: *Annotated bibliographies in combinatorial optimization,* M. Dell'Amico, F. Maffioli and S. Martello, eds, Wiley, 1997, 241–259.

[27] J. Lehel and Zs. Tuza, Triangle-free partial graphs and edge-covering theorems. *Discrete Math.,* **30** (1982), 59–63.

[28] S. C. Locke, Maximum $k$-colourable subgraphs, *J. Graph Theory* **6** (1982), 123–132.

[29] M. Mahajan and V. Raman, Parametrizing above guaranteed values: MaxSat and MaxCut, *J. Algorithms* **31** (1999), 335–354.

[30] Nguyen Van Ngoc and Zs. Tuza, Linear time approximation algorithms for the Max Cut problem, *Combinatorics, Probability and Computing* **2** (1993), 201–210.

[31] S. Poljak and D. Turzik, A polynomial time heuristic for certain subgraph problems with guaranteed worst case bound, *Discrete Math.,* **58** (1986), 99–104.

[32] S. Poljak and Zs. Tuza, Maximum cuts and largest bipartite subgraphs, in: *Combinatorial optimization, DIMACS Ser. in Discrete Mathematics and Theoretical Computer Science,* volume 20, W. Cook, L. Lovász and P. Seymour, eds, 1995, xi+4 41pp., 181–244.

[33] S. Poljak and D. Turzik, A polynomial algorithm for constructing a large bipartite subgraph, with an application to a satisfiability problem, *Canadian J. Math.,* **34** (1982), 519–524.

[34] H. Saran and V. V. Vazirani, Finding $k$-cuts within twice the optimal, in: *Proc. 32nd Annual IEEE Symposium on Foundations of Computer Science* (1991).

[35] Zs. Tuza, Maximum cuts: improvements and local algorithmic analogues of the Edwards-Erdős inequality, *Discrete Math.,* **194** (1999), 39–58.

B. Bollobás
*Trinity College*
*Cambridge CB2 1TQ*

and

*Department of Mathematical Sciences*
*University of Memphis*
*Memphis TN38152*

e-mail:
bollobas@ibex.msci.memphis.edu

A. D. Scott
*Department of Mathematics*
*University College London*
*Gower Street*
*London WC1E 6BT*

e-mail: scott@math.ucl.ac.uk

and

*Department of Mathematical Sciences*
*University of Memphis*
*Memphis TN38152*

BOLYAI SOCIETY
MATHEMATICAL STUDIES, 10

**Contemporary Combinatorics**
**Budapest, 2002,** pp. 247–276.

# Random Colorings of a Cayley Tree

G. R. BRIGHTWELL and P. WINKLER

Probability measures on the space of proper colorings of a Cayley tree (that is, an infinite regular connected graph with no cycles) are of interest not only in combinatorics but also in statistical physics, as states of the antiferromagnetic Potts model at zero temperature, on the "Bethe lattice".

We concentrate on a particularly nice class of such measures which remain invariant under parity-preserving automorphisms of the tree. Making use of a correspondence with branching random walks on certain bipartite graphs, we determine when more than one such measure exists. The case of "uniform" measures is particularly interesting, and as it turns out, plays a special role.

Some of the results herein are deducible from previous work of the authors and by members of the statistical physics community, but many are new. We hope that this work will serve as a helpful glimpse into the rapidly expanding intersection of combinatorics and statistical physics.

## 1. Introduction

### 1.1. Uniform Proper Colorings

The *Cayley tree* $T_r$ is uniquely defined by being connected, $r+1$-regular and cycle-free; thus $T_1$ is the two-way infinite path and $T_2$ the complete binary tree. A (proper) $q$-coloring of $T_r$ can be thought of as a homomorphism $\varphi$ from $T_r$ to the complete graph $K_q$ on $q$ nodes, which we label $\{1, 2, \ldots, q\}$. (We call the nodes of $T_r$ "sites" in accordance with physics tradition and to distinguish them from the nodes of the target graph $K_q$.)

What would it mean to choose such a coloring $\varphi$ *uniformly* at random? If we were instead coloring a finite graph, there would be no difficulty in selecting $\varphi$ with equal probability from the finitely many possibilities. Even on $T_r$, if the coloring were not constrained to be proper, uniformity would be easily achievable by choosing the color of each site uniformly and independently.

In the case at hand, however, we are reduced to agreeing on what properties a "uniform" probability measure on the space of colorings should have, and then asking whether there is a unique measure with these properties.

We will begin by listing every property we can think of that makes sense for the infinite tree $T_r$ and holds for the uniform distribution on colorings of any finite graph. Letting $\mu$ be a measure on proper colorings of $T_r$ by $q$ colors, we demand:

1. $\mu$ is a *Gibbs measure.* This means that for any finite subtree $U$ of $T_r$, and any proper coloring $\psi$ of $U$, the probability that $\varphi$ agrees with $\psi$ on $U$, given that $\varphi = \psi$ on the boundary of $U$, is the same for $\varphi$ drawn from $\mu$ as it is for $\varphi$ drawn uniformly from the set of all proper colorings of $U$. (The boundary $\partial U$ of $U$ is the set of those points in $U$ which are adjacent to some point of $T_r \setminus U$.) For much more information about Gibbs measures, in far more general settings, the reader is referred to the excellent texts by Baxter and Georgii [1, 9].

2. $\mu$ is *simple.* This means that if we draw $\varphi$ from $\mu$, conditioned on any particular site $u$ being colored by any fixed color $j$, then the restrictions of $\varphi$ to each connected component of $T_r \setminus \{u\}$ are independent.

3. $\mu$ is *invariant.* This means that for any automorphism $\theta$ of $T_r$, $\mu \circ \theta = \mu$; in other words, if $A$ is any measurable set of colorings, then $\mu\big(\{\varphi \circ \theta : \varphi \in A\}\big) = \mu(A)$.

In statistical physics, the Cayley tree $T_r$ is frequently called a "Bethe lattice". Coloring by $q$ colors with symmetric interactions produces the "$q$-state Potts model"; the model is "antiferromagnetic" if adjacent sites are discouraged (and "at zero temperature", forbidden) from having the same color. The Bethe lattice is not a statistical physicist's first choice as a setting for theorems on model behavior — a lattice such as a cubic grid, in some Euclidean space, is preferred — but it has yielded many useful insights. Although variational methods used on more "amenable" lattices do not always work on the Bethe lattice (see e.g. [6]), the absence of cycles permits direct methods for analysis such as those we employ below.

## 1.2. Constraint Graphs and Activities

We now generalize the setting in two ways. First, let us replace the target graph $K_q$ by an arbitrary, finite "constraint graph" $H$. We can think of $H$ as representing a generalized set of coloring constraints, namely that adjacent sites of $T_r$ may receive colors $i$ and $j$ if and only if $i$ and $j$ are adjacent in $H$; if $H$ has a loop at node $i$ then color $i$ may be used on both adjacent sites. Such a coloring will be called a *coloring by H* when confusion with ordinary proper coloring is threatened. The use of $H$ is intended to model physical systems with hard constraints.

There are constraint graphs other than $K_q$ of natural combinatorial interest, for instance the graph $J$ consisting of nodes 0 and 1 with an edge connecting the two nodes and a loop at node 0. In any $J$-coloring of $T_r$ the sites colored "1" form an independent set, and conversely. We are thus talking now about random independent sets instead of random proper colorings, and there is again a name for this model in statistical physics: the (discrete) *hard-core lattice gas model,* or "hard-core model" for short. The intuition is that the sites colored "1" are occupied by a gas molecule, and forces between molecules prevent occupation of two neighboring sites. (An elegant and accessible treatment of the discrete hard-core model can be found in [2].)

The second generalization replaces the uniform distribution with a "multiplicative" distribution, defined by positive reals $\lambda_1, \dots, \lambda_n$ associated with the nodes of $H$. The *activity* $\lambda_i$ of node $i$ reflects the relative preference given to color $i$; in particular, if the graph $G$ to be colored by $H$ is finite, the probability of a particular coloring $\varphi$ is proportional to

$$\prod_{u \in G} \lambda_{\varphi(u)}.$$

This implies in particular that if the colors of the neighbors of a site $u$ are fixed in such a way that colors $j_1, \dots, j_k$ are permissible at $u$, then the conditional probability that $u$ is colored $j_1$ is $\lambda_{j_1}/(\lambda_{j_1} + \cdots + \lambda_{j_k})$. This condition then applies also when $G$ is infinite, as a special case of the Gibbs condition.

To see that the multiplicative measures really are a natural extension of the uniform, consider for example the constraint graph which consists of a path on three nodes with a loop at the center node. Coloring uniformly by this $H$ is equivalent to selecting an independent set and then coloring the nodes of that set by two colors; the distribution thus induced on uncolored independent sets is the same as that given by coloring by $J$, with activities satisfying $\lambda_1/\lambda_0 = 2$.

For the hard-core model, activity can be thought of as the result of applying pressure (positive or negative) to the gas, and is often called "fugacity".

For the Potts model, the activity of any single color is controlled by an "external field". The effect of the field, however, is understood by physicists to become infinite when the temperature reaches 0; thus non-uniform activity sets fall outside their framework at zero temperature. The result is that many of the questions we attack, motivated by our combinatorial point of view, are not explored e.g. in the analysis of the antiferromagnetic Potts model on the Bethe lattice found in Peruggi et al. [14, 15, 16].

A physicist might even argue that there is nothing special at all about the uniform distribution, for example in the hard-core model. However, if the constraint graph is symmetric then the uniform distribution (where all the activities are equal) preserves this symmetry; and in fact we will see that in critical cases, the uniform set of activities stands apart from all others.

### 1.3. Semi-invariance

In much more general settings than ours, a result of Dobrushin [7] states that there is always at least one Gibbs measure for a given "specification". The problem of determining when there is more than one such Gibbs measure is a central problem of statistical mechanics, often called the *DLR problem* after Dobrushin, Lanford and Ruelle (see e.g. [17]); when more than one Gibbs measure exists, there is said (by some) to be a *phase transition*.

The DLR problem for uniformly random colorings of $T_r$ can be put entirely in finite combinatorial terms. Let $T_r^n$ be an $n$-level $r$-branching tree, and choose a proper $q$-coloring of $T_r^n$ uniformly at random. Let $p_n$ be the maximum probability that the root is colored "red" given the colors of the leaves; then $p_n \to 1/q$ if and only if there is a unique uniform Gibbs measure for $q$-colorings of $T_r$. Roughly speaking, if there is a phase transition, then there is "long-range order" meaning that the color of a site can be influenced by the colors of sites arbitrarily far away. We give more details in Section 3.4.

We are particularly interested in Gibbs measures satisfying the nice conditions given in Section 1.1, but we would like sometimes to relax one of our conditions slightly as follows. A measure is said to be *semi-invariant* if it is invariant under *even* automorphisms of $T_r$, i.e. those automorphisms which preserve the parity of a site. (For specificity, we choose a root $w$ of $T_r$ which is deemed to be of parity 0 mod 2, the parity of all other points being equivalent mod 2 to their distance to $w$.)

The hard-core model on $T_r$ has been extensively studied (see e.g. [10, 9, 4, 3]); for any $r$ and any $\lambda := \lambda_1/\lambda_0$, there is a unique simple invariant Gibbs measure. However, for $\lambda > r^r/(r-1)^{r+1}$ there are two additional simple semi-invariant Gibbs measures, one with most of its independent set on even sites, the other mostly on odd.

We will see that the situation for proper colorings is similar in two respects: there is always a unique simple invariant Gibbs measure, and sometimes multiple simple semi-invariant Gibbs measures. For instance, in the case $q = r + 1$ ($q$-coloring the $q$-regular tree), it will turn out that there is a unique simple semi-invariant Gibbs measure *only* when $\lambda_1 = \lambda_2 = \cdots = \lambda_q$.

Our results come from a very useful correspondence between invariant (and semi-invariant) simple Gibbs measures, and node-weighted branching random walks on a constraint graph.

## 2. Gibbs Measures

### 2.1. Homomorphisms

As noted $T_r$ will always stand for the $(r+1)$-regular Cayley tree and $H$ for a finite constraint graph, possibly with loops. If $G$ is any finite or infinite graph (the "board") we denote the space of homomorphisms from $G$ to $H$ by $\mathrm{Hom}\,(G,H)$. Adjacency between sites of $G$ or nodes of $H$ is denoted by $\sim$, so that $\varphi \in \mathrm{Hom}\,(G,H)$ and $u \sim v$ in $G$ imply $\varphi(u) \sim \varphi(v)$ in $H$.

Assume that a positive real activity $\lambda_i$ is associated with each node $i$ of $H$. If the board $G$ is finite, we denote the resulting multiplicative probability distribution on $\mathrm{Hom}\,(G,H)$ by $m_G$; thus

$$m_G(\{\psi\}) := \frac{\prod_{u\in G}\lambda_{\psi(u)}}{\sum_{\varphi\in \mathrm{Hom}\,(G,H)}\prod_{u\in G}\lambda_{\varphi(u)}}.$$

When the board is infinite, there is a bit more work to be done in determining when a measure on $\mathrm{Hom}\,(G,H)$ meets the "specifications" given by the set of activities on $H$. To begin with, a finite set $U$ of sites of $G$ will be called a *patch* and we will deliberately confuse $U$ with the subgraph of $G$ induced by $U$. As before, the *boundary* $\partial U$ of $U$ will be the set of sites in $U$ which are adjacent to at least one site of $G\setminus U$.

If $U$ is a patch and $\varphi\in\mathrm{Hom}\,(G,H)$, we denote by $\varphi\restriction U$ the restriction of $\varphi$ to $U$; thus $\varphi\restriction U\in\mathrm{Hom}\,(U,H)$. If $A$ is an event of the form

$$A=\{\varphi\in\mathrm{Hom}\,(G,H)\ :\ \varphi\restriction U\in F\}$$

for some patch $U$ and some $F\subset\mathrm{Hom}\,(U,H)$, then we call $A$ a "patch event". We equip $\mathrm{Hom}\,(G,H)$ with the $\sigma$-field (denoted by $\mathcal{F}$) generated by the patch events, and consider henceforth only measures $\mu$ on $\langle\ \mathrm{Hom}\,(G,H),\mathcal{F}\rangle$ such that $\mu(\ \mathrm{Hom}\,(G,H))\ =1$.

Let $U_1\subset U_2\subset\cdots$ be a nested sequence of patches whose union is $G$. We can define a compact topology on the space $\mathcal{M}$ of all probability measures on $\mathrm{Hom}\,(G,H)$ via the metric $\rho$ defined as follows:

$$\rho(\mu,\nu)=\sum_{i=1}^{\infty}2^{-i}\|\mu\restriction U_i,\nu\restriction U_i\|$$

where $\mu\restriction U$ is the measure induced by $\mu$ on $\mathrm{Hom}\,(U,H)$, and $\|\cdot\|$ is the total variation metric between probability distributions (a.k.a. half the $\ell^1$ distance). Of course this topology is not sensitive to the choice of the $U_i$'s, but it is useful to fix them anyway.

A measure $\mu$ on $\mathrm{Hom}(G,H)$ is said to be a *Gibbs measure* if for any patch $U \subset G$, and almost every $\psi \in \mathrm{Hom}(G,H)$,

$$\Pr_{\mu}\left(\varphi \restriction U = \psi \restriction U \,|\, \varphi \restriction ((G \setminus U) \cup \partial U) = \psi \restriction ((G \setminus U) \cup \partial U)\right)$$

$$= \Pr_{m_U}\left(\varphi \restriction U = \psi \restriction U \,|\, \varphi \restriction \partial U = \psi \restriction \partial U\right).$$

It suffices to verify the Gibbs condition for the $U_i$'s in our nested sequence; e.g. in the case $G = T_r$, for $U_i$ equal to the set of sites at distance at most $i$ from the root.

It may be possible for a measure to satisfy the Gibbs condition in a trivial and somewhat unsatisfactory way. For example, suppose we are $q$-coloring $T_r$ (with root $w$) for some $q \le r+1$, and let $\psi$ be any fixed coloring in which the children of every node exhibit all colors other than the color of the parent. Let $\mu$ be the measure which assigns probability 1 to $\psi$. Then the colors of $\psi$ on any $\partial U_i$ force the colors on $\partial U_{i-1}$ and so on all the way to the root, thus matching $m_U$ regardless of the activities. Thus $\mu$ is a ("frozen") Gibbs measure, and is also vacuously simple — but not invariant or semi-invariant. A frozen state of $\mathrm{Hom}(T_2, K_3)$ is illustrated in Fig. 1; for more about frozen Gibbs measures the reader is referred to [5]. (Note that frozen Gibbs measures can only occur when adjacent sites are absolutely forbidden to have the same color; that is, appropriately, at zero temperature. This accounts for their absence from most of the statistical physics literature.)

Normally the activities affect the definition of Gibbs measure via their tacit role in $m_U$. The remaining properties of measures that we use — simplicity, invariance and semi-invariance — are defined in the previous section for the case $G = T_r$ (which is all we shall need) and do not depend on the activities.

## 2.2. Simple Invariant Gibbs Measures

Simple invariant Gibbs measures on $\mathrm{Hom}(T_r, H)$ for arbitrary $H$ were studied in [4], and the results can be applied directly to the case $H = K_q$. We need only this half of the main theorem of [4]:

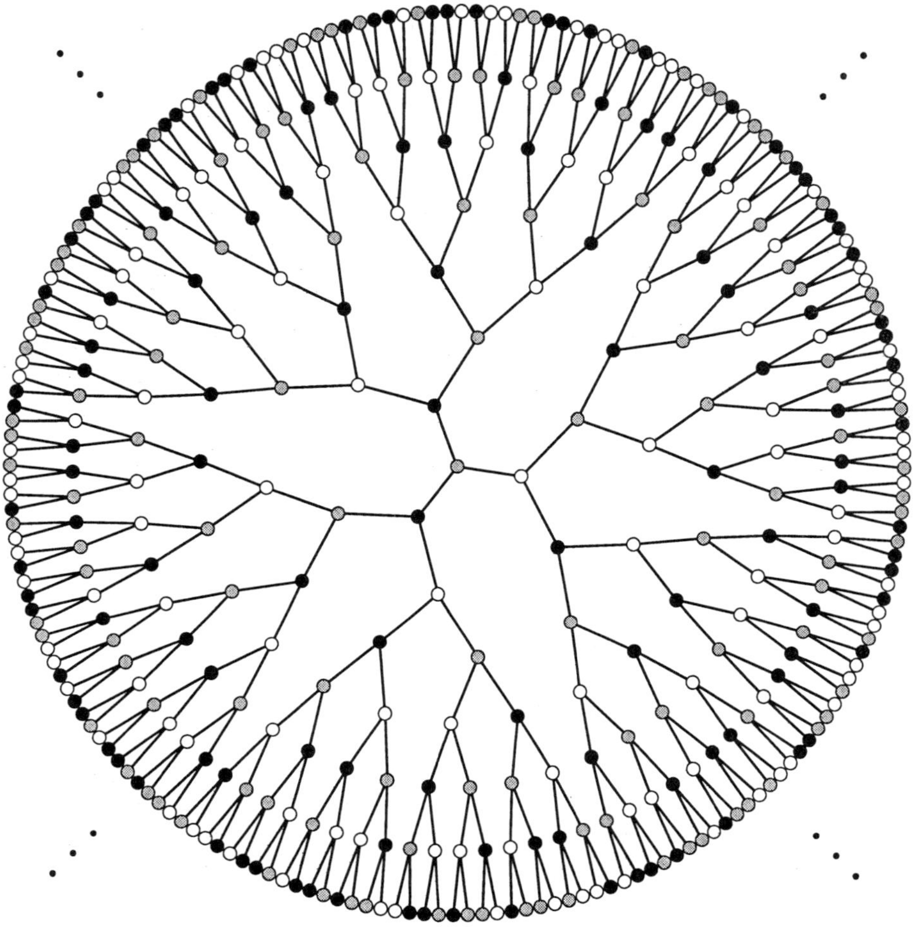

*Fig. 1.* A frozen 3-coloring of the complete binary tree

**Theorem 2.1.** *Let $H$ be a connected graph which satisfies the following two conditions:*

(a) *Every looped node of $H$ is adjacent to all other nodes of $H$;*

(b) *With its loops deleted, $H$ is a complete multipartite graph.*

*Then for every set of activities on $H$, and every $r \geq 1$, there is a unique simple invariant Gibbs measure on the space* $\mathrm{Hom}\,(T_r, H)$.

Since the graphs $K_q$ (as well as $J$) satisfy the two conditions, we conclude that for the scope of this paper, we always have a unique simple invariant Gibbs measure.

The proof of the main theorem in [4] relied strongly on the fact that simple invariant Gibbs measures on $\mathrm{Hom}(T_r, H)$ are relatively tangible objects, obtainable by branching random walks on $H$. The measures are obtained as follows.

Let $H$ be arbitrary but connected, and give each node $i$ of $H$ a positive real weight $w_i$. Let the probability distribution $\pi$ on the nodes of $H$ be defined by letting $\pi_i$ be proportional to $w_i \sum_{j\sim i} w_j$ ($\pi$ is then the stationary distribution for a random walk on $H$ weighted by the $w_i$). We can (randomly) determine a coloring $\varphi$ of $T_r$ by $H$ as follows:

First, drop an amoeba onto $H$ which lands on node $i$ with probability $\pi_i$; this $i$ becomes the color of the root $w$ of $T_r$.

Second, the amoeba divides into $r+1$ baby amoebas and each takes an independent random step to a neighbor $j$ of $i$, with probabilities proportional to $w_j$. These $j$'s become the colors of $w$'s $r+1$ children.

Next, each baby amoeba divides into $r$ new amoebas, and all take independent random steps to neighbors, always in accordance with the weights. These grand-babies determine the colors of the $r(r+1)$ grandchildren of $w$.

The process continues, each new amoeba splitting $r$ ways at the next step, and every site of $T_r$ is thus colored. The following theorem from [4] is critical, though quite straightforward to prove.

**Theorem 2.2.** *The above process produces a simple invariant Gibbs measure on* $\mathrm{Hom}(T_r, H)$ *for activities given by:*

$$\lambda_i = \frac{w_i}{\left(\sum_{j\sim i} w_j\right)^r}.$$

*Moreover, every simple invariant Gibbs measure on* $\mathrm{Hom}(T_r, H)$ *can be obtained in this way.*

Results similar to, but not quite the same as, Theorem 2.2 had earlier been proved by Peruggi [13] and Zachary [21].

The map which transforms the weights into activities is surjective and, in the cases considered in this paper, injective as well. We will see that things get much more interesting when the invariance condition is relaxed to semi-invariance.

## 2.3. Constraint Graphs and Their Doubles

A graph $H$ is said to be *ergodic* if it is connected and not bipartite (note that a graph containing a loop cannot be bipartite). Thus the graphs $K_q$ for $q > 2$, and the graph $J$ above associated with the hard-core model, are ergodic.

Given an ergodic graph $H$ on nodes $1, 2, \ldots, q$ we form its bipartite "double", denoted $2H$, as follows: the nodes of $2H$ are $\{1, 2 \ldots, q\} \cup \{-1, -2, \ldots, -q\}$ with an edge between $i$ and $j$ just when $i \sim -j$ or $-i \sim j$ in $H$. Note that $2H$ is loopless; a loop at node $i$ in $H$ becomes the edge $\{-i, i\}$ in $2H$.

A coloring $\psi$ of $T_r$ by $2H$ induces a coloring $|\psi|$ by $H$ via $|\psi|(v) = |\psi(v)|$. In the reverse direction, a coloring $\varphi$ of $T_r$ by $H$ may be transformed to a coloring $\bar{\varphi}$ by $2H$, by putting $\bar{\varphi}(v) = \varphi(v)$ for even sites $v \in T_r$ and $\bar{\varphi}(v) = -\varphi(v)$ for odd $v$.

Let $\lambda = (\lambda_1, \ldots, \lambda_n)$ be a set of activities for $H$ and suppose that $\mu$ is a simple invariant Gibbs measure on Hom$(T_r, H)$ corresponding to $\lambda$. ¿From $\mu$ we can obtain a simple invariant Gibbs measure $\bar{\mu}$ on Hom$(T_r, 2H)$ by selecting $\varphi$ from $\mu$, and flipping a fair coin to decide between $\bar{\varphi}$ (as defined above) and $-\bar{\varphi}$. Obviously $\bar{\mu}$ yields the activity set $\bar{\lambda}$ on $2H$ given by $\bar{\lambda}_i = \lambda_{|i|}$. Furthermore, the weights on $H$ which produce $\mu$ extend to $2H$ by $w_{-i} = w_i$.

Conversely, suppose $\nu$ is a simple invariant Gibbs measure on Hom$(T_r, 2H)$ whose activity set satisfies $\lambda_{-i} = \lambda_i$ for each $i$. Then the measure $|\nu|$, obtained by choosing $\psi$ from $\nu$ and taking its absolute value, is certainly an invariant Gibbs measure on Hom$(T_r, H)$ for $\lambda \restriction \{1, \ldots, q\}$, but is it simple?

In fact, if the weights on $2H$ which produce $\nu$ do not satisfy $w_{-i} = cw_i$, then $|\nu|$ will fail to be simple. To see this, observe that if the weights are not proportional then there are nodes $i \sim j$ of $H$ such that $p_{-i,-j} \neq p_{i,j}$ in the random walk on $2H$. Suppose that $|\psi|$ is conditioned on the color of the root $w$ of $T_r$ being fixed at $i$, and let $x$ and $y$ be distinct neighbors of

$w$. Set $\alpha = \Pr\left(\psi(w) = i \,\middle|\, |\psi(w)| = i\right)$. Then

$$\Pr\left(|\psi|(x) = j\right) = (1-\alpha)p_{-i,-j} + \alpha p_{i,j}$$

but

$$\Pr\left(|\psi|(x) = j \wedge |\psi|(y) = j\right) = (1-\alpha)p^2_{-i,-j} + \alpha p^2_{i,j} > \Pr\left(|\psi|(x) = j\right)^2$$

so the colors of $x$ and $y$ are not independent given $|\psi|(w)$.

However, we can recover simplicity at the expense of one bit worth of symmetry. Let $\nu^+$ be $\nu$ conditioned on $\psi(u) > 0$, and define $\nu^-$ similarly. Then $|\nu^+|$ and $|\nu^-|$ are essentially the same as $\nu^+$ and $\nu^-$, respectively, and all are simple; but these measures are only semi-invariant.

On the other hand, suppose $\mu$ is a simple semi-invariant Gibbs measure on $\mathrm{Hom}\,(T_r, H)$. Let $\theta$ be a parity-reversing automorphism of $T_r$ and define $\mu' := \mu \circ \theta$, so that $\frac{1}{2}\mu + \frac{1}{2}\mu'$ is fully invariant (but generally no longer simple). However, $\nu := \frac{1}{2}\bar{\mu} + \frac{1}{2}(-\overline{\mu'})$ is a simple *and* invariant Gibbs measure on $\mathrm{Hom}\,(T_r, 2H)$, thus given by a node-weighted random walk on $2H$. We can recover $\mu$ as $\nu_+$, hence:

**Theorem 2.3.** *Every simple semi-invariant Gibbs measure on* $\mathrm{Hom}\,(T_r, H)$ *is obtainable from a node-weighted branching random walk on* $2H$*, with its initial state drawn from the stationary distribution on positive nodes of* $2H$*.*

Suppose, instead of beginning with a measure, we start by weighting the nodes of $2H$ and creating a Gibbs measure as in Theorem 2.3. Suppose the activities of the measure are $\{\lambda_i : i = \pm 1, \ldots, \pm q\}$. By identifying color $-i$ with $i$ for each $i > 0$, we create a measure on $H$-colorings, but this will not be a Gibbs measure unless it happens that $\lambda_{-1}, \ldots, \lambda_{-q}$ are proportional to $\lambda_1, \ldots, \lambda_q$.

We could assure this easily enough by making the weights proportional as well, e.g. by $w_{-i} = w_i$; then the resulting measure on $\mathrm{Hom}\,(G, H)$ could have been obtained directly by applying these weights to $H$, and is thus a fully invariant simple Gibbs measure. To get new, semi-invariant Gibbs measures on $\mathrm{Hom}\,(G, H)$, we must somehow devise weights for $2H$ such that $w_{-i} \not\propto w_i$ yet $\lambda_{-i} \propto \lambda_i$.

Restated with slightly different notation, simple semi-invariant Gibbs measures are in 1–1 correspondence with solutions to the "fundamental equations"

$$\lambda_i = \frac{u_i}{\left(\sum_{j\sim i} v_j\right)^r} = \frac{v_i}{\left(\sum_{j\sim i} u_j\right)^r}$$

for $i = 1, \ldots, q$. Such a solution will be invariant if $u_i = v_i$ for each $i$.

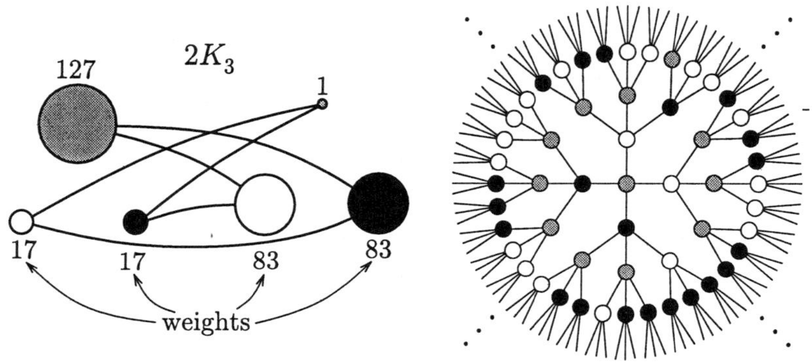


*Fig. 2.* An asymmetric phase for 3-coloring a 3-branching tree

Figure 2 illustrates a semi-invariant, but not invariant, simple Gibbs measure for uniform 3-colorings of $T_3$. Approximate weights of the nodes of $2H = 2K_3$ are given along with part of a sample coloring drawn from this measure. Additional measures may be obtained by permuting the colors or by making all the weights equal (invariant case).

## 3. Results

### 3.1. Semi-invariant Measures: General $r$ and $q$

We are now in a position to begin answering the key question of this paper: for which $r$ and $q$, and which sets of activities, is there a unique simple semi-invariant Gibbs measure on $\mathrm{Hom}(T_r, K_q)$?

Let us dispose of the less interesting cases first. If $q = 2$ there are only two possible colorings of $T_r$, and every measure on $\mathrm{Hom}(T_r, K_2)$ is a convex combination of the two; all of these are simple semi-invariant Gibbs measures but only the $\frac{1}{2}$–$\frac{1}{2}$ mix is fully invariant.

The Cayley tree $T_1$ is a doubly-infinite path, known for not exhibiting phase transitions even in much more general circumstances than ours. For all $q > 2$ and all sets of activities for $K_q$, the only Gibbs measure of any kind on $\mathrm{Hom}(T_1, K_q)$ is the one given by a (non-branching) stationary node-weighted random walk on $K_q$. The weights are uniquely determined by the activities (see [4] for a direct proof).

Henceforth we will assume that $r \geq 2$ and $q \geq 3$. When $q < r + 1$, we will see that all choices of activities including the uniform set yield multiple simple semi-invariant Gibbs measures.

When $q > r + 1$, there is only one simple semi-invariant Gibbs measure for the uniform set of activities, but multiple simple semi-invariant Gibbs measures for some other choices of activities.

The critical case is at $q = r + 1$, that is, when the number of colors is equal to the degree of the Cayley tree. Here we will show that there are multiple simple semi-invariant Gibbs measures for all activities *except* the uniform case, where there is just one.

If we consider all possible Gibbs measures, we find frozen measures (such as the ones mentioned above) and others whenever $q \leq r + 1$. We believe that when $q > r + 1$ and the activities are equal, the unique simple semi-invariant Gibbs measure is in fact the only Gibbs measure of any kind; we are able to prove this, however, only for $q > cr$, with fixed $c > 1$.

### 3.2. Multiple Simple Semi-invariant Gibbs Measures

Our next task is to show the existence of multiple simple semi-invariant Gibbs measures for certain values of $q$ and $r$ and certain ranges of the activities.

**Theorem 3.1.** *There are multiple simple semi-invariant Gibbs measures for* $\mathrm{Hom}\,(T_r, K_q)$ *in the following cases:*

- $q < r + 1$, *for all sets of activities;*
- $q \leq r + 1$, *for all non-uniform sets of activities;*
- *any* $q, r > 1$, *for some non-uniform sets of activities.*

The uniform case with $q < r + 1$ is contained in the work of Peruggi, di Liberto and Monroy [16], where they find phase transitions of several types at zero temperature and zero external field. The other cases are in line with their work but, as we mentioned earlier, the case of hard constraints and finite variation from uniform activities does not have an exact counterpart in the physicists' setting.

**Proof.** We proceed indirectly, by showing that, in certain circumstances, the solution to a certain optimization problem is an "asymmetric" solution to the fundamental equations, and so corresponds to a simple Gibbs measure that is semi-invariant but not invariant. There will also always be a simple invariant Gibbs measure, and of course another simple semi-invariant Gibbs measure can be obtained by switching the parity. The quantities involved in our optimization problem do not seem to be equivalent to any of the usual physical parameters such as pressure or mean free energy.

Given any non-negative real vectors $\lambda = (\lambda_1, \ldots, \lambda_q)$, $\mathbf{u} = (u_1, \ldots, u_q)$ and $\mathbf{v} = (v_1, \ldots, v_q)$, set

$$G(\mathbf{u}, \mathbf{v}) = \sum_{j \neq k} u_j v_k; \quad H_\lambda(\mathbf{u}) = \sum_{j=1}^{q} \frac{u_j^{1+1/r}}{\lambda_j^{1/r}}.$$

For any fixed $q \geq 3$, $r \geq 2$ and vector $\lambda > 0$, consider the problem of maximizing $G(\mathbf{u}, \mathbf{v})$ over non-negative $\mathbf{u}$, $\mathbf{v}$, subject to the constraints $H_\lambda(\mathbf{u}) \leq 1$, $H_\lambda(\mathbf{v}) \leq 1$.

Note that the constraints define a compact region of $\mathbb{R}^{2q}$, and that $G$ is continuous over this region, so the maximum is attained.

Suppose that some variable $u_j$ or $v_j$ is zero at the maximum. Then there are some $k$, $l$ distinct from $j$ with $u_j = 0$, $v_k > 0$, and $u_l > 0$ (or similarly with the roles of $\mathbf{u}$ and $\mathbf{v}$ interchanged). Then $\partial G/\partial u_j \geq v_k > 0$, $\partial H_\lambda(\mathbf{u})/\partial u_j = 0$, while $\partial H_\lambda(\mathbf{u})/\partial u_l > 0$ at the supposed maximum. Hence it is possible, by increasing $u_j$ and decreasing $u_l$, to increase $G$ while holding $H_\lambda(\mathbf{u})$ fixed — a contradiction.

Therefore all the variables are strictly positive at the maximum, and clearly we have equality in the constraints $H_\lambda(\mathbf{u})$, $H_\lambda(\mathbf{v}) \leq 1$. This means that the maximum will be a stationary point of the Lagrangian

$$L(\mathbf{u}, \mathbf{v}, \mu_1, \mu_2) = G(\mathbf{u}, \mathbf{v}) - \mu_1\big(H_\lambda(\mathbf{u}) - 1\big) - \mu_2\big(H_\lambda(\mathbf{v}) - 1\big).$$

Setting the partial derivatives of $L$ equal to 0 gives us:

$$\sum_{k \neq j} v_k = \mu_1 \frac{r+1}{r} \left(\frac{u_j}{\lambda_j}\right)^{1/r} \qquad (j = 1, \ldots, q),$$

$$\sum_{k \neq j} u_k = \mu_2 \frac{r+1}{r} \left(\frac{v_j}{\lambda_j}\right)^{1/r} \qquad (j = 1, \ldots, q),$$

together with the constraints $H_\lambda(\mathbf{u}) = H_\lambda(\mathbf{v}) = 1$. Solutions to this system of equations are in 1–1 correspondence with solutions to the fundamental equations; indeed, if $(\mathbf{u}, \mathbf{v})$ is a solution to the equations above, then $(c_1\mathbf{u}, c_2\mathbf{v})$ is a solution to the fundamental equations, for appropriately chosen constants $c_1$, $c_2$ — conversely, any solution to the fundamental equations can be scaled to satisfy the constraints $H_\lambda(\mathbf{u}) = H_\lambda(\mathbf{v}) = 1$ as well as the equations above.

Therefore stationary points of the Lagrangian are in 1–1 correspondence with simple semi-invariant Gibbs measures for the set of activities $\lambda$.

We know that there is one invariant solution, and therefore a stationary point with $u_j = v_j$ for all $j$ and $\mu_1 = \mu_2 = \mu$. Let $w_j$ be the common value of $u_j$ and $v_j$ at this point. If this is the only solution to the fundamental equations, then this must be the solution to our optimization problem.

Suppose that $w_1 \geq w_2 \geq \cdots \geq w_q$ (this is equivalent to $\lambda_1 \geq \lambda_2 \geq \cdots \geq \lambda_q$). Suppose also that either $q < r+1$ or $q = r+1$ and the $\lambda_j$ (hence the $w_j$) are not all equal. Observe then that, in either case, $\sum_{j\neq 1} w_j \leq (q-1)w_2 \leq rw_2$, and $\sum_{j\neq 2} w_j \leq (q-1)w_1 \leq rw_1$. Furthermore, equality is not possible in the latter inequality, and we can choose $\gamma > 0$ such that

$$2 > \frac{(1+\gamma)}{r}\left(\frac{\sum_{j\neq 2} w_j}{w_1} + \frac{\sum_{j\neq 1} w_j}{w_2}\right).$$

In the case $q > r+1$, we can find weights $w_j$, and $\gamma > 0$, satisfying the above inequality. Such weights define a set of activities, and it is this set for which we claim that there are multiple simple semi-invariant Gibbs measures.

Now choose a small $\delta > 0$, set

$$\delta_1 = \delta\left(\frac{w_2}{\lambda_2}\right)^{1/r}, \quad \delta_2 = \delta\left(\frac{w_1}{\lambda_1}\right)^{1/r},$$

and consider the effect of the following small changes to the variables:

$$u_1 = w_1 + \delta_1 - \frac{\delta_1^2(1+\gamma)}{2rw_1};$$

$$u_2 = w_2 - \delta_2 - \frac{\delta_2^2(1+\gamma)}{2rw_2};$$

$$v_1 = w_1 - \delta_1 - \frac{\delta_1^2(1+\gamma)}{2rw_1};$$

$$v_2 = w_2 + \delta_2 - \frac{\delta_2^2(1+\gamma)}{2rw_2}.$$

For $j > 2$, we keep $u_j = v_j = w_j$.

We have that

$$H_\lambda(\mathbf{u}) =$$

$$= H_\lambda(\mathbf{w}) + \frac{w_1^{1+1/r}}{\lambda_1^{1/r}}\left(\left(\frac{u_1}{w_1}\right)^{1+1/r} - 1\right) + \frac{w_2^{1+1/r}}{\lambda_2^{1/r}}\left(\left(\frac{u_2}{w_2}\right)^{1+1/r} - 1\right)$$

$$= H_\lambda(\mathbf{w}) + \frac{w_1^{1+1/r}}{\lambda_1^{1/r}}\left(\frac{\delta_1}{w_1}(1+1/r) - \left(\frac{\delta_1}{w_1}\right)^2 \frac{\gamma(r+1)}{2r^2} + O\left(\frac{\delta_1}{w_1}\right)^3\right)$$

$$- \frac{w_2^{1+1/r}}{\lambda_2^{1/r}}\left(\frac{\delta_2}{w_2}(1+1/r) + \left(\frac{\delta_2}{w_2}\right)^2 \frac{\gamma(r+1)}{2r^2} + O\left(\frac{\delta_2}{w_2}\right)^3\right).$$

By choice of $\delta_1$ and $\delta_2$, the first order terms cancel. Thus, for $\delta$ sufficiently small, we have $H_\lambda(\mathbf{u}) \leq H_\lambda(\mathbf{w}) = 1$. Similarly $H_\lambda(\mathbf{v}) \leq H_\lambda(\mathbf{w}) = 1$ for small $\delta$, so our new choice is feasible.

Now let us determine the effect on $G(\mathbf{u}, \mathbf{v})$. To simplify matters, define $x_j = (w_j/\lambda_j)^{1/r}$ for each $j$, and $\nu = \mu(1+1/r)$, so that $\sum_{k \neq j} w_k = \nu x_j$ for $j = 1, 2$, and $\delta_1 = \delta x_2$, $\delta_2 = \delta x_1$.

Then we have

$$G(\mathbf{u}, \mathbf{v}) - G(\mathbf{w}, \mathbf{w}) =$$

$$= \nu\big((u_1 - w_1)x_1 + (u_2 - w_2)x_2 + (v_1 - w_1)x_1 + (v_2 - w_2)x_2\big)$$

$$+ (u_1 - w_1)(v_2 - w_2) + (u_2 - w_2)(v_1 - w_1)$$

$$= \nu\delta x_1 x_2\Bigg[\left(1 - \frac{\delta x_2(1+\gamma)}{2rw_1}\right) + \left(-1 - \frac{\delta x_1(1+\gamma)}{2rw_2}\right)$$

$$+ \left(-1 - \frac{\delta x_2(1+\gamma)}{2rw_1}\right) + \left(1 - \frac{\delta x_1(1+\gamma)}{2rw_2}\right)\Bigg] + 2\delta^2 x_1 x_2 + O(\delta^3)$$

$$= \delta^2 x_1 x_2 \left(2 - \frac{\nu(1+\gamma)}{r}\left(\frac{x_2}{w_1} + \frac{x_1}{w_2}\right)\right) + O(\delta^3)$$

$$= \delta^2 x_1 x_2 \left(2 - \frac{1+\gamma}{r}\left(\frac{\sum_{k\neq 2} w_k}{w_1} + \frac{\sum_{k\neq 1} w_k}{w_2}\right)\right) + O(\delta^3).$$

By choice of $\gamma$, this is positive for $\delta$ sufficiently small.

In conclusion, making the indicated change to **u** and **v** keeps the solution feasible while increasing the objective function, thus proving that the choice **w** is not a maximum. ■

### 3.3. Unique Simple Semi-invariant Gibbs Measures

We saw in the previous section that for $q < r+1$, there are multiple simple semi-invariant Gibbs measures for all activities, including uniform. Here we show that in the uniform case, this threshold is exact.

**Theorem 3.2.** *For $q \geq r+1$, there is only one simple semi-invariant Gibbs measure on* $\mathrm{Hom}\,(T_r, K_q)$ *with uniform activities.*

**Proof.** Suppose that there are weights $(u_1, u_2, \ldots, u_q, v_1, v_2, \ldots, v_q)$ satisfying the fundamental equations with $\lambda_1 = \cdots = \lambda_q = 1$, so that

$$u_j = \left(\sum_{k\neq j} v_k\right)^r; \quad v_j = \left(\sum_{k\neq j} u_k\right)^r \quad (j = 1, \ldots, q).$$

Suppose that $u_1 \geq u_2 \geq \ldots \geq u_q$, which implies that $v_1 \leq v_2 \leq \ldots \leq v_q$. Set

$$\alpha = u_1/u_q; \quad \beta = \frac{u_1 + u_2 + \cdots + u_{q-1}}{(q-1)u_q},$$

so that $\alpha \geq \beta \geq 1$.

We know that all the weights are equal at the unique invariant solution, so what we are claiming is that there is no solution to the equations above with $\beta > 1$.

We have

$$\alpha^{1/r} = \left(\frac{u_1}{u_q}\right)^{1/r} = 1 + \frac{v_q - v_1}{\sum_{k=1}^{q-1} v_k}.$$

Note that

$$\sum_{k=1}^{q-1} v_k = \sum_{k=1}^{q-1} \big(u_q + (q-2)\beta u_q + (\beta u_q - u_k)\big)^r \geq (q-1)\big(u_q + (q-2)\beta u_q\big)^r,$$

since the terms $(\beta u_q - u_k)$ sum to zero. Hence, using also that $v_1 = \big((q-1)\beta + 1 - \alpha\big)^r u_q^r$, $v_q = \big((q-1)\beta\big)^r u_q^r$, we see that

$$\alpha^{1/r} \leq 1 + \frac{\big((q-1)\beta\big)^r - \big((q-1)\beta + 1 - \alpha\big)^r}{(q-1)\big(1 + (q-2)\beta\big)^r}$$

$$= 1 + \frac{1 - \left(1 - \frac{\alpha - 1}{(q-1)\beta}\right)^r}{(q-1)\left(1 - \frac{\beta - 1}{(q-1)\beta}\right)^r}.$$

The right hand side here is decreasing in $q$, so for $q \geq r+1$ we have

$$0 \leq F_r(\alpha, \beta) = -\alpha^{1/r} + 1 + \frac{1 - \left(1 - \frac{\alpha-1}{r\beta}\right)^r}{r\left(1 - \frac{\beta-1}{r\beta}\right)^r}.$$

We next claim that $F_r(\alpha, \beta)$ is decreasing in $\alpha$. To see this, we start by observing that

$$\frac{\partial F_r}{\partial \alpha} = \frac{1}{r}\left(-\alpha^{-(r-1)/r} + \frac{1}{\beta}\frac{\big(1 - (\alpha - 1)/r\beta\big)^{r-1}}{\big(1 - (\beta - 1)/r\beta\big)^r}\right)$$

$$= \alpha^{-(r-1)/r}\left(-1 + \frac{G(\alpha)^{(r-1)/r}}{G(\beta)}\right),$$

where the function $G(x) = x\big(1 - (x-1)/r\beta\big)^r$ is decreasing for $x \geq \beta$, and satisfies $G(\beta) \geq 1$. Thus

$$G(\beta) \geq G(\beta)^{(r-1)/r} \geq G(\alpha)^{(r-1)/r},$$

and so indeed $F_r$ is decreasing in $\alpha$. Therefore we have

$$0 \leq F_r(\alpha, \beta) \leq F_r(\beta, \beta) = H_r(\beta) = -\beta^{1/r} + 1 + \frac{1}{r}\left(\left(1 - \frac{\beta - 1}{r\beta}\right)^{-r} - 1\right).$$

Now we claim that $H_r(\beta)$ is decreasing in $\beta$. Indeed, its derivative, multiplied by $r$, is

$$\frac{-1}{\beta^{(r-1)/r}} + \frac{\beta^{r-1}}{\left(1 + \frac{r-1}{r}(\beta - 1)\right)^{r+1}}.$$

This is negative since $1 + \frac{r-1}{r}(\beta - 1) \geq \big(1 + (\beta - 1)\big)^{(r-1)/r} = \beta^{(r-1)/r}$, and also we have strict inequality if $\beta > 1$, so $H_r(\beta) \leq H_r(1) = 0$, with strict inequality, and therefore a contradiction, unless $\beta = 1$.

In conclusion, in the case we are considering, there is no solution to the fundamental equations unless all the weights are equal, which is the required result. ∎

### 3.4. Uniqueness of Gibbs Measures

In this subsection we will show that when $q$ is sufficiently large relative to $r$, there is just one Gibbs measure for Hom $(T_r, K_q)$ with uniform activities. Salas and Sokal [18], and apparently also Kotecký as cited in [9], were able to show that with any board of maximum degree $r+1$, there is a unique Gibbs measure in the uniform case as long as $q > 2r + 2$. We can do somewhat better in our special case.

Our approach is to show that there is, indeed, no long-range order; that is, the restriction of a random coloring $\varphi$ to far-away sites has diminishing effect on the values of $\varphi$ on a fixed patch of $T_r$.

Given $r < q$ probability vectors $\mathbf{p}_1 = (p_{11}, \ldots, p_{1q})$, $\ldots$, $\mathbf{p}_r = (p_{r1}, \ldots, p_{rq})$ in $[0,1]^q$, each with entries summing to 1, we form their *Potts product* $\mathbf{p}_1 * \cdots * \mathbf{p}_r$ as the probability vector proportional to $(p'_{11}p'_{21}\cdots p'_{r1}, \ldots, p'_{1q}p'_{2q}\cdots p'_{rq})$, where $p'_{ij} = 1 - p_{ij}$, for $i = 1, \ldots, r$, and $j = 1, \ldots, q$. Note that, since $q > r$, this vector cannot be identically zero, and hence normalization to a probability vector is always possible.

Given $q$ and $r$, and a positive integer $n$, consider a finite portion $T_r^n$ of $T_r$, defined from $T_r$ by deleting one of the $r+1$ branches from the root, and taking for $T_r^n$ all sites at distance at most $n$ from the root. Now, for every assignment of a probability vector $\mathbf{p}(u)$ to each leaf $u$ of $T_r^n$, we recursively define a probability vector $\mathbf{p}(x)$ at each internal site $x$ of $T_r^n$ as the Potts product $\mathbf{p}(x_1) * \ldots * \mathbf{p}(x_r)$, where $x_1, \ldots, x_r$ are the children of $x$. In particular, any assignment of a probability vector to each leaf gives rise to a probability vector $\mathbf{p}(w)$ at the root $w$.

Suppose that each probability vector assigned to a leaf is a 0–1 vector, to be thought of as specifying the color of that leaf. For any site $x$ of $T_r^n$, let $T(x)$ denote the portion of $T_r^n$ above $x$, i.e., the subtree obtained from $T_r^n$ by deleting the branch from $x$ that includes the root $w$. We claim that, for any site $x$ in $T_r^n$, the $j^{\text{th}}$ coordinate of $\mathbf{p}(x)$ is proportional to the number of proper $q$-colorings of $T(x)$ in which $x$ receives color $j$, and the leaves of $T(x)$ receive the colors specified. Indeed, this is clearly seen by induction, working in from the leaves, noting that each $T(x)$ is a copy of $T_r^m$ for some $r$.

In particular, the $j^{\text{th}}$ coordinate of the vector $\mathbf{p}(w)$ is the probability that the root $w$ has color $j$, in a uniform random proper $q$-coloring of $T_r^n$ consistent with the give leaf colorings.

(This approach is particular to the case of uniform activities. In a more general setting, one could define a version of the Potts product taking account of the activities, and then interpret the entries of the vector $\mathbf{p}(x)$ as representing the "effective activity", summarizing the effect of the entire subtree above $x$.)

We would like to find conditions (on $q$ and $r$) guaranteeing that, as $n \to \infty$, the root vector $\mathbf{p}(w)$ tends to the uniform vector $(1/q, \ldots, 1/q)$, regardless of the leaf vectors. As we shall see, this implies that there is a unique Gibbs measure in this case.

To get some feel for this, consider the case $q = r+1$, take a frozen $q$-coloring $\varphi$ of the tree $T_r$, and delete a branch from the root corresponding to a repeated color among the children of the root. Now the colors on sites $u$ at distance $n$ from the root produce 0–1 probability vectors $\mathbf{p}(u)$. For any internal vertex $x$, we claim that the derived probability vector $\mathbf{p}(x)$ is also the 0–1 vector corresponding to $\varphi(x)$, which we may take to be the last color $q$. Indeed, note that the $r$ children of $x$ are each given one of the colors $1, 2, \ldots, q-1$, so we may assume for an induction that their assigned vectors are $(1,0,\ldots,0)$, $(0,1,0,\ldots,0)$, $\ldots$, $(0,\ldots,0,1,0)$. The Potts product of these vectors is $(0,\ldots,0,1)$, which indeed corresponds to the assignment of color $q$ to $x$. Hence the root vector $\mathbf{p}(w)$ does not necessarily converge to uniform as $n \to \infty$ in the case $q = r+1$.

Define the *eccentricity* $E(\mathbf{x})$ of a non-negative vector $\mathbf{x} = (x_1, \ldots, x_q)$ to be

$$\sum_{j \neq k} (x_j - x_k)^2 \Big/ \Big( \sum_{j=1}^{q} x_j \Big)^2.$$

Note that the denominator is 1 for a probability vector, while the eccentricity is invariant under scaling: $E(t\mathbf{x}) = E(\mathbf{x})$ for all scalars $t$ and vectors $\mathbf{x}$.

Let $M(q, r; n)$ be the maximum eccentricity of $\mathbf{p}(w)$, over all assignments of probability vectors $\mathbf{p}(u)$ to the leaves $u$ of $T_r^n$.

**Theorem 3.3.** *Suppose that $q$ and $r$ are such that $M(q, r; n) \to 0$ as $n \to \infty$. Then there is a unique Gibbs measure (namely, the simple invariant Gibbs measure generated by the node-weighted branching random walk) for* $\mathrm{Hom}\,(T_r, K_q)$ *with uniform activities.*

**Proof.** Let $\mu$ be any Gibbs measure; we aim to show that $\mu$ is equal to the measure $\nu$ defined from the $r$-branching random walk on $K_q$, with all nodes weighted equally.

To show that the two measures are equal, take any patch $U$, any proper coloring $\varphi$ of $U$, and any $\varepsilon > 0$. We need to show that $\big| \Pr_\mu(\psi = \varphi) - \Pr_\nu(\psi = \varphi) \big| < \varepsilon$. We may assume (by taking a larger patch if necessary) that $U$ is connected (so it is a subtree of $T_r$), and that all its elements have degree 1 or $r+1$. Note that $\Pr_\nu(\psi = \varphi) = 1/q(q-1)^{|U|-1}$, since all of the $q(q-1)^{|U|-1}$ proper $q$-colorings of $U$ are equally likely under $\nu$.

Suppose that $U$ has $\ell$ leaves. Now choose $n$ sufficiently large (to be specified later), and let $V$ denote the set of all sites of $T_r$ at distance at most $n$ from our patch $U$. Note that $(V \setminus U) \cup \partial U$ consists of $\ell$ disjoint copies of $T_r^k$, one rooted at each leaf of $U$.

We generate a random sample from the Gibbs measure $\mu$ according to the following procedure: first take a random sample $\psi$ from $\mu$, then uncolor all sites in $V \setminus \partial V$. Now choose, uniformly at random, a proper coloring of $V \setminus \partial V$ consistent with the coloring of $\partial V$. By the Gibbs property, this is a valid way of constructing a sample coloring $\psi'$.

Now take any coloring $\varphi$ of $U$, and observe that the probability that $\psi' \upharpoonright U = \varphi$ is equal to the sum, over all colorings $\rho$ of $\partial V$, of $\Pr_\mu(\psi \upharpoonright \partial V = \rho) \Pr(\psi' \upharpoonright U = \varphi \,|\, \psi \upharpoonright \partial V = \rho)$. The second probability here refers only to the recoloring phase, and so is proportional to the number of proper colorings of $V \setminus (U \cup \partial V)$ consistent with both the coloring $\rho$ of $\partial V$ and the coloring $\varphi \upharpoonright \partial U$ of the leaves of $U$. This is, by the earlier discussion, proportional to the product, over all the leaves $u$ of $U$, of the $\varphi(u)$-entry of the vector $\mathbf{p}(u)$ generated from the 0–1 probability vectors set (by $\rho$) at the leaves of the copy of $T_r^n$ rooted at $u$. This vector $\mathbf{p}(u)$ has eccentricity at most $M = M(q, r; n)$, by definition, and so each entry is within $\sqrt{M}$ of $1/q$. Hence the probability $\Pr(\psi' \upharpoonright U = \varphi \,|\, \psi \upharpoonright \partial V = \rho)$ is proportional to a number lying between $\left(1/q - \sqrt{M}\right)^\ell$ and $\left(1/q + \sqrt{M}\right)^\ell$, for every $\varphi$ and $\rho$. Therefore

$$\left(\frac{1/q - \sqrt{M}}{1/q + \sqrt{M}}\right)^\ell \frac{1}{q(q-1)^{|U|-1}} \le \Pr(\psi' \upharpoonright U = \varphi)$$

$$\le \left(\frac{1/q + \sqrt{M}}{1/q - \sqrt{M}}\right)^\ell \frac{1}{q(q-1)^{|U|-1}},$$

so we can ensure that $\Pr(\psi' \upharpoonright U = \varphi)$ is within $\varepsilon$ of $1/q(q-1)^{|U|-1}$ by choosing $n$ to make $M$ small enough. ∎

To prove that $M(q, r; n) \to 0$, the obvious approach is to show that, if $\mathbf{p}_1, \ldots, \mathbf{p}_r$ are any probability vectors with eccentricities at most $m$, then the eccentricity of $\mathbf{p}_1 * \ldots * \mathbf{p}_r$ is at most $tm$ for some $t < 1$. It turns out to be worthwhile to refine this approach somewhat.

We start with a lemma dealing with the normalization involved in the definition of the Potts product.

**Lemma 3.4.** *Suppose* $\mathbf{p}_1 = (p_{11}, \dots, p_{1q}), \dots, \mathbf{p}_r = (p_{r1}, \dots, p_{rq})$ *are probability vectors, all of whose entries are at most* $U$. *Then*

$$\sum_{j=1}^{q} p'_{1j} \cdots p'_{rj} \geq \max\left(q - r, q(1-U)^{r/qU}\right).$$

*and so in particular every entry of* $\mathbf{p}_1 * \ldots * \mathbf{p}_r$ *is at most* $1/(q-r)$.

**Proof.** Consider the problem of minimizing $\sum_{j=1}^{q} p'_{1j} \cdots p'_{rj}$, subject to the conditions that all the $p'_{ij}$ are (to start with) between 0 and 1, and $\sum_{j=1}^{q} p'_{ij} = q - 1$ for all $i = 1, \dots, r$. For any fixed assignment of values to all the variables except $p'_{i1}, \dots, p'_{iq}$ (for some fixed $i$), it is clear that the minimum is obtained by setting $q-1$ of these variables to 1, and the other (that value $p'_{ij}$ whose multiplier $\prod_{h \neq i} p'_{hj}$ is maximum) to 0. Hence, at the optimum, exactly $r$ of the $qr$ variables $p'_{ij}$ are set to 0, making $r$ of the terms $p'_{1j} \cdots p'_{rj}$ equal to 0, and the other $q - r$ terms equal to 1. Therefore

$$\sum_{j=1}^{q} p'_{1j} \cdots p'_{rj} \geq q - r.$$

To see the other bound, we start by using the arithmetic-geometric mean inequality to find that

$$\sum_{j=1}^{q} \prod_{i=1}^{r} p'_{ij} \geq q\left(\prod_{j=1}^{q} \prod_{i=1}^{r} p'_{ij}\right)^{1/q}.$$

Now each of the $qr$ terms $p'_{ij}$ can be written as $1 - \lambda_{ij} U$, where $0 \leq \lambda_{ij} \leq 1$. Since the sum of all the $qr$ terms $p'_{ij}$ is $r(q-1)$, the sum of all the $\lambda_{ij}$ is equal to $r/U$. Now since

$$p'_{ij} = 1 - \lambda_{ij} U \geq (1-U)^{\lambda_{ij}},$$

we have

$$\sum_{j=1}^{q} \prod_{i=1}^{r} p'_{ij} \geq q\left((1-U)^{r/U}\right)^{1/q},$$

as desired.

The final statement follows because an entry of $\mathbf{p}_1 * \ldots * \mathbf{p}_r$ is obtained by dividing a number at most 1 by a normalizing factor of at least $q-r$. ∎

Note: We can also see informally that the entries of $\mathbf{p}_1 * \ldots * \mathbf{p}_r$ are at least $1/(q-r)$, by interpreting $p_{ij}$ as the probability that the $i$th child of a site $u$ has color $j$; since any possible coloring of the children permits at least $q-r$ colors at $u$, the probability of any one color at $u$ is at most $1/(q-r)$.

Thus, if we take any probability vectors at the leaves of our $T_r^n$, the entries of their Potts products, one layer in, are all at most $U = 1/(q-r)$. This in turn implies a lower bound on the entries two layers in, namely $L = \left(1 - 1/(q-r)\right)^r/(q-1)$, since $q-1$ is an easy upper bound on the normalization constant. Better bounds could be found by iterating this procedure.

To prove that $M(q,r;n) \to 0$ as $n \to \infty$, it thus suffices to show that, if $\mathbf{p}_1 = (p_{11}, \ldots, p_{1q}), \ldots, \mathbf{p}_r = (p_{r1}, \ldots, p_{rq})$ are probability vectors with eccentricity at most $m$ and entries between $L$ and $U$, then $E(\mathbf{p}_1 * \cdots * \mathbf{p}_r) \leq tm$ for some $t < 1$.

Accordingly, take such vectors $\mathbf{p}_1, \ldots, \mathbf{p}_r$, and consider

$$E(\mathbf{p}_1 * \cdots * \mathbf{p}_r) = \frac{\sum_{k \neq j} (p'_{1j} \cdots p'_{rj} - p'_{1k} \cdots p'_{rk})^2}{\left(\sum_{j=1}^{q} p'_{1j} \cdots p'_{rj}\right)^2}.$$

We need an upper bound on the numerator of this expression, which we can then combine with the lower bound on the denominator given by Lemma 3.4.

By telescoping, we see that

$$p'_{1j} \cdots p'_{rj} - p'_{1k} \cdots p'_{rk} = \sum_{i=1}^{r} (p'_{ij} - p'_{ik}) \prod_{h<i} p'_{hj} \prod_{h>i} p'_{hk}.$$

Since all the $p'_{ij}$ are at most $1-L$, while $p'_{ij} - p'_{ik} = p_{ik} - p_{ij}$, we find that

$$|p'_{1j} \cdots p'_{rj} - p'_{1k} \cdots p'_{rk}| \leq (1-L)^{r-1} \sum_{i=1}^{r} |p_{ik} - p_{ij}|.$$

Thus we have

$$\sum_{j\neq k}(p'_{1j}\cdots p'_{rj}-p'_{1k}\cdots p'_{rk})^2 \le (1-L)^{2(r-1)}\sum_{j\neq k}\left(\sum_{i=1}^{r}|p_{ik}-p_{ij}|\right)^2$$

$$\le (1-L)^{2(r-1)}r\sum_{i=1}^{r}\sum_{j\neq k}(p_{ik}-p_{ij})^2$$

$$\le r^2(1-L)^{2(r-1)}m.$$

Combining this bound with that from Lemma 3.4, we see that

$$E(\mathbf{p}_1 * \ldots * \mathbf{p}_r) \le \min\left\{\left(\frac{r(1-L)^{r-1}}{q-r}\right)^2, \left(\frac{r(1-L)^{r-1}}{q(1-U)^{r/qU}}\right)^2\right\}\cdot m.$$

Hence if $q$ and $r$ are such that either of the two terms is less than 1, for the values of $L$ and $U$ given earlier, then there is a unique Gibbs measure for Hom $(T_r, K_q)$, with uniform activities.

In particular, if $q \ge 2r$, then $r(1-L)^{r-1}/(q-r) < 1$ and we have uniqueness. Also, if $r$ and $q$ are both large, with $q = (d+1)r$, then $U = 1/dr$, $L \approx e^{-1/d}/(d+1)r$, and

$$\frac{r(1-L)^{r-1}}{q(1-U)^{r/qU}} \approx \frac{1}{d+1}\exp\left(\frac{1-e^{-1/d}}{d+1}\right).$$

This quantity is less than 1 for $d > 0.6296$. Therefore we have the following result.

**Theorem 3.5.** *Suppose that $q \ge 2r$, or that $q \ge 1.6296\,r$ and $r$ is sufficiently large. Then there is a unique Gibbs measure for Hom $(T_r, K_q)$, with uniform activities.*

The reader will have no difficulty in improving the constant 1.6296 in the above theorem. Our own calculations give also more specific bounds, namely that we have a unique uniform Gibbs measure when $r \geq 5$ and $q > \left(\frac{5}{3}\right) r$, or indeed when $q \geq \frac{5}{3}(r+1) - 1$; for comparison purposes, note that $r+1$ is the maximum degree of the tree $T_r$. However, what we believe is that there is a unique Gibbs measure even for $q = r+2$, and this method does not seem able to give any result of this type. Even the case $q = 5$, $r = 3$ remains a challenge, although it is not impossible that the method used here can be improved to cover this case.

## 4. Remarks and Conjectures

We have pretty much settled the question of when there is more than one simple semi-invariant Gibbs measure, but have left some important open questions concerning general Gibbs measures.

We have shown that, when $q \geq r+1$, there is only one simple semi-invariant Gibbs measure in the uniform case, but multiple such measures for some non-uniform activity sets. We suspect that when $q > r+1$, there is only one Gibbs measure of any kind for the uniform set of activities. We have been able to prove this for $q = 4$ and $r = 2$, and generally when $q \geq 2r$ and, for large $r$, when $q \geq 1.6296\, r$.

For the critical case $q = r+1$, all non-uniform sets of activities produce multiple simple semi-invariant Gibbs measures. There are other Gibbs measures, however, regardless of the activity set, and it appears to be a daunting challenge to classify them — even in the ground case $q = 3$, $r = 2$.

When $q < r+1$, we know that there are multiple simple semi-invariant Gibbs measures for every set of activities (plus frozen Gibbs measures and more).

The state of affairs is roughly illustrated in Figure 3.

There are many related issues which we have not considered in this work. For example, the question of whether our simple semi-invariant Gibbs measures are *extremal* (that is, not mixes of other Gibbs measures) is complex, and linked to the issue of whether information about the starting state of a branching random walk is available from states reached at a much later time (see e.g. [5, 8, 11, 12]). (We can show that for $q \geq r+1$, the simple invariant Gibbs measure for the uniform set of activities is indeed extremal.)

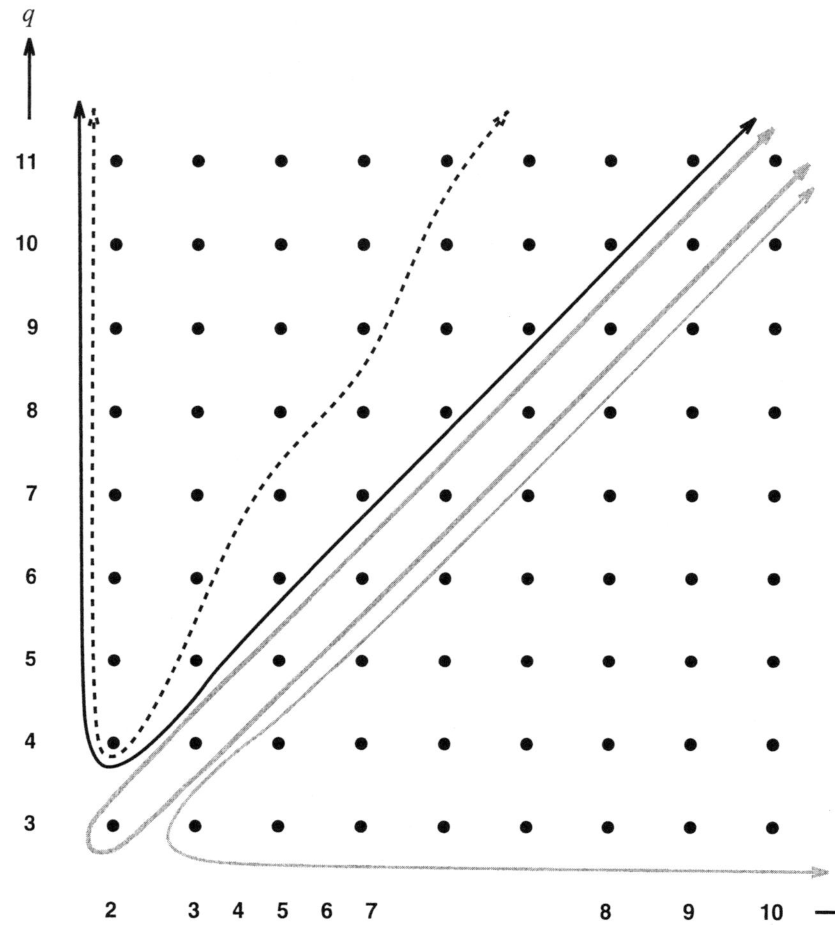


unique Gibbs measure for uniform activities (conjectured), multiple simple semi-invariant Gibbs measures for some non-uniform activities

unique Gibbs measure for uniform activities (proven)

unique simple semi-invariant Gibbs measure for uniform activities, multiple simple semi-invariant Gibbs measures for all non-uniform activities, non-semi-invariant Gibbs measures for all activities

multiple simple semi-invariant and non-semi-invariant Gibbs measures for all activities

*Fig. 3.* Simple semi-invariant, and general, Gibbs measures for $q$-coloring the $r$-branching Cayley tree

The DLR problem (uniqueness of Gibbs measures) is itself closely allied with the problem of whether "heat bath" dynamics, which make local changes to colorings of a large finite board, constitute rapidly mixing Markov chains. For example, Vigoda [20] has shown that when the number of colors is at least 11/6 times the maximum degree of the board, rapid mixing ensues; it is thus arguably noteworthy that our constant 1.6296 undercuts 11/6. See also Sokal [19] for related results and conjectures.

**Acknowledgment.** We would like to thank Alan Sokal, and also an anonymous referee, for several useful comments that have improved the presentation of this paper.

## References

[1] R. J. Baxter, *Exactly Solved Models in Statistical Mechanics,* Academic Press (London, 1982).

[2] J. van den Berg and J. E. Steif, Percolation and the hard-core lattice gas model, *Stochastic Proc. and their Appls.,* **49** (1994), 179–197.

[3] G. R. Brightwell, O. Häggström and P. Winkler, Nonmonotonic behavior in hard-core and Widom–Rowlinson models, *J. Stat. Physics,* **94** (1999), 415–435.

[4] G. R. Brightwell and P. Winkler, Graph homomorphisms and phase transitions, *J. Comb. Theory (Series B),* **77** (1999), 221–262.

[5] G. R. Brightwell and P. Winkler, Gibbs measures and dismantlable graphs, *J. Comb. Theory (Series B),* to appear (2000).

[6] R. M. Burton, C.-E. Pfister and J. E. Steif, The variational principle for Gibbs states fails on trees, *Markov Process. Related Fields,* **1** #3 (1995), 387–406.

[7] R. L. Dobrushin, The description of a random field by means of conditional probabilities and conditions of its regularity, *Theory Probab. Appl.,* **13** #2 (1968), 197–224.

[8] W. Evans, C. Kenyon, Y. Peres and L. J. Schulman, Broadcasting on trees and the Ising model, *Ann. Appl. Probab.,* to appear (2000).

[9] H.-O. Georgii, *Gibbs Measures and Phase Transitions,* de Gruyter (Berlin, 1988).

[10] F. P. Kelly, Stochastic models of computer communication systems, *J. R.Statist. Soc. B,* **47** #3 (1985), 379–395.

[11] E. Mossel, Recursive reconstruction on periodic trees, *Random Structures & Algorithms,* **13** #1 (1998), 81–97.

[12] E. Mossel, Reconstruction on trees: Beating the second eigenvalue, to appear in *Ann. Appl. Probab.*

[13] F. Peruggi, Probability measures and Hamiltonian models on Bethe lattices. I. Properties and construction of MRT probability measures, *J. Math. Phys.,* **25** (1984), 3303–3315.

[14] F. Peruggi, F. di Liberto and G. Monroy, The Potts model on Bethe lattices. I. General results, *J. Phys. A: Math. Gen.,* **16** (1983), 811–828.

[15] F. Peruggi, F. di Liberto and G. Monroy, Critical behaviour in three-state Potts antiferromagnets on a Bethe lattice, *Physica,* **131A** (1985), 300–310.

[16] F. Peruggi, F. di Liberto and G. Monroy, Phase diagrams of the $q$-state Potts model on Bethe lattices, *Physica,* **141A** (1987), 151–186.

[17] B. Prum and J. C. Fort, *Stochastic Processes on a Lattice and Gibbs Measures,* Kluwer (Dordrecht, 1991).

[18] J. Salas and A. D. Sokal, Absence of phase transition for antiferromagnetic Potts models via the Dobrushin uniqueness theorem, *J. Stat. Phys.,* **86** (1997), 551–579.

[19] A. D. Sokal, A personal list of unsolved problems concerning Potts models and lattice gases, *Markov Proc. Rel. Fields,* to appear (available from xxx.lanl.gov as cond-mat/0004231).

[20] E. Vigoda, Improved bounds for sampling colorings, *Proc. 40th Symp. on Foundations of Comp. Sci.,* I.E.E.E. Computer Society, Los Alamitos CA (1999), 51–59.

[21] S. Zachary, Countable state space Markov random fields and Markov chains on trees, *Ann. Probab.,* **11** (1983), 894–903.

Graham R. Brightwell
*Department of Mathematics*
*London School of Economics*
*Houghton St.*
*London WC2A 2AE*
*England*

e-mail: g.r.brightwell@lse.ac.uk

Peter Winkler
*Bell Labs 2C-365*
*Lucent Technologies*
*700 Mountain Ave.*
*Murray Hill, NJ 07974*
*U.S.A.*

e-mail: pw@lucent.com

BOLYAI SOCIETY
MATHEMATICAL STUDIES, 10

**Contemporary Combinatorics**
**Budapest, 2002,** pp. 277–288.

# On the Expected Incremental Cost of a Minimum Assignment

D. COPPERSMITH and G. B. SORKIN

For a complete bipartite graph with edge weights chosen randomly from some distribution, the random assignment problem is to determine the expected cost of a minimum-cost matching. For edge weights chosen from the exponential distribution with parameter 1, it is conjectured that the expected cost of a minimum-cost perfect matching in a complete $n \times n$ bipartite graph is exactly $\sum_{i=1}^{n} 1/i^2$, tending to $\pi^2/6$ in the limit. It is further conjectured that the expected cost of a minimum-cost matching of cardinality $k$ in a complete $m \times n$ bipartite graph is $F(k,m,n) := \sum_{i,j \geq 0,\ i+j<k} \frac{1}{(m-i)(n-j)}$.

In this note we show that, under certain hypotheses, the cost of augmenting a minimum $(m-1)$-assignment in an $(m-1) \times n$ bipartite graph, to a minimum $m$-assignment in an $m \times n$ bipartite graph, is equal to $F(m,m,n) - F(m-1,m-1,n)$. However, the hypotheses required are such that this result, while intriguing, does not provide a proof of the conjectures.

## 1. Motivation

Let $A$ be a complete $m \times n$ bipartite graph whose edge weights (or costs; we use the terms interchangeably) are chosen independently from the exponential distribution with mean 1. Equivalently, we may regard $A$ as an $m \times n$ matrix with i.i.d. exponential entries. A $k$-assignment is a choice of $k$ elements of the matrix $A$, all lying in different rows and different columns; a minimum $k$-assignment is one in which the sum of the chosen elements is minimized.

Where $\mathbb{E}A^\star(k,m,n)$ denotes the expectation of a minimum-cost $k$-assignment, Mézard and Parisi made the remarkable conjecture that as $n \to \infty$, $\mathbb{E}A^\star(n,n,n) \to \pi^2/6$ [13, 14]; an equally remarkable proof has just been announced by Aldous [1]. In the interim, however, Parisi developed a stronger conjecture [15], that

$$\mathbb{E}A^\star(n,n,n) = \sum_{i=1}^{n} 1/i^2.$$

In [4] we made the following conjecture, and showed that it agrees with Parisi's when $k = m = n$.

**Conjecture 1.**

$$\mathbb{E}A^\star(k,m,n) = \sum_{i,j\geq 0,\ i+j<k} \frac{1}{(m-i)(n-j)}.$$

We proved the conjecture in the cases $k \leq 3$ and $k = m = n = 4$, and it has since been proved for $k \leq 4$, $k = m = 5$, and $k = m = n = 6$ [2], and now for $k = 6$ and $k = m = n = 7$ [12].

The purpose of this note, a follow-up to [4], is to demonstrate a construction which might be used to give an inductive proof of Conjecture 1 when $k = m$. However, since we have not been able to make this connection (because we cannot see a way to enforce the hypotheses of Theorem 3), the present work is a hammer in search of a nail.

**Lemma 2.** *The minimum assignment problem corresponds to the linear program, and its dual, shown in Figure* 1.

**Proof.** The LP is a relaxation in which the selection variables $x_{ij}$ are permitted to come from the interval $[0,1]$ rather than the set $\{0,1\}$, so we need only show that it has an integral optimal solution. For the usual assignment problem, with $k = m = n$, one argues that this is so because the LP is a special case of the transportation problem, but this is not true of our LP.

| **Primal:** | **Dual:** |
|---|---|
| minimize $\sum_{i=1}^{m}\sum_{j=1}^{n} a_{ij}x_{ij}$ | maximize $kz - \sum_{i=1}^{m} u_i - \sum_{j=1}^{n} v_j$ |
| subject to | subject to |
| $x_{ij} \geq 0 \ (i = 1 \ldots m,\ j = 1 \ldots n)$ | $z \geq 0;\ u_i \geq 0 \ (i = 1 \ldots m);\ v_j \geq 0 \ (j = 1 \ldots n)$ |
| $\sum_{i=1}^{m}\sum_{j=1}^{n} x_{ij} \geq k$ | $a_{ij} - z + u_i + v_j \geq 0 \ (i = 1 \ldots m,\ j = 1 \ldots n)$ |
| $\sum_{j=1}^{n} x_{ij} \leq 1 \ (i = 1 \ldots m)$ | |
| $\sum_{i=1}^{m} x_{ij} \leq 1 \ (j = 1 \ldots n)$ | |

*Fig. 1.* The (primal) linear program describing the assignment problem for a matrix $A$, and its dual LP. The variables $x_{ij}$ in basic optimal solutions to the primal LP will have values of 0 or 1 (not fractional), $x_{ij} = 1$ indicating assignment of row $i$ to column $j$. We call the variables $u_i$ and $v_j$ row duals and column duals.

However, it is still true here that the extreme points are integral, because the constraint matrix is unimodular. The constraint matrix has the form

$$\begin{pmatrix} 1 & 1 & 1 & 1 & 1 & 1 & 1 & 1 & 1 & 1 & 1 & 1 \\ 1 & 1 & 1 & 1 & & & & & & & & \\ & & & & 1 & 1 & 1 & 1 & & & & \\ & & & & & & & & 1 & 1 & 1 & 1 \\ & & I & & & & I & & & & I & \end{pmatrix}$$

consisting of a first row with $mn$ 1's, a second group of $m$ rows each with $n$ 1's, and a third group of rows formed from $m$ identity matrices, each $n \times n$.

It is a classical result that a matrix is totally unimodular if every group of rows can be partitioned into two, such that when the rows within each group are summed, the difference between the sums is a vector whose entries are all in $\{-1,0,1\}$ [7]. Here there are just two cases. If the first row is *not* selected, then partition the selected rows into those of the second type and those of the third type; each sum is a 0–1 vector, and the difference is in $\{-1,0,1\}^{mn}$. If the first row *is* selected, partition into the first row and all other rows; the sum over the second partition is in $\{0,1,2\}^{mn}$, and subtracting the first gives a vector in $\{-1,0,1\}^{mn}$.

The dual LP is obtained from the primal by a standard, mechanical, transformation. ■

The (flawed) motivation for the main result described in this note is the following. Given a matrix $A$ and a corresponding optimal dual solution $(\mathbf{z},\mathbf{u},\mathbf{v}) = (z, u_1, \ldots, u_m, v_1, \ldots, v_n)$, consider the matrix $A'$ with the duals "subtracted out", $a'_{ij} = a_{ij} - z + u_i + v_j$. For convenience of phrasing, assume that $A$ is "in general position": that it has a unique minimum $k$-assignment and no events of probability 0 occur. (But note that even under these assumptions, an optimal *dual* is not usually unique.) Then $A'$ is elementwise non-negative, and its zero elements $a'_{ij}$ include the elements $a_{ij}$ of the (unique) minimum $k$-assignment.

For the case $k = m = n$, subtraction of a canonical optimal dual was the basis of Karp's upper bound of 2 for the expected cost of a minimum assignment [9]. Again when $k = m = n$, for any collection of feasible dual variables $(\mathbf{z},\mathbf{u},\mathbf{v})$, an optimal $n$-assignment for a matrix $A$ is also optimal for the corresponding residual matrix $A'$, and vice-versa; the problem instances $A$ and $A'$ are equivalent up to the additive constant $nz - \sum_{i=1}^n u_i - \sum_{j=1}^n v_j$. This was exploited in Lazarus's 1979 lower bound for random assignment [11]. Recent works including [4, 2, 3, 12] have relied on a similar equivalence between $A$ and a residual $A'$, where $A'$ is obtained from $A$ in ways different from but reminiscent of the subtraction of duals.

Such methods are most useful if the nonzero entries of the reduced matrix $A'$ are also independent, mean-1 exponentials. In fact, as per our Theorem 3 and Corollary 4, $A'$ is particularly amenable to analysis if it contains exactly $k$ zeros (corresponding to edges in the matching) and other entries which are independent, mean-1 exponentials. Unfortunately, the matrix $A'$ obtained from subtracting an optimal dual from a random matrix $A$ does not have these properties. For one thing, for a dual corresponding to an optimal $k$-assignment in $A$, the residual $A'$ may have as many as $2k+1$ zeros, with nonzero probability. Moreover, the remaining entries of $A'$ are not exponentially distributed, because the "subtracted" dual quanties $z - u_i - v_j$ can in fact be negative. For example, consider the matrix shown below, where the entries not shown numerically are large. A minimum 3-assignment, consisting of the diagonal $a_{11}$, $a_{22}$, $a_{33}$, has corresponding optimal row duals $u_i$ shown along its left side, column duals $v_j$ along the top edge, and $z$ at the corner.

$$\begin{array}{c|ccc} 7 & 7 & 4 & 0 \\ \hline 0 & 0 & a_{12} & a_{13} \\ 2 & a_{21} & 1 & 5 \\ 0 & a_{31} & a_{32} & 7 \end{array}$$

This gives $a'_{21} = a_{21} - 7 + 7 + 2 = a_{21} + 2$. For any fixed $c \geq 0$, and an exponential random variable $X$, conditional upon $X - c \geq 0$, the random variable $X - c$ is again exponentially distributed; but of course this is not true if $c < 0$. When $z - u_i - v_j$ *adds* to a value $a_{ij}$, it seems impossible to preserve the exponential distribution.

Nonetheless, such considerations suggest that the following theorem might be of use.

## 2. THE COST OF AN ALTERNATING PATH

**Theorem 3.** *Let $A$ be an $m \times n$ bipartite graph whose edge weights $a_{11}, a_{22}, \ldots, a_{kk}$ are all 0, and whose other edge weights are i.i.d. exponential rv's. Then the expected cost of extending the 0-cost $k$-assignment to a minimum $(k+1)$-assignment in $A$ is*

$$\Delta(k, m, n) = \sum_{r=0}^{k} \frac{\binom{k}{r}}{\binom{n}{r}} \cdot \frac{1}{(m - k + r)(n - r)}.$$

**Proof.** Recall that with respect to a given matching $\mathcal{M}$, an alternating path is a path that alternates between edges in $\mathcal{M}$ ("matching edges") and edges not in $\mathcal{M}$ ("non-matching edges"). If the first and last vertices of an alternating path are both unmatched, the path is called "augmenting": its symmetric difference with $\mathcal{M}$ is a new matching $\mathcal{M}'$ whose cardinality is one larger than $\mathcal{M}$'s. The cost of an alternating path is the sum of its non-matching edge weights, minus the sum of its matching edge weights; in this context the matching edge weights are 0.

It is a standard result in matching theory that an optimal $k$-matching may be augmented to an optimal $(k+1)$-matching via a shortest alternating path from any of the $m-k$ previously unmatched "left vertices" (rows) to any of the $n-k$ previously unmatched right vertices (columns). (The result is elementary by consideration of the symmetric difference between optimal $k$- and $(k+1)$-matchings, but discussions may be found in [3, 10].)

An augmenting, alternating path is uniquely described by its non-matching edges, from $j_0 \in \{k+1, \ldots, m\}$ to some $j_1$, from $j_1$ to $j_2$, and so forth, until an (unmatched) $j \in \{k+1, \ldots, n\}$ is reached. If the path has length $2\ell+1$, its cost is $\sum_{r=0}^{\ell} a(j_r, j_{r+1})$.

Let $D$ be a directed non-bipartite analogue of $A$, having self-loops with weights $d_{ii} = a_{ii}$ for $i = k+1, \ldots, m$, having edges with weights $d_{ij} = a_{ij}$ for $i \in [m]$ and $j \in [n]$, $i \neq j$, and having no other edges (in particular no out-edges from $i \in \{m+1 \ldots n\}$). In $D$, distinguish vertices $k+1, \ldots, m$ as "sources" and vertices $k+1, \ldots, n$ as "terminals". (Some sources are also terminals.) Then each augmenting path in $A$ corresponds to a simple path in $D$, and a minimum-cost augmenting path in $A$ corresponds to a shortest non-trivial path from a source to a terminal in $D$.

Find such a path using Dijkstra's algorithm [5], as follows.[1] Initialize $S = S(0)$ to be the sources $k+1, \ldots, m$ and $T = T(0)$ to be $1, \ldots, n$. For $i \in S$, let $\operatorname{dist}(i)$ be the weight of the shortest path from any source to $i$, and let $\operatorname{dist}(S) = \max_{i \in S} \operatorname{dist}(i)$. Initially, for any $i \in \{k+1, \ldots, m\}$, $\operatorname{dist}(i) = \operatorname{dist}\big(S(0)\big) = 0$. We assert that for every $r$, the weights $a_{ij}$ of edges $ij$, $i \in S(r)$, $j \in T(r)$, are i.i.d. exponential rv's conditioned upon

$$\operatorname{dist}(i) + a_{ij} \geq \operatorname{dist}\big(S(r)\big). \tag{1}$$

At step 0 this means they are not conditioned, which is true by hypothesis. We will show inductively that the assertion (1) continues to hold.

Consider taking step $r$, which adds a vertex to $S(r)$ and removes the same vertex from $T(r)$, giving $S(r+1)$ and $T(r+1)$. Before this step, by

[1] The argument from this point through the paragraph following (2) also appears in [8], in the language of percolation in lieu of Dijkstra's algorithm.

the inductive hypothesis, the weights $a_{ij}$ of edges $ij$, $i \in S(r)$, $j \in T(r)$, are i.i.d. exponential rv's conditioned upon $\text{dist}(i) + a_{ij} \geq \text{dist}(S)$. (Also, if $\text{dist}(i) + a_{ij}$ were less than $\text{dist}(S)$, vertex $j$ would already be in $S$, not in $T$.) Equivalently, the random variables

$$X_{ij} = \text{dist}(i) + a_{ij} - \text{dist}(S)$$

form an i.i.d. exponential collection.

In step $r$, Dijkstra's algorithm adds to $S(r)$ the vertex $j(r+1) \in T(r)$ which satisfies

$$\text{(2)} \qquad \begin{aligned} \text{dist}\big(j(r+1)\big) &= \min_{i \in S,\ j \in T} \{\, \text{dist}(i) + a_{ij}\} \\ &= \text{dist}(S) + \min_{i \in S,\ j \in T} \{X_{ij}\} \end{aligned}$$

By definition, $\text{dist}\big(S(r+1)\big) = \text{dist}\big(j(r+1)\big)$.

The asserted conditioning (1) now holds for edges $uv \in S(r+1) \times T(r+1)$. If $u \in S(r)$, then $uv \in S(r) \times T(r)$ was one of the edges considered in (2), therefore $\text{dist}(u) + a_{uv} \geq \text{dist}\big(S(r+1)\big)$. Otherwise $u = j(r+1)$, and the edges out of $u$, never having been examined before, are unconditioned, or equivalently, "conditioned" by $a_{uv} \geq 0$; since $\text{dist}(u) = \text{dist}\big(j(r+1)\big) = \text{dist}\big(S(r+1)\big)$, this is equivalent to (1).

If $j$ is any terminal $k+1, \ldots, n$, then Dijkstra's algorithm terminates. What is the expected distance from the sources to this terminal?

In step $r$, when the $m-k+r+1$'st vertex is about to be added to $S(r)$, $S(r)$ contains $m-k+r$ vertices (including the original sources $k+1, \ldots, m$), and $T(r)$ contains $n-r$ vertices (the original $1, \ldots, n$ less the vertices shifted into $S$). Thus, the minimum in (2) is the minimum of $(m-k+r)(n-r)$ i.i.d. exponential rv's, and has expectation $1/\big[(m-k+r)(n-r)\big]$. It follows that

$$\mathbb{E}\big(\text{dist}\big(S(r+1)\big)\big) - \mathbb{E}\big(\text{dist}\big(S(r)\big)\big) = 1/\big[(m-k+r)(n-r)\big].$$

Dijkstra's algorithm continues until a terminal is hit. Since the vertices added are by construction distinct, but otherwise random (in (2), the residual edge costs $X_{ij}$ out of $S$ are i.i.d.), the probability that all the vertices hit in $r$ steps are among the non-terminals $1, \ldots, k$ (and Dijkstra's algorithm needs to be continued) is $\binom{k}{r}/\binom{n}{r}$. Note that after $k+1$ steps, failure is impossible, as there are only $k$ non-terminal vertices, and this is reflected in $\binom{k}{k+1} = 0$.

The expected distance to the terminal hit can be computed by summing, over $r$, the probability that a terminal has not yet been hit, times the cost to add the next vertex in this event. The sum is

$$\sum_{r=0}^{k} \frac{\binom{k}{r}}{\binom{n}{r}} \cdot \frac{1}{(m-k+r)(n-r)}. \qquad \blacksquare$$

The next corollary may be interpreted as saying that the expected cost of augmenting an $(m-1)$-assignment to an $m$-assignment, in an $m \times n$ bipartite graph, and under the conditions of Theorem 3, is in agreement with Conjecture 1.

**Corollary 4.** *With* $F(k,m,n) := \sum_{i,j \geq 0,\ i+j<k} \frac{1}{(m-i)(n-j)}$ *per Conjecture 1, and* $\Delta(m-1,m,n)$ *as given by Theorem 3,* $\Delta(m-1,m,n) = F(m,m,n) - F(m-1,m-1,n)$.

**Proof.** Pairing off corresponding terms, $F(m,m,n) - F(m-1,m-1,n) = \frac{1}{m}\sum_{i=0}^{m-1}\frac{1}{n-i}$. To show that

$$\sum_{r=0}^{m-1} \frac{\binom{m-1}{r}}{\binom{n}{r}} \frac{1}{(r+1)(n-r)} = \frac{1}{m}\sum_{i=0}^{m-1}\frac{1}{n-i},$$

it suffices to show equality after multiplication by $n(n-1)\cdots(n-m+1)$, that is, to show that

$$\sum_{r=0}^{m-1} \frac{(m-1)!(n-r-1)!n!}{(m-1-r)!n!(r+1)(n-m)!}$$

$$= \frac{1}{m}\sum_{i=0}^{m-1} n(n-1)\cdots\widehat{(n-i)}\cdots(n-m+1),$$

where $\widehat{(n-i)}$ indicates absence of this term from the product. Equivalently, and writing $N$ in lieu of $n$ to emphasize the dependence on this parameter, we wish to show that

$$\sum_{r=0}^{m-1} \frac{(m-1)!(N-r-1)\cdots(N-m+1)}{(m-1-r)!(r+1)}$$

$$= \frac{1}{m}\sum_{i=0}^{m-1} N(N-1)\cdots\widehat{(N-i)}\cdots(N-m+1).$$

Now observe that as functions of $N$, both sides of the would-be equality above are simply polynomials of degree $m-1$. Thus it suffices to prove equality at any $m$ distinct values of $N$, and we shall use $N = 0, \ldots, m-1$. (So now $N < m$, whereas $n \geq m$ in the original context.) In this range, terms of the left side vanish when $r \leq N-1$, and terms of the right side vanish except when $i = N$, so the desired equality may be rewritten as

$$\text{(3)} \qquad \sum_{r=N}^{m-1} \frac{(m-1)!(-1)^{m-r-1}(m-1-N)\cdots(r+1-N)}{(m-1-r)!(r+1)}$$

$$= (-1)^{m-1-N}\frac{1}{m}N!(m-1-N)!$$

Attacking the left side,

$$\sum_{r=N}^{m-1} \frac{(m-1)!(-1)^{m-r-1}(m-1-N)\cdots(r+1-N)}{(m-1-r)!(r+1)}$$

$$= \sum_{r=N}^{m-1} \frac{(m-1)!(-1)^{m-r-1}(m-1-N)!}{(m-1-r)!(r+1)(r-N)!},$$

which, substituting $s = r - N$,

$$= \sum_{s=0}^{m-1-N} \frac{(m-1)!(-1)^{m-s-N-1}(m-1-N)!}{(m-1-s-N)!(s+N+1)s!}$$

$$= (m-1)! \sum_{s=0}^{m-1-N} \binom{m-1-N}{s} \frac{(-1)^{m-s-N-1}}{s+N+1},$$

which, employing

$$\sum \binom{n}{k} \frac{1}{k+c+1}(-1)^{n-k} = \sum \binom{n}{k} (\int_0^1 x^{k+c} dx)(-1)^{n-k}$$
$$= \int_0^1 x^c \sum \binom{n}{k} x^k (-1)^{n-k} dx = \int_0^1 x^c (x-1)^n dx,$$

leads to

$$= (m-1)! \int_0^1 (1-x)^{m-1-N} x^N dx (-1)^{m-1-N},$$

and by iterated integration by parts

$$= (m-1)! \frac{(m-1-N)!N!}{m!}(-1)^{m-1-N}$$

$$= \frac{1}{m} N!(m-1-N)!(-1)^{m-1-N},$$

matching the right side of (3) as desired. ■

## 3. Conclusions

On the negative side, Corollary 4, which refers to an increment $\Delta(m-1, m, n)$, fails to generalize to arbitrary increments $\Delta(k, m, n)$. For example, as per the Corollary, we know that $\Delta(3, 4, 6)$ is equal to $F(4, 4, 6) - F(3, 3, 6)$.

On the other hand, $\Delta(3,5,6)$ is not equal to $F(k,m,n) - F(k',m',n')$ for any values $k,m,n,k',m',n' \in \{0,\ldots,7\}$.

It is an open question whether Corollary 4 is merely an algebraic coincidence, or whether it might be used to prove the correctness of Conjecture 1 in the case $k = m$. Such a proof might require producing the unbiased edges of Theorem 3 from the biased edges that result when optimal dual variables are subtracted from the original edge weights, or from the biases introduced by growing a minimum $k$-assignment through successive augmentations.

Alternatively, it is possible that biases could be tolerated, using Theorem 3 to produce a bound on the expected assignment cost, rather than its exact value.

Recently, Dotsenko [6] exploited an equivalent of Corollary 4 (independently derived) for a provocative but unfortunately incomplete proof of Parisi's conjecture.

**Acknowledgments.** We thank Jon Lee and David Williamson for helpful discussions regarding integrality of the LP and the construction of its dual. We also appreciate the comments of an anonymous referee.

## References

[1] D. Aldous, *The $\zeta(2)$ limit in the random assignment problem,* currently available as http://www.stat.berkeley.edu/~aldous/me94.ps.Z, September 2000.

[2] S. E. Alm and G. B. Sorkin, On the expectation and variance of a minimum assignment, *Random Structures and Algorithms,* to appear.

[3] M. W. Buck, C. S. Chan, and D. P. Robbins, *On the expected value of the minimum assignment,* Mathematics arXiv, http://arxiv.org/ps/math/0004175v2.

[4] D. Coppersmith and G. B. Sorkin, Constructive bounds and exact expectations for the random assignment problem, *Random Structures and Algorithms,* **15** (1999), 133–144.

[5] E. W. Dijkstra, A note on two problems in connexion with graphs, *Numerische Mathematik,* **1** (1959), 269–271.

[6] V. S. Dotsenko, Exact solution of the random bipartite matching model, *J. Phys. A,* **33** (2000), 2015–2030.

[7] Alain Ghouila-Houri, Caractérisation des matrices totalement unimodulaires, *C. R. Acad. Sci. Paris,* **254** (1962), 1192–1194.

[8] S. Janson, One, two and three times $\log n/n$ for paths in a complete graph with random weights, *Combinatorics, Probability, and Computing,* **8** (1999), 347–361.

[9] Richard M. Karp, *An upper bound on the expected cost of an optimal assignment,* Discrete Algorithms and Complexity: Proceedings of the Japan-US Joint Seminar (David S. Johnson et al., eds.), Perspectives in Computing, vol. 15, Academic Press, 1987, pp. 1–4.

[10] E. L. Lawler, *Combinatorial optimization: networks and matroids,* Holt, Rinehart and Winston (New York, 1976).

[11] A. J. Lazarus, Certain expected values in the random assignment problem, *Oper. Res. Lett.,* **14** (1993), no. 4, 207–214.

[12] S. Linusson and J. Wästlund, *A generalization of the random assignment problem,* Mathematics arXiv, http://arxiv.org/ps/math/0006146.

[13] M. Mézard and G. Parisi, Replicas and optimization, *J. Physique Lettres,* **46** (1985), 771–778.

[14] M. Mézard and G. Parisi, On the solution of the random link matching problems, *J. Physique Lettres,* **48** (1987), 1451–1459.

[15] G. Parisi, *A conjecture on random bipartite matching,* Physics e-Print archive, http://xxx.lanl.gov/ps/cond-mat/9801176, January 1998.

Don Coppersmith
*IBM T.J. Watson Research Center*
*Department of Mathematical Sciences*
*Yorktown Heights NY 10598*
e-mail: copper@watson.ibm.com

Gregory B. Sorkin
*IBM T.J. Watson Research Center*
*Department of Mathematical Sciences*
*Yorktown Heights NY 10598*
e-mail: sorkin@watson.ibm.com

BOLYAI SOCIETY
MATHEMATICAL STUDIES, 10

**Contemporary Combinatorics**
**Budapest, 2002,** pp. 289–301.

# On Concentration of Probability

S. JANSON

We give a survey of several methods to obtain sharp concentration results, typically with exponentially small error probabilities, for random variables occuring in combinatorial probability.

## Introduction

In probabilistic combinatorics, it is often important to show that a random variable is sharply concentrated about its mean. A well-known, simple, but rather weak result of this type is Chebyshev's inequality

$$\mathbb{P}\big(|X - \mathbb{E}\,X| \geq t\big) \leq \operatorname{Var}(X)/t^2,$$

which holds for any random variable $X$ with finite variance and any $t > 0$. This inequality is very general (and very useful), but is in many cases too weak. The purpose of this paper is to give a survey of some more or less recent stronger inequalities, which under suitable assumptions yield estimates that decrease exponentially as $t \to \infty$.

This paper is based on lectures at the workshop "Probabilistic combinatorics" at the Paul Erdős center in Budapest 1998.

## 1. Sums of independent variables

An important case is when the variable $X$ can be written as a sum $\sum_1^n X_i$ of independent random variables. We consider here only the case when each $X_i$ is an indicator variable; thus $X_i \in \mathrm{Be}(p_i)$ where $p_i = \mathbb{P}(X_i = 1) = \mathbb{E}\, X_i$. (The results hold more generally, assuming only that $0 \le X_i \le 1$.) Let $\lambda = \mathbb{E}\, X = \sum_1^n p_i$.

Consider first the case of a binomially distributed variable $X \in \mathrm{Bi}(n,p)$; this is of the type above with all $p_i = p$, and thus $\lambda = np$. Applying Markov's inequality to $e^{uX}$, one finds

$$\text{(1)} \quad \mathbb{P}(X \ge \lambda + t) \le e^{-u(\lambda+t)}\, \mathbb{E}\, e^{uX} = e^{-u(\lambda+t)}(1 - p + pe^u)^n, \quad u \ge 0.$$

The right hand side attains its minimum at $e^u = (\lambda + t)(1-p)/(n - \lambda - t)p$, assuming $0 < \lambda + t < n$, which yields

$$\text{(2)} \quad \mathbb{P}(X \ge \mathbb{E}\, X + t) \le \left(\frac{\lambda}{\lambda + t}\right)^{\lambda+t} \left(\frac{n-\lambda}{n-\lambda-t}\right)^{n-\lambda-t}, \quad 0 \le t \le n - \lambda;$$

for $t > n - \lambda$ the probability is 0. This bound is implicit in Chernoff [8] and appears explicitly in Okamoto [19]. For applications, it is usually convenient to replace the right hand side of (2) by a larger but simpler bound. Our favourites are the following, which follow from (2) by some calculus.

**Theorem 1.** *If $X \in \mathrm{Bi}(n,p)$ and $\lambda = np$, then*

$$\text{(3)} \qquad \mathbb{P}(X \ge \mathbb{E}\, X + t) \le \exp\left(-\frac{t^2}{2(\lambda + t/3)}\right), \qquad t \ge 0;$$

$$\text{(4)} \qquad \mathbb{P}(X \le \mathbb{E}\, X - t) \le \exp\left(-\frac{t^2}{2\lambda}\right), \qquad t \ge 0.$$

**Remark.** The inequality (3) would not hold in general without the term $t/3$ in the denominator.

We now return to the case when $X_i \in \mathrm{Be}(p_i)$ with different $p_i$. Using Jensen's inequality, it is easily seen that (1) holds in this case too, with

$p = \lambda/n = \sum p_i/n$. Consequently, the following generalization of Theorem 1 holds.

**Theorem 2.** *If $X_i \in \mathrm{Be}(p_i)$, $i = 1, \dots, n$, are independent and $X = \sum_1^n X_i$, then* (2), (3) *and* (4) *hold, with $\lambda = \mathbb{E}\, X$.*

Many other similar bounds, often sharper than (3) and (4), have been proved under these, or more general, assumptions by the same method (which goes back at least to Bernstein [4]); see e.g. [3], [11] and [1, Appendix A].

**Remark.** Much more precise estimates of the tail probabilities for sums of independent, but not necessarily identically distributed, random variables were obtained by Feller [10] using different methods. Feller's result implies, for the cases above and with $\sigma^2 = \operatorname{Var} X$,

$$\mathbb{P}(X \geq \mathbb{E}\, X + t) = e^{\theta_1 \frac{t}{\sigma^2} x^2} \left(1 - \Phi(x) + \frac{\theta_2}{\sigma} e^{-x^2/2}\right), \qquad 0 < t < \sigma^2/12,$$

where $x = t/\sigma$, $\Phi$ is the normal distribution function, $|\theta_1| \leq 6/7(1 - 12t/\sigma^2)$ and $|\theta_2| < 9$.

Very precise asymptotic results for the binomial distribution are given by Littlewood [16].

**Some related distributions.** The Poisson distribution $\mathrm{Po}(\lambda)$ is a limit of binomial distributions, and thus Theorem 1 implies the same bounds for the Poisson case.

**Theorem 3.** *If $X \in \mathrm{Po}(\lambda)$, then* (3) *and* (4) *hold.*

Furthermore, as a special case of a result in [22], a hypergeometric distribution is the distribution of a certain sum of independent indicator variables. (The proof is algebraic; there is no known probabilistic interpretation of these indicators, which in general have irrational expectations.) Consequently we can apply Theorem 2.

**Theorem 4.** *Let $X$ have a hypergeometric distribution with parameters $N$, $n$ and $m$. Then (3) and (4) hold, with $\lambda = \mathbb{E}\, X = mn/N$.*

(This was proved more directly by Hoeffding [11].)

## 2. Sums of dependent variables

In this section we discuss random variables of the type $X = \sum_{\alpha \in \mathcal{A}} I_\alpha$, where $I_\alpha \in \mathrm{Be}\,(p_\alpha)$ are indicator variables as above, but now they may be dependent. (However, the inequalities below are useful only in cases when "most" variables are independent.) In order to quantify the dependence, we say that a graph $\Gamma$ is a *dependency graph* for the family $\{I_\alpha\}_{\alpha \in \mathcal{A}}$ if the vertex set $V(\Gamma) = \mathcal{A}$ and two subfamilies $\{I_\alpha\}_{\alpha \in A}$ and $\{I_\alpha\}_{\alpha \in B}$ are independent whenever $A$ and $B$ are two disjoint subsets of $\mathcal{A}$ such that there is no edge in $\Gamma$ connecting $A$ and $B$.

**Remark.** A slightly weaker definition of dependency graph is commonly used for the Lovasz local lemma [1].

We concentrate for simplicity on bounds for $\mathbb{P}\,(X = 0)$, i.e., on the probability that $I_\alpha = 0$ for every $\alpha \in \mathcal{A}$. Similar bounds exist for the lower tail probabilities $\mathbb{P}\,(X \leq \mathbb{E}\, X - t)$, $t \geq 0$, but not for the upper tail [12, 13].

A special case, which has had many applications in the last decade, is when there is an underlying family $\{J_j\}_{j \in \mathcal{J}}$ of *independent* random indicator variables, each index $\alpha$ is a (nonempty) subset of $\mathcal{J}$ and $I_\alpha = \prod_{j \in \alpha} J_j$. In other words, if $W$ denotes the random set $\{j \in \mathcal{J} : J_j = 1\}$, then $I_\alpha$ is the indicator of the event $\alpha \subseteq W$, where $\alpha \in \mathcal{A}$, an arbitrary family of nonempty subsets of $\mathcal{J}$. In this case, we define a dependency graph $\Gamma$ by taking $V(\Gamma) = \mathcal{A}$ and letting the edge set consist of the edges $\alpha\beta$ for all pairs $\alpha, \beta \in \mathcal{A}$ with $\alpha \neq \beta$ and $\alpha \cap \beta \neq \emptyset$.

We fix a dependency graph $\Gamma$ for $\{I_\alpha\}$ and write $\alpha \sim \beta$ if $\alpha, \beta \in \mathcal{A}$ and there is an edge in $\Gamma$ between $\alpha$ and $\beta$. (In particular, then $\alpha \neq \beta$.) We further write $\alpha \sim \{\beta, \gamma\}$ if $\alpha \sim \beta$ or $\alpha \sim \gamma$, and define

- $\lambda = \mathbb{E}\, X = \sum_{\alpha \in \mathcal{A}} p_\alpha$

- $\Delta = \sum_{\{\alpha,\beta\}\,:\,\alpha\sim\beta} \mathbb{E}\,(I_\alpha I_\beta)$, summing over *unordered* pairs $\{\alpha, \beta\}$, i.e. over the edges in $\Gamma$. As a sum over ordered pairs, $\Delta = \frac{1}{2}\sum_{\alpha\in\mathcal{A}}\sum_{\beta\sim\alpha} \mathbb{E}\,(I_\alpha I_\beta)$.
- $\delta_\alpha = \sum_{\beta\sim\alpha} p_\beta$.
- $\delta = \max_{\alpha\in\mathcal{A}} \delta_\alpha$.
- $\varepsilon = \max_{\alpha\in\mathcal{A}} p_\alpha$.

In the special case we thus have $\Delta = \frac{1}{2}\sum\sum_{\alpha\neq\beta,\,\alpha\cap\beta\neq\emptyset} \mathbb{E}\,(I_\alpha I_\beta)$. Then, the following bounds hold [14].

**Theorem 5.** *In the special case, with $X = \sum I_A$, $\lambda = \mathbb{E}\,X$ and $\Delta$ as above,*

$$\text{(5)} \qquad \mathbb{P}\,(X = 0) \le \exp\,(-\lambda + \Delta);$$

$$\text{(6)} \qquad \mathbb{P}\,(X = 0) \le \exp\left(-\frac{\lambda^2}{\lambda + 2\Delta}\right) = \exp\left(-\frac{\lambda^2}{\sum\sum_{\alpha\cap\beta\neq\emptyset} \mathbb{E}\,(I_\alpha I_\beta)}\right).$$

*The first bound is best for $\Delta \le \lambda/2$, while the second is better for larger $\Delta$.*

**Remark.** Boppana and Spencer [6] proved a slightly different version of (5), viz.

$$\text{(7)} \qquad \mathbb{P}\,(X = 0) \le e^{\Delta/(1-\varepsilon)} \prod_{\alpha\in\mathcal{A}} (1 - p_\alpha),$$

which has the advantage of being exact when $\Delta = 0$ (i.e. when the variables $I_\alpha$ are independent), but otherwise seems slightly less convenient for applications. Although the upper bounds (5) and (7) are quite close when $\varepsilon = \max p_\alpha$ is small, neither of them dominates the other. It is intriguing to note that the conceivable common improvement $e^{\Delta}\prod_\alpha(1 - p_\alpha)$ fails to be an upper bound to $\mathbb{P}\,(X = 0)$; this is seen by the simple example where $X = I_1 + I_2$ with $I_1 = I_2 \in \mathrm{Be}\,(p)$, for which $\Delta = p$ and $\mathbb{P}\,(X = 0) = 1 - p > e^p(1 - p)^2$, cf. [13].

Note also that in the special case, by the FKG inequality, we have the lower bound

$$\mathbb{P}(X=0) \geq \prod_{A\in\mathcal{A}} (1-p_A) \geq \exp\left(-\frac{\lambda}{1-\varepsilon}\right),$$

which shows that the bounds above are quite sharp when $\Delta$ and $\varepsilon$ are small.

The inequalities (5)–(7) have been widely used, although they apply only to the sum of indicator variables with a very special structure. For example, they apply to the number of copies of a given graph in $\mathbb{G}(n,p)$, but they do not apply to the number of induced copies.

In a paper which has too long been neglected, Suen [20] proved, at about the same time as [14, 6] and independently of them, a similar inequality for the general case. The Suen inequality and later versions of it [13] yield almost as good bounds as (5)–(7), but are thus much more widely applicable.

We do not give Suen's original inequality [20] here, but rather a slight improvement of it (8) followed by two consequences of it [13], which are more convenient for applications and closely resemble (5) and (6) for the special case above.

**Theorem 6.** *Let $X=\sum_\alpha I_\alpha$, where $\{I_\alpha\}_{\alpha\in\mathcal{A}}$ is a finite family of Bernoulli random variables having a dependency graph $\Gamma$. Then, with the notation above,*

$$(8)\quad \mathbb{P}(X=0) \leq \exp\left(\sum_{\{\alpha,\beta\}\,:\,\alpha\sim\beta} \mathbb{E}(I_\alpha I_\beta) \prod_{\gamma\sim\{\alpha,\beta\}} (1-p_\gamma)^{-1}\right) \prod_{\nu\in\mathcal{A}} (1-p_\nu);$$

$$(9)\quad \mathbb{P}(X=0) \leq \exp(-\lambda+\Delta e^{2\delta});$$

$$(10)\quad \mathbb{P}(X=0) \leq \exp\left(-\min\left(\frac{\lambda^2}{8\Delta},\frac{\lambda}{2},\frac{\lambda}{6\delta}\right)\right) = e^{-\lambda^2/\max(8\Delta,2\lambda,6\delta\lambda)}.$$

The bounds (9) and (10) are similar to the ones in (5) and (6), but with extra terms involving $\delta$ and somewhat worse constants in (10). In applications, $\delta$ is usually small and the bounds (9) and (10) are often as useful as (5) and (6). Nevertheless, in at least one application (a recent bound for the 3-SAT problem by Kirousis, Kranakis, Krizanc and Stamatiou [15], where $\delta \approx 0.089$), the extra factor $e^{2\delta}$ in (9) affects the final result significantly, and it is desirable to reduce it as much as possible. It is shown in [13], using a proof by Spencer (personal communication), that if $0 \leq \delta+\varepsilon$, this factor $e^{2\delta}$ can be replaced by the smallest root $\varphi$ of $\varphi = e^{(\delta+\varepsilon)\varphi}$; this is the version used in [15].

It is an open problem whether the factor $e^{2\delta}$ in (9) can be eliminated completely, i.e. whether (5) (or (6), (7)) holds in the general case too.

## 3. Azuma's inequality

The final inequalitites that we consider apply to random variables of the form $X = f(Z_1, \dots, Z_N)$, where $Z_1, \dots, Z_N$ are independent random variables; here the underlying random variables $Z_k$ can take values in arbitrary (and possibly different) spaces $\Lambda_k$. Our basic assumption is the following Lipschitz condition, for some numbers $c_1, \dots, c_N$. (In most applications, $c_k = 1$.)

(L) If the vectors $z,\ z' \in \prod_1^N \Lambda_i$ differ only in the $k$th coordinate, then $\left|f(z) - f(z')\right| \leq c_k$, $k = 1, \dots, N$.

In combinatorial applications, one often considers a random subset $W$ of a set $A = \{a_1, \dots, a_N\}$, and sets $Z_k = \mathbf{1}[a_k \in W]$, the indicator variable of the event $\{a_k \in W\}$.

More generally, we may partition the ground set $A$ into blocks $B_k$ and let $Z_k$ be the vector of the indicator variables $\big(\mathbf{1}[a_i \in W]\big)_{i \in B_k}$ (thus $Z_k$ takes values in the product space $\{0,1\}^{B_k}$). A common instance of this construction is "vertex exposure" for a random graph $\mathbb{G}(n,p)$, where $A = \big\{\{i,j\} : 1 \leq i < j \leq n\big\}$ and $B_k = \big\{\{i,k\} : 1 \leq i < k\big\}$, $k = 1, \dots, n$.

The first result for this situation is based on the following martingale inequality, known as Azuma's inequality [2], although it was earlier proved by Hoeffding [11].

**Theorem 7.** *If $(X_k)_0^N$ is a martingale, and there exist constants $c_k \geq 0$ such that $|X_k - X_{k-1}| \leq c_k$, $k = 1, \ldots, N$, then*

$$\mathbb{P}(X_N \geq X_0 + t) \leq \exp\Bigl(-t^2/2\sum_1^N c_k^2\Bigr), \qquad t \geq 0.$$

For applications to our situation, let $X_k = \mathbb{E}(X \mid Z_1, \ldots, Z_k)$; then the assumptions of Theorem 7 are satisfied, and thus

$$\mathbb{P}(X \geq \mathbb{E}X + t) \leq \exp\Bigl(-t^2/2\sum_1^N c_k^2\Bigr).$$

In fact, the exponent here (but not in Theorem 7) may be improved by a factor 4 [17].

**Theorem 8.** *Let $Z_1, \ldots, Z_N$ be independent random variables, with $Z_k$ taking values in a set $\Lambda_k$. Assume that a (measurable) function $f : \Lambda_1 \times \Lambda_2 \times \ldots \times \Lambda_N \to \mathbb{R}$ satisfies the Lipschitz condition* (L). *Then, the random variable $X = f(Z_1, \ldots, Z_N)$ satisfies, for any $t \geq 0$,*

$$\mathbb{P}(X \geq \mathbb{E}X + t) \leq \exp\Bigl(-2t^2\sum_1^N c_k^2\Bigr), \tag{11}$$

$$\mathbb{P}(X \leq \mathbb{E}X - t) \leq \exp\Bigl(-2t^2\sum_1^N c_k^2\Bigr). \tag{12}$$

For further related results, including an extension to unbounded martingale differences, see the survey by Bollobás [5]. For another extension (weakening (L) by a truncation), see [7].

## 4. Talagrand's inequality

Talagrand [21] has given several inequalities on concentration of measure, with various applications. We will here treat one of his inequalities only; it yields results similar to those obtained by Azuma's inequality, although often much stronger, and it has already had important combinatorial applications.

**Remark.** The interested reader should study also the other inequalities by Talagrand in [21] and subsequent papers; it seems reasonable to guess that some of them too will be useful in combinatorial applications. Moreover, there are other proofs of the results below, with extensions in different directions, by Marton [18] and Dembo [9].

We continue with the assumptions of the preceding section, in particular condition (L), and assume furthermore the following, for some function $\psi$ and the same $c_k$ as in (L).

(C) If $z = (z_k)_1^N \in \prod_1^N \Lambda_k$ and $r \in \mathbb{R}$ with $f(z) \geq r$, then there exists a set $J \subseteq \{1, \ldots, N\}$ with $\sum_{k \in J} c_k^2 \leq \psi(r)$, such that for all $y = (y_k)_1^N \in \prod_1^N \Lambda_k$ with $y_k = z_k$ when $k \in J$, we have $f(y) \geq r$.

In other words, there exists a vector $(z_k)_{k \in J}$ (called a *certificate*), which forces $f \geq r$, such that the index set $J$ is not too large. (The set $J$ generally depends on $z$ and $r$.)

Every function $f$ trivially satisfies (C) with $\psi(r) = \sum_1^N c_k^2$ for all $r$; just take $J = \{1, \ldots, N\}$. This choice of $\psi$ in the theorem below yields an inequality similar to the one in Theorem 8, but generally somewhat weaker. In many interesting applications, however, (C) holds with a much smaller $\psi$; this leads to stronger estimates that significantly surpass Azuma's inequality. (It is immaterial that we obtain inequalities for the deviation from the median instead of the mean.)

**Example.** Let $X$ be the order of the largest independent set in the random graph $\mathbb{G}(n, p)$. Then, using vertex exposure as described above, (L) holds with $c_k = 1$, and (C) holds with $\psi(r) = \lceil r \rceil$ for $r \geq 0$; the certificate is just any independent set of order $\lceil r \rceil$.

Talagrand's inequality may then be stated as follows.

**Theorem 9.** *Suppose that $Z_1, \dots, Z_N$ are independent random variables taking their values in some sets $\Lambda_1, \dots, \Lambda_N$, respectively. Suppose further that $X = f(Z_1, \dots, Z_N)$, where $f : \Lambda_1 \times \cdots \times \Lambda_N \to \mathbb{R}$ is a (measurable) function such that* (L) *and* (C) *hold for some constants $c_k$, $k = 1, \dots, N$, and some function $\psi$. Then, for every $r \in \mathbb{R}$ and $t \geq 0$,*

$$\mathbb{P}(X \leq r - t)\,\mathbb{P}(X \geq r) \leq e^{-t^2/4\psi(r)}. \tag{13}$$

*In particular, if $m$ is a median of $X$, then for every $t \geq 0$,*

$$\mathbb{P}(X \leq m - t) \leq 2e^{-t^2/4\psi(m)} \tag{14}$$

*and*

$$\mathbb{P}(X \geq m + t) \leq 2e^{-t^2/4\psi(m+t)}. \tag{15}$$

Actually, the theorem above is a simple corollary of the following more general inequality due to Talagrand [21].

**Theorem 10.** *Suppose that $A$ and $B$ are two (measurable) subsets of $\prod_1^N \Lambda_k$ such that for some $t \geq 0$ the following separation condition holds: for every $z \in B$, there exists a non-zero vector $\alpha = (\alpha_i)_1^N \in \mathbb{R}^N$ such that for every $y \in A$,*

$$\sum_{i\,:\,y_i \neq z_i} \alpha_i \geq t\Big(\sum_1^N \alpha_i^2\Big)^{1/2}.$$

*Then*

$$\mathbb{P}(A)\,\mathbb{P}(B) \leq e^{-t^2/4}.$$

**Remark.** Talagrand [21, Corollary 4.2.5] further showed that the conclusion of Theorem 10 can be improved to

$$\sqrt{\log\left(1/\,\mathbb{P}(A)\right)} + \sqrt{\log\left(1/\,\mathbb{P}(B)\right)} \geq t/\sqrt{2},$$

which implies for example that (14) can be improved to the smaller, but more complicated, bound

$$\mathbb{P}(X \leq m - t) \leq \exp\left(-\frac{1}{2\psi(m)}\left(t - \sqrt{2\log 2\,\psi(m)}\right)^2\right),$$

$$t \geq \sqrt{2\log 2\,\psi(m)}.$$

**Acknowledgement.** This survey is based on a chapter in a forthcoming book on random graphs that I am writing together with Tomasz Łuczak and Andrzej Ruciński. I thank these my coauthors for helpful discussions, which have contributed to this paper too.

## References

[1] N. Alon and J. Spencer, *The Probabilistic Method,* Wiley (New York, 1992).

[2] K. Azuma, Weighted sums of certain dependent variables, *Tôhoku Math. J.,* **3** (1967), 357–367.

[3] G. Bennett, Probability inequalities for the sum of independent random variables, *J. Amer. Statist. Assoc.,* **57** (1962), 33–45.

[4] S. Bernstein, On a modification of Chebyshev's inequality and of the error formula of Laplace, *Ann. Sci. Inst. Savantes Ukraine, Sect. Math.,* **1** (1924), 38–49 (Russian).

[5] B. Bollobás, Martingales, isoperimetric inequalities and random graphs, in: *Combinatorics* (Eger 1987), Colloq. Math. Soc. János Bolyai, 52, North-Holland, Amsterdam 1988, 113–139.

[6] R. Boppana and J. Spencer, A useful elementary correlation inequality, *J. Combin. Th. Ser. A.,* **50** (1989), 305–307.

[7] T. K. Chalker, A. P. Godbole, P. Hitczenko, J. Radcliff and O. G. Ruehr, On the size of a random sphere of influence graph, *Adv. Appl. Probab.,* **31** (1999), 596–609.

[8] H. Chernoff, A measure of asymptotic efficiency for tests of a hypothesis based on the sum of observations, *Ann. Math. Statist.*, **23** (1952), 493–507.

[9] A. Dembo, Information inequalities and concentration of measure, *Ann. Probab.*, **25** (1997), 927–939.

[10] W. Feller, Generalization of a probability limit theorem of Cramér, *Trans. Amer. Math. Soc.*, **54** (1943), 361–372.

[11] W. Hoeffding, Probability inequalities for sums of bounded random variables, *J. Amer. Statist. Assoc.*, **58** (1963), 13–30.

[12] S. Janson, Poisson approximation for large deviations, *Random Struct. Alg.*, **1** (1990), 221–229.

[13] S. Janson, New versions of Suen's correlation inequality, *Random Struct. Alg.*, **13** (1998), 467–483.

[14] S. Janson, T. Łuczak and A. Ruciński, An exponential bound for the probability of nonexistence of a specified subgraph in a random graph, in: *Random Graphs '87* (Poznań 1987), Wiley, Chichester 1990, 73–87.

[15] L. M. Kirousis, E. Kranakis, D. Krizanc and Y. C. Stamatiou, Approximating the unsatisfiability threshold of random formulas, *Random Struct. Alg.*, **12** (1998), 253–269.

[16] J. E. Littlewood, On the probability in the tail of a binomial distribution, *Adv. Appl. Probab.*, **1** (1969), 43–72.

[17] C. McDiarmid, On the method of bounded differences, in: *Surveys in Combinatorics* (Norwich 1989), London Math. Soc. Lect. Not. 141, Cambridge Univ. Press, Cambridge 1989, 148–188.

[18] K. Marton, A measure concentration inequality for contracting Markov chains, *Geom. Funct. Anal.*, **6** (1996), 556–571. Erratum, *Geom. Funct. Anal.*, **7** (1997), 609–613.

[19] M. Okamoto, Some inequalities relating to the partial sum of binomial probabilities, *Ann. Inst. Statist. Math.*, **10** (1958), 29–35.

[20] W. C. S. Suen, A correlation inequality and a Poisson limit theorem for nonoverlapping balanced subgraphs of a random graph, *Random Struct. Alg.*, **1** (1990), 231–242.

[21] M. Talagrand, Concentration of measure and isoperimetric inequalities in product spaces, *Inst. Hautes Études Sci. Publ. Math.*, **81** (1995), 73–205.

[22] V. A. Vatutin and V. G. Mikhaĭlov, Limit theorems for the number of empty cells in an equiprobable scheme for group allocation of particles, *Teor. Veroyatnost. i Primenen.*, **27** (1982), 684–692 (Russian); English transl.: *Theor. Probab. Appl.*, **27** (1982), 734–743.

Svante Janson

*Department of Mathematics*
*Uppsala University*
*PO Box 480*
*S-751 06 Uppsala*
*Sweden*

e-mail: `svante.janson@math.uu.se`